ESPRIT Basic Research Series

W0232121

Edited in cooperation with
the Commission of the European Communities, DG III

Editors: P. Aigrain   F. Aldana   H. G. Danielmeyer
O. Faugeras   H. Gallaire   R. A. Kowalski   J. M. Lehn
G. Levi   G. Metakides   B. Oakley   J. Rasmussen   J. Tribolet
D. Tsichritzis   R. Van Overstraeten   G. Wrixon

Ph. Lalanne P. Chavel (Eds.)

Perspectives for Parallel Optical Interconnects

Springer-Verlag Berlin Heidelberg GmbH

Volume Editors

Philippe Lalanne
Pierre Chavel
Institut d'Optique Théorique et Appliquée
Centre Scientifique d'Orsay
B.P. 147
F-91403 Orsay Cedex, France

CR Subject Classification (1991): B.4, C.1-2

ISBN 978-3-642-49266-2   ISBN 978-3-642-49264-8 (eBook)
DOI 10.1007/978-3-642-49264-8

Publication No. EUR 15134 EN of the Commission of the European Communities,
Dissemination of Scientific and Technical Knowledge Unit, Directorate-General Information Technologies and Industries, and Telecommunications, Luxembourg
Neither the Commission of the European Communities nor any person acting on behalf
of the Commission is responsible for the use which might be made of the following
information.

Typesetting: Camera-ready by authors
45/3140 - 5 4 3 2 1 0 - Printed on acid-free paper

Foreword

This volume is a monograph on parallel optical interconnects. It presents not only the state-of-the-art in this domain but also the necessary physical and chemical background. It also provides a discussion of the potential for future devices.

Both experts and newcomers to the area will appreciate the authors' proficiency in providing the complete picture of this rapidly growing field. Optical interconnects are already established in telecommunications and should eventually find their way being applied to chip and even gate level connections in integrated systems.

The inspiring environment of the Basic Research Working Group on Optical Information Technology WOIT (3199), together with the excellent and complementary skills of its participants, make this contribution highly worthwhile.

G. Metakides

Table of contents

Part 1.2 Active interconnect components

Second Section: Interconnection schemes and systems
Part 2.1 Parallel schemes

13 Interconnects with optically thick elements 297

Gilles Pauliat and Gérard Roosen

1 Perspectives for parallel optical interconnects: introduction

P. Chavel and Ph. Lalanne
Institut d'Optique, CNRS, Orsay

1.1 Optical interconnects and ESPRIT B.R.A. WOIT (3199)

This volume is the result of the interactions between the partners of the "Optical Interconnects" subgroup of WOIT. WOIT is the acronym for "Workshop on Optical Information Technology", which is also ESPRIT Basic Research Actions number 3199. It was supported by the European Community during the period 1989-1992.

Among the 24 partners of WOIT, eight were active in the Optical Interconnects subgroup:

- Heriot-Watt University, Physics Department, Edinburgh;
- Institut d'Electronique Fondamentale (IEF), Université de Paris-Sud, an associated laboratory of Centre National de la Recherche Scientifique (CNRS), Orsay;
- Institut d'Optique Théorique et Appliquée (IOTA), an associated laboratory of Centre National de la Recherche Scientifique (CNRS), Orsay;
- King's College London, Department of Physics;
- Laboratoire d'Automatique et d'Analyse des Systèmes (LAAS), an laboratory of Centre National de la Recherche Scientifique (CNRS), Toulouse;
- Office National d'Etudes et de Recherches Aérospatiales, Centre d'Etudes et de Recherches de Toulouse (ONERA CERT);
- the Swiss Working group on Optical Interconnects (SWOI), which itself consists of three partners:
    o Centre Suisse d'Electronique et de Microtechniques (CSEM), Neuchâtel;
    o Paul Scherrer Institut, Zürich;
    o Université de Neuchâtel, Institut des Microtechniques;
- University of Erlangen-Nürnberg, Institut für Angewandte Optik.

In addition to this introduction, the book is organised in 15 chapters, all written by scientists from those eight partners.

1.2 What are optical interconnects?

There are two converging approaches to optical interconnects : one arises from the domain of optical telecommunications and the other from the field of optical processing.

Guided wave optics has been used for about 20 years in optical telecommunications : fibre optics links exist in the whole range of distances, from city networks to intercontinental communications and are now routinely operating at bandwidths larger than 1 GHz. Fibre amplifiers have rapidly emerged as a mature technology and its application to repeaters without the need for light detection and reemission is becoming a meaningful industrial perspective. Through fibre optics as well as through planar guided wave optics, photonics is playing an increasing role in switching.

Optical local area networks connecting computers also exist and may develop more widely if the cost of some critical components can be reduced. The next step for optics is therefore to go inside the computer. In fact, fibre connections between boards and holographic backplanes already exist and within the ESPRIT programme, an important project, OLIVES [1], has been devoted to the development of optical interconnects at various levels within a computer cabinet and of the related technologies.

A slightly different approach of the role of optics in processors is the domain called optical computing, or optical processing. Optical Fourier transforms and optical correlations have been used for many years in specialised processors; optical logic devices and optical bistables now exist and are available commercially at least in the form of AT&T Bell Laboratories' SEED. The global purpose of the WOIT collaboration was to investigate the physics and the technology of such devices and to develop concepts for "optical architectures" that use them for the implementation of more or less specialised processors. Such architectures may embody various concepts, such as neural networks, low level image processing machines, matrix vector product coprocessors, cellular automata. In all cases, imaging plays a central role and the connections between the optical gates or memories are implemented by a suitable modification of plain geometric imaging. That is where the need for a special study of optical interconnects arises within WOIT and this explains the viewpoint taken by the Interconnects subgroup in its book.

One can pinpoint the distinction between those two ways of looking at optical interconnects as follows : the one emphasises beam propagation and encodes information in time, the other utilises the spatial domain as well as the time domain to data encoding and is necessitates therefore imaging components, which implies devices with a large number of spatial modes - as opposed to most fibres and guided optics devices nowadays; in contrast with multimode fibres where it is quite difficult to propagate independent data on the different modes, the imaging optical interconnections that form the core of the book may carry up to one independent information per mode.

The two viewpoints can easily be presented as complementary in a perspective presented by J.W. Goodman [2] : optics is being used for shorter and shorter connections. A very long connection is called a telecommunication channel and optics has succeeded there already. It is now going down to cabinet, backplane, and perhaps chip level interconnections; the next step down might be gate level interconnections : if optics can ever be technically and commercially competitive at the gate level, then the use of optical gates will gain in relevance, imaging optics will be needed to interconnect them and processors will become all-optoelectronic if not all optical.

1.3 Optical interconnects : how ?

There are both components and systems aspects to our subject. Optical interconnects necessitate
- specific passive devices
- specific active devices
- the development of appropriate systems using those devices for the implementation specific interconnection schemes.

The organisation of the book, that will now be examined in some detail below, reflects that division, with an additional part devoted to the discussion of theoretical and practical limits. All the chapters do not present the same technical level of description. For example, chapters 4, 5, 6 on the design and control of passive components, chapters 12 and 14 on architectures provide a thorough description containing original experimental results obtained by the author's laboratories. The other chapters are more general or introductory. We expect that this marriage makes the book relevant for both newcomers and specialists.

1.3.1 Passive devices

By passive device, we mean a device whose function, or state of transmission, or reflection, or refraction, or polarisation, is constant during the operation of the processor and therefore during the use of the interconnection that it realises. This includes a number of purely optical devices, such as lenses or gratings, as well as electrical or optoelectronic devices, such as fixed sources or detectors. There is little specificity associated to sources and detectors for optical interconnect applications, so that only purely optical passive components are considered.

Chapter 2 and chapter 3 are by the same authors as this introduction. The former contain general considerations about the various families of passive optical components — refractive and diffractive —, the diffraction limit, and the basic optical setups that can be used in systems. The latter reviews the basics of light refraction and of imaging by lenses as appropriate for their application in the field of optical interconnects. The necessity of miniature lens arrays is introduced, and the effect of downscaling on aberrations, i.e. on image quality, is investigated on a test case. A review of the state of the art in microlens array fabrication concludes the chapter.

The next two chapters, respectively by H.P. Herzig and R. Dändliker, and by H.P. Herzig, M.T. Gale, H.W. Lehmann and R. Morf, all of the Swiss Working group on Optical Interconnects, cover diffractive optical interconnection components in more details. These are divided into two groups : natural holograms and computer generated elements. Chapter 4 reviews holographic fanout elements, holographic aberrations, and the basis of coupled wave analysis of diffraction in thick holographic gratings; recent original results by the authors on the comparison between sequential and simultaneous recording of holographic lenslet arrays are covered in detail, with theory and experiments, and with conclusions on the importance of phase control in simultaneous recording and on optimal conditions to minimise crossmodulations and to maximise diffraction efficiency. The technological means for the fabrication of computer generated diffractive optics improve. Chapter 5 examines the design of multilevel diffractive lenses and fanout elements and includes a novel algorithm for a rigorous calculation of diffraction by lamellar gratings. Useful practical information on available techniques for the fabrication of such elements concludes the chapter.

Chapter 6, by J. Schwider of Erlangen, addresses the question of adapting testing and control methods to ever increasing miniaturisation of the components. Automated operation of interference microscopes are the main approach here, but for every type of component a special setup must be conceived. The difficulty is even greater with multicomponent and multifunctional devices.

1.3.2 Active devices

In active devices, a control signal, that may be in electronic or in optical form, is used to modify the interconnect function. Progress in the physics and in the technology of active devices has been the triggering event in the present development of optoelectronics in general and of optical interconnects in particular.

Chapter 7, by F. Lozes-Dupuy, H. Martinot and S. Bonnefont of LAAS, gives an introduction to the impact of semiconductor technology to optoelectronic devices. It includes elements of growth techniques and processes that can be used to fabricate the present family of devices. These devices are based on heterostructures — the composition of the semiconductor varies in space, either continuously or abruptly. Illustrative examples, based on an extensive bibliography, emphasise the SEED devices, already mentioned, the photothyristor arrays that combine an optical or electronic command with triggered emission of light, and the various families of microlaser arrays that have appeared in the very recent past. Other devices designed for the modulation of light can be grouped under the common denomination of Spatial Light Modulators. They are discussed by A.C. Walker of Heriot-Watt University in chapter 8. The electro-optic, magneto-optic and photorefractive effects are exploited in the liquid crystals or ordinary, insulating or semiconductor, crystals devices that have been developed for several years; some of them are commercially available. Mechanical effects in microstructures are also mentioned, and a special section is devoted to the optical disk — presently the most widespread spatial light modulator. Chapter 9, by D. Bize of ONERA CERT, is a short review of photorefractive effects and materials. Chapter 10, by J. Turunen of Heriot-Watt University, is devoted to acousto-optic devices and has the form of a digest of the physical effects and parameters relevant for the design and performance of those components.

1.3.3 Schemes for parallel optical interconnects

Before specific setups based on the various components described in chapters 2-10 are described, one chapter is devoted to the study of density in parallel optical interconnects. While the theory of diffraction allows to introduce a number of degrees of freedom for any optical beam, thus setting an ultimate limit to the density of any interconnection scheme, the theoretical limit is out of reach for two reasons, as shown by N. Streibl of Erlangen in chapter 11 :
- light detection is performed by photodiodes with some finite integration area, and because of that there is always some slight spillover on any channel from its neighbours; a maximal tolerable amount of crosstalk must then be specified to guarantee correct detection with a given energy and accuracy or bit error rate, and this in turns determines a maximal practical density, the ratio between the theoretical and the practical limit depends on the beam profile and geometry;
- if the geometrical relation between sources and detectors can be described as a simple imaging — a "space invariant" scheme, then no other restriction applies. If, however, a space variant geometry is required, which is useful in particular in the case of randomly reconfigurable connections where any source can send data at will to any channel, the maximal number of channels is basically reduced to the square root of its space invariant value.

Chapter 12, by A.G. Kirk and T.J. Hall of King's College London, and chapter 13, by G. Pauliat and G. Roosen of IOTA, are devoted to a systematic description of interconnection schemes, the former with optically thin elements and the latter with optical thick elements. The difference lies precisely in the amount of space variance : in an optically thin element, beams from two point sources arrive at slightly different angles but undergo basically the same modification, thus forming two identical, angularly separated, output beams; in an optically thick element, they essentially "see" two different structures. Chapter 12 covers free space imaging and convolution geometries and their recently developed glass-filled free space counterparts, matrix vector products, and their application to various cases such as optical switching, neural networks, cellular automata, fibre to fibre interconnects. Chapter 13 derives the interconnection capacity parameters from the physics of thick gratings; illustrations cover in particular photorefractive materials used for correlations, optical memories, self-aligning interconnections.

In chapter 14, D. Fey and W. Stork of Erlangen give a tutorial review of the most important switching network, with the idea that multistage networks may be an important field of application for optics. The fine control of wavelength that is now possible with diode lasers may be one way to devise real implementations. Experimental results are given for an optical perfect shuffle and a butterfly with special purpose holograms developed in the author's laboratory, and wavelength multiplexing architectures are discussed.

1.3.4 Limits of optical interconnects

Chapter 15, by K. Zürl from Erlangen discusses the speed-spatial density trade off that generally comes up when a given data throughput per area is required. Based on minimal power consumption, it provides interesting conclusions on the relevant architectures to be chosen.

Chapter 16, co-authored by Th. Maurin of IEF and the authors of this introduction, include a summary of the constraints and performances of electrical interconnects in computer systems and conclude with a quantitative comparison of the optoelectronic and the all-electronic implementation of some simple interconnection schemes between chips and boards.

1.4 Optical interconnects : why ?

Various reasons are usually cited for developing optical interconnects and predicting their future success. In this section, we review some of these reasons and comment briefly on them.

The first keyword here is "speed". However, care should be exercised to define what is meant by "speed". If it designates the usual physical quantity obtained by dividing a distance by the time it takes to cover it, than an optical signal travels in a given medium at the (group) velocity of light in that medium, which is usually slightly slower that the velocity c of light in vacuum. However, the same is true for any electromagnetic signal : even without optics, signals in a computer travel at a speed close to velocity c and very little can be expected from optics. Strictly speaking, Brillouin precursors always travel at exactly velocity c in any medium, irrespective of frequency. However, the word "speed" in computer science is often used in a different meaning and designates a time delay : namely, the time it takes for a signal emitted by a source before it is detected at the receiver end. This means that it must reach a predetermined threshold, fixed by the technology and for that some minimal amount of energy from the source must reach the detector. With purely electrical interconnects, this implies loading a number of conducting lines, i.e. loading capacitors, whence an RC delay; because the propagation modes of optical signals does not imply loading capacitor, the RC delay in optoelectronic connections is restricted to the source and detector capacitances and can be significantly reduced. Hence, there is a delay advantage in using optics even there is no velocity advantage. This advantage depends on many technological rather than fundamental physical factors.

The second keyword is "fan-in". Goodman [3] has shown that there is no energy advantage related with optical fan-in (or fan-out). It may nevertheless be a significant advantage to use optics to send a large number of signals to one given detector without the problem of imposing the same electrical potential to all sources; optical bit to bit comparison between two words on one single threshold detector has been used as the basic principle of a general purpose computer [4]. Here again, there is no fundamental advantage, but there is a practical advantage for some processor architectures.

Fan-in should not be confused with space density. "Time density", i.e. bandwidth, is a fundamental advantage of optics just because the carrier frequency of optics is in the hundreds of terahertz range, considerably higher than that of microwaves for example, and optical telecommunication take benefit of it. This book is mainly devoted to the space density of optical interconnects, related to the fact that the optical wavelength is small — in the micrometer range or slightly less —; this is also a fundamental advantage and has still to be fully exploited in interconnection applications.

Our final keyword is reprogrammability. Naively speaking, reprogramming a set of interconnections may mean changing a processor of one type into a processor of another type at well without wasting the energy to always implement both sets of interconnects and using only one set. More technically, this may mean changing a set of interconnection in a multistage network. This is potentially a useful goal. Optoelectronic or photorefractive devices described in this book may allow to reach it.

Concluding the activity of the WOIT subgroup on optical interconnects, we like to express the wish that our research is just a little upstream from that developed in the precompetitive part of the ESPRIT programme, or indeed of already competitive developments in industry, and that optical interconnects using the time modulation of large images rather than a few individual time modulated laser beams will allow to increase the performances of computer systems in a few years.

Acknowledgements

The authors wish to thank the CEC-ESPRIT Basic Research for giving them the opportunity to undertake this work as well as the necessary support.

They thank all chapter authors for their contribution and for their efforts in preparing these overviews of the state of the art in the field. A procedure of refereeing similar to that usually employed for journals was used and made their work heavier. In addition to some of the authors of other chapters, more than ten anonymous reviewers from outside the WOIT partnership were used to help in the preparation of this volume : their help and advice deserves a special acknowledgement here.

References

1. Esprit II Project 2289 (OLIVES), final report, April 1992.
2. Goodman, J.W.: Optical Interconnects. Oral paper at ICO XV, Garmisch-Partenkirchen, August 5-10 1990.
3. Goodman, J.W.: Fan-in and fan-out with optical interconnections. Opt. Acta 32, 1489 (1985).
4. Guilfoyle, P.S.: Digital optical computing fundamentals, implementation, and ultimate limits. In Digital Optical Computing, SPIE Proc. CR35, 288-309 (1990).

First section: Components

Part 1.1 Passive interconnect components

2 Free space interconnects

Ph. Lalanne and P. Chavel
Institut d'Optique, CNRS, Orsay

2.1 Introduction: 3D optical interconnects

Optical fibers and planar integrated waveguides are now both used for communications when extremely large bandwidths of up to 100 GHz are required. In monomode operation, these structures confine the optical waves in cross sections lying between 1 and 10 μm. Geometrically, these guided interconnections are analogous to electrical ones. Fibers and planar integrated waveguides are respectively alternatives for 1D electrical wires and for electrical stipes engraved on 2D integrated circuits.

In contrast to these serial processing concepts is the increasing interest in the possibility offered by optics for providing highly parallel and massive connectivity between optoelectronic processors equipped with detectors or modulators in array format. This approach has no electrical counterpart; we will call it free space optics or by opposition to the previous description, 3D optics. Its main interest lies in the fact that optical beams (or optical channels) may cross in 3D space without crosstalk. This property comes from the nature of photons, that do not interact with each other. This is the basic justification for the existence of this book.

The first and simple approach to free space optical interconnects is illustrated on figure 2.1: a signal emitted by a source is broadcast to a large number of photodiodes. However, this simple scheme is too crude an approach for efficient broadcasting through free space. Because the amplitude of an elementary spherical wavelet issuing from one point source decays as 1/d when propagating, our simple approach of figure 2.1 is not energetically efficient.

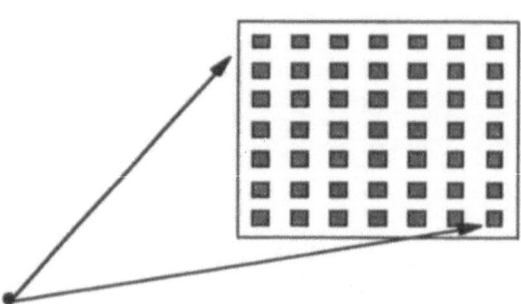

Fig. 2.1 One to many interconnects through free space.

To provide a better use of energy and to realise specific interconnects, generally speaking, we need to add passive components in order both to conjugate optical sources and detectors and to decode the different wavefronts propagating simultaneously through the same free space (each implementing an independent channel, as does an electrical wire). Part 1.1 of the present volume describes these passive components while part 1.2 examines active interconnect components such as spatial light modulators or photorefractive crystals. *This chapter is here just to introduce the set of fixed interconnect elements needed for focusing, decoding, managing free space optical channels.*

At the beginning of the following section (section 2.2.1), the degree of parallelism provided by free space interconnects is reflected in term of the number of resolvable points available in an optical line according to the concept first formulated by D. Gabor. Section 2.2 presents several types of architectures generally used for optical interconnects. Space variant and invariant schemes are provided. While section 2.1 deals with theoretical limitations, section 2.3 presents a few practical limitations that need to be taken into account: 1) through two examples giving the volume required for an optical channel in a space variant or invariant scheme, space invariant interconnects are claimed to realise larger space-bandwidth product, 2) through two recently proposed concepts, the alignment tolerances of optical setups are briefly pointed out.

2.2 Optical free space channels and their implementations

2.2.1 Diffraction and degrees of freedom

The question we address here is how many independent channels can be theoretically carried through free space. We call it the number of degrees of freedom available in an optical line. That concept has been first formulated by D. Gabor [1].

This number is directly related to the number of resolvable spots accessible in an image plane.

Figure 2.2 shows a focused beam limited by a square pupil of size a and located at distance D from the focus. Due to diffraction, a $sinc^2(x)$ pattern results in the image plane. Its spread is about $\lambda D/a$. This effect corresponds to a basic physical limitation that is best summarised by using the concept known to opticists as "étendue": the étendue of an optical beam is defined as the product of the cross section by the square of the numerical aperture, or equivalently, by the product of the image and pupil areae divided by their distance squared. Generally speaking, the étendue is a constant of any lossless optical beam transformation: it is conserved in reflection and refraction. Because of diffraction, the spherical beam of figure 2.2 has an étendue of λ^2, and no beam can ever have a smaller étendue.

Gabor has shown that an optical beam can only transmit a finite amount of data. This amount, also known as the Gabor number, is the number of degrees of freedom usable for any polarisation in any lossless linear optical wave transformation and is equal to the étendue of any optical beam divided by the wavelength squared. More recent argumentations can be found in reference [2].

Such a density is an ultimate limit still out of sight since off-axis aberrations often result in spot sizes much larger than the diffraction limit.

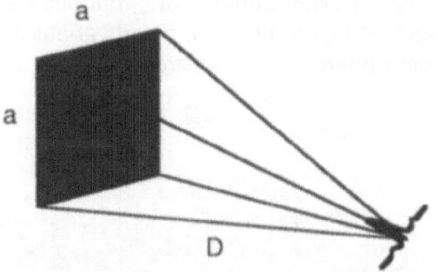

Fig.2.2 Diffraction by a square aperture.

2.2.2 Two basic interconnect setups

A use of energy better than in figure 2.1 can be achieved simply by using a lens to conjugate the two planes of sources and detectors. However, as figure 2.3a shows, no broadcasting is achieved and only many one-to-one connecting channels exist.

A much more compact but equivalent approach consists in replacing the macrolens by a lenslet array, each lenslet conjugating one source and one detector (fig.2.3b). In a 2f-2f configuration using a macrolens, the volume required to image NxN sources whose spacing is p is proportional to $4f(Np)^2$. Replacing f by NpN_0, using the definition the f-number of the lens, leads to a volume of $4NpN_0(Np)^2$. When one microlens is assigned per source, the miniature focal length is then equal to pN_0. Therefore, the volume reduction is proportional to the square root of the number of lenslets.

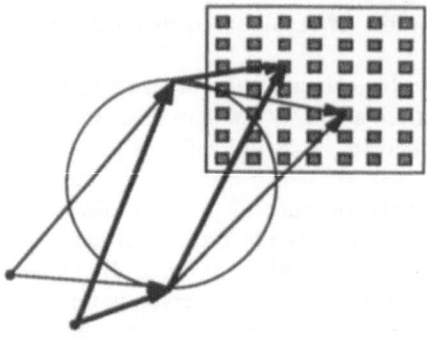

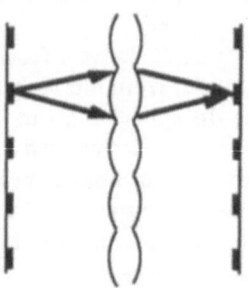

Fig.2. 3a Many 1 to 1 with a lens   **3b** with a lenslet array.

A compromise between energy conservation and the broadcasting ability shown in figure 2.1 can also be implemented with a lenslet array. As suggested in figure 2.3c, the source is conjugated with all the detectors while most of the emitted light is received correctly. This kind of optical element capable of converting a plane or spherical wave into many concentrated spots of equal intensity and with little loss is called an array illuminator. The regularity we gave to figure 2.3c is arbitrary. It is just here to introduce the concept of array illuminators, that has received much attention in the literature [3]. Chapters 4 and 5 will give details about their fabrications while chapter 12 will illustrate their interests in some architectures.

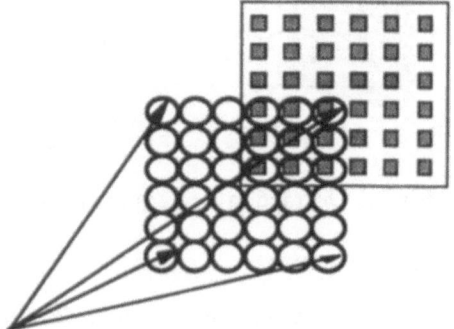

Fig.2.3c One to many interconnects with a lenslet array.

A slightly more general interconnect scheme can be conceptually achieved by replacing the periodic lenslet array by a combination of lenses and prisms of various angles. Figure 2.3d shows a lens whose area is divided into four quadrants, each quadrant containing its own prism. It results in a four-point impulse response which provides the interconnect pattern. The imaging properties of the lens are maintained, as suggested in the figure where two impulse responses are provided through two sources. As the prism operation consists in deviating the beam, geometrical optics straightfordwardly shows that such a combination of lenses and prisms is equivalent to using only lenses, some of them being off axis. While the scheme of figure 2.3d allows an additional flexibility, the manufacturing of such components is a problem. In practice, as suggested by figure 2.3e, a hologram is used to replace both lenses and prisms. The imaging properties still remain.

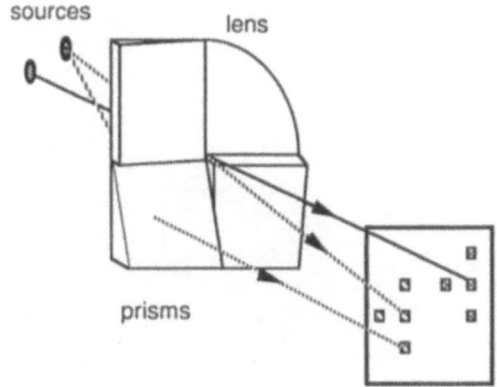

Fig.2.3d Many to Many interconnects with a combination of prims and lens hologram

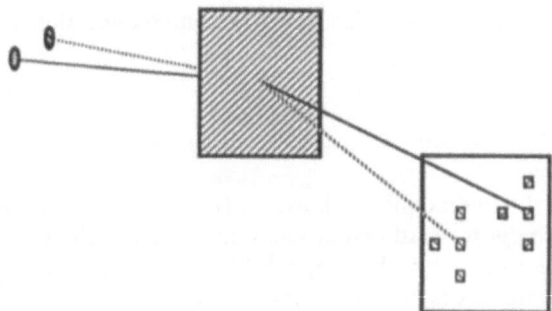

Fig.2.3e The same interconnection with a hologram.

As shown in figure 2.4, the most general interconnect scheme consists in providing an arbitrary space variant interconnection between each of NxN transmitting ports and one or more of NxN receiving ports. According to the application, these connections may need to be reprogrammable or not. In the so-called crossbar architecture, 2D spatial light modulators can be used as switching arrays for 1D port arrays, see chapters 8 and 12. Volume holograms using the Bragg effect have been studied for dense interconnects requiring very large fan-outs [4]. The Bragg effect, that has been first introduced for the interpretation of X-ray diffraction by crystal lattices, corresponds to the reinforcement of constructive interferences for particular directions of incident beams onto volume holograms. Between two NxN two-dimensional surfaces, there are N^4 possible connections; however, a 3D hologram can only provide a number of interconnections proportional to N^3. As is suggested in reference 4, a possible compromise is to map $N^{3/2}$ to $N^{3/2}$ ports, using a fractal layout. Chapter 13 is more specifically centred on interconnects using volume holograms.

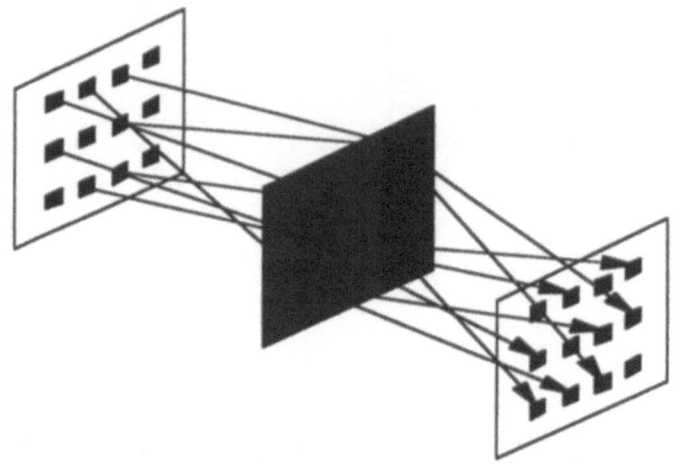

Fig.2.4 Arbitrary space variant interconnections.

2.3. Limitations

At different levels, the interconnects shown in figures 2.3 and 2.4 present practical limitations. Here, we put forward two of them: the volume required to realise the set of expected connections and the alignability of the system of components. Concerning the compactness of the system, we are going to compare the mean volume required to realise one channel in two different architectures. A more general approach can be found in reference 5.

Alignability and compactness are not the only important limitations encountered in optical systems. For instance, noise, especially in coherent illumination is also crucial. However, alignability and compactness seem more appropriate for such an introduction chapter because of their primary importance for the substitution of optical interconnects for electrical interconnects in future systems.

2.3.1 Compactness

In this section, we show that the mean volume required by an interconnection largely depends on the scheme investigated. Two setups are studied. The first one is space variant and is generally used in switching networks described in chapter 14. The second is space invariant. For the sake of clarity, we consider a simple plane to plane imaging setup.

Space variant interconnects. The mean volume required for realising one channel is estimated in the case of space-variant interconnections under the paraxial approximation. In figure 2.5, we denote by a the maximal transverse distance between two points linked by the interconnection. This parameter is a very important characteristics of the interconnect since it quantifies its complexity: as a increases, aberrations may increase, longer communications are achieved and more sophisticated components are required.

Figure 2.5 shows a double hologram structure. The transmitter illuminates the first hologram which is located just to its right. This hologram deflects the beam to the second grating, that in turn deflects and focuses the incident beam onto the receiver. We choose to establish the volume requirement in this double pass structure because it satisfies quite well the requirements encountered in dense and space-variant optical interconnects [6]. Moreover, it seems quite promising for implementing optical switching networks and chip to chip interconnects (see chapters 14 and 16).

We denote by i the maximum deflection angle (see figure 2.5). This angle i can be limited by the spatial bandwidth of the hologram or also by the designer in order to control the optical system aberrations. Consequently, the distance D between the transmitter and the receiver holograms is at least equal to:

$$D > a/tg(i) \tag{2.1}$$

Because of the symmetry, the two microholograms in the receiver and emitter plane are both disks with the same diameter h. Under the assumption that the first microhologram is uniformly illuminated, the beam which results in an Airy disk diffraction pattern in the receiver hologram plane has a spread in the x axis of figure 2.5 equal to:

$$1.22 \; \lambda D/[2h(\cos^3 i)] \tag{2.2}$$

The term in $\cos^3 i$ takes into account the inclination between the beam and the hologram plane. In the direction perpendicular to the figure, the cosine is not cubed except if an arbitrary interconnection (not in the plane of figure 2.5) is investigated. Equalling this spread with the diameter h ensures a crosstalk less than 16 percent, we find:

$$h^2 = 0.64 \; (\lambda D)/\cos^3(i) \tag{2.3}$$

The question now arises to infer the interconnect density. The distance between two adjacent emitters or receivers being h, the volume required per interconnection is Dh^2. The interconnect density N/V per unit volume is equal to $1/[Dh^2]$. Replacing h^2 given in eq. 2.3 in the previous expression, we obtain an interconnect density equal to $1.6 \cos^3 i/\lambda D^2$. According to equation 2.1, it is then inferior to:

$$N/V < 1.6 \cos(i).\sin^2(i)/\lambda.a^2 \qquad (2.4)$$

That means that the mean value of an interconnect volume is proportional to the square of the maximal transversal distance of the link. For a deviation angle i equal to 20° and a wavelength of 1μm, N/V (cm^{-3}) is equal to $1700/a^2$ with a in cm. For a=10cm, this density is as low as to 17 interconnections per cm^3.

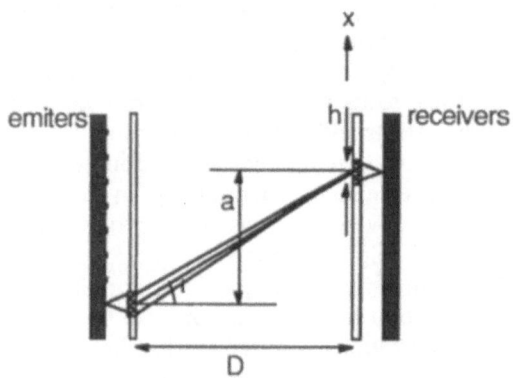

Fig.2.5 Plane to plane interconnects scheme.

Space invariant interconnects. In the case of space invariant optical interconnects, let us estimate the interconnect density for a very simple case consisting in the imaging of one plane onto another one. Under the paraxial approximation, the lens is supposed to be diffraction limited. Let us denote by r and f its radius and focal length. Then:
  - the volume required to conjugate the two planes is $4\pi r^2 f$ in a 2f-2f configuration,
  - the spread of each resolved point is $\lambda f/r$.
Therefore, if off-axis aberrations restrict the object area to a fraction k of the lens area, the number of links provided per unit volume is equal to:

$$\pi k r^2/4\pi\lambda^2 f^3 \qquad (2.5)$$

For an interconnected surface $\pi k r^2$ equal to 1 cm^2 and for a focal length f equal to 3r, we find 70,000 interconnections per cm^3.

Compared with the previous results with a=1 cm (this also corresponds to a 1 cm^2 interconnected surface), the density is significantly increased. Note that this result has been obtained with miniature components in case of the variant scheme while a conventional lens has been supposed in the invariant configuration.

2.3.2 Alignability

Development and fabrication of optical interconnects depend on the easiness with which optical components and light beams can be aligned in the system. Difficult or time-consuming alignment in an interconnection system is obviously a penalty for its development since it will probably be expensive and poorly reliable. Over the last few years, a significant effort for assigning a general framework to optical interconnect systems has been made in research laboratories [7]. Although it is yet an unsolved issue, several approaches are worth paying attention to.

3D stacked planar optics. The first approach has been proposed in 1982 by Iga [8] and coworkers. The concept of stacked planar optics is to construct 2D arrays of lightwave components which can be stacked all together to provide a necessary function. All components must have the same two dimensional spatial periodicity. As 2D arrays of components, they can be fabricated by the planar technology of microoptics. Figure 2.6 shows for instance a stacked planar optics of multifiber components. Other configurations realising different processing functions have been proposed: For instance, configurations with active optical components such as surface emitting laser arrays or SEED arrays [9] or with multichannel processing through optical correlation [10] have also been investigated. With this concept, a simultaneous alignment of a large number of optical components can be expected through mass production.

3D confined interconnects. The second approach is more recent and is sketched in figure 2.7. It consists in implementing free space interconnects by folding the beams within a glass plate. The upper surface can be etched so as to implement gratings, whereas the bottom surface is simply used in a total internal reflection mode. Figure 2.7 shows a f-f-f-f configuration: the input and output planes are respectively the object and image focal planes of the two Fresnel zone lenses, while the grating realises an arbitrary space variant interconnection in the Fourier domain. Applications to optical backplane connections are under study in the ESPRIT OLIVES project [11] (see also chapter 16) through image transportation while filtering has been investigated by AT&T [12].

The main result of this approach is the reduction of the expensive mechanical alignment since this planar positioning technique, that can easily achieve an accuracy of 0.1 µm reduces the problem of controlling the six degrees of freedom of each optical component. The alignment is made with the same methods used in microelectronic technology masking; therefore, it is not more critical or less accurate.

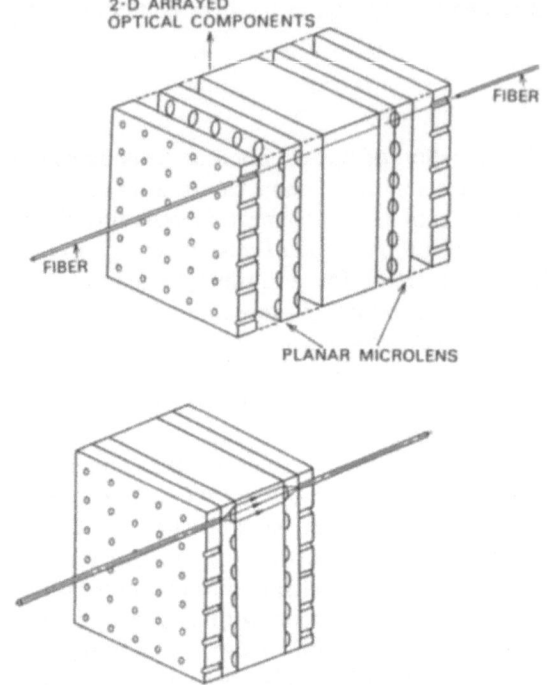

Fig.2.6 Stacked planar optics of multifiber components.

This glass-filled free space interconnection system also presents little environmental sensitivities such as those due to mechanical constraints or dust deposition.

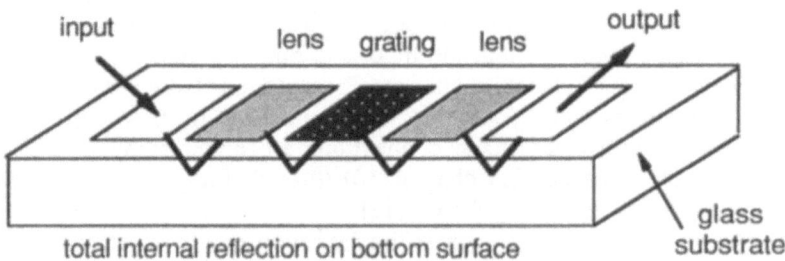

Fig.2.7 4f filtering system within a glass plate.

To build complex systems consisting of many such modules, a standardization "towards a digital optical platform" is needed [13]. As shown on figure 2.8, grooves can be etched into the substrates during the masking process to make a precise alignment of the module [12]. An index matching fluid might be used to ensure that the travelling light is not disturbed at the module interface.

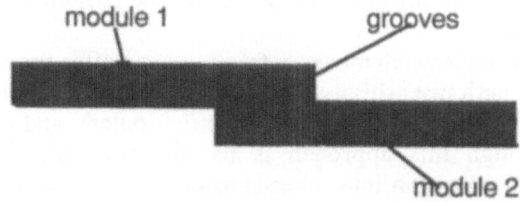

Fig.2.8 Alignment of two modules by etched grooves.

Fig. 2.9 Monolithic integration of microprisms and microlenses on one substrate. (Photograph provided by the courtesy of Maria and Stephan Kufner).

An other technique for integrating refractive components such as microprisms, microbeamsplitters and microlenses has been recently proposed [14]. It consists in irradiating polymethyl methacrylate (PMMA) with deep protons. The prisms are fabricated with an orientation of 45° relative to the substrate surface, and can be used for deflection when total internal reflection is considered, or for splitting when two PMMA blocks are stacked by adhesive techniques. Microlenses can be produced by diffusion of monomer vapour into irradiated domains that generate a volume growth. First experiments with a 500 μm microlens have shown resolutions as high as 256 line pairs per millimetre. Figure 2.9 shows a monolithic integration of microprisms and microlenses on opposite sides. The lens diameters and the height of the prisms are 500 μm.

Compared to diffractive elements, refractive ones offer the advantage of reduced angle and wavelength insensitivity (for instance, insensitivity to wavelength shift of laser diodes operating without thermal electric cooler), and offer large numerical apertures. Although this approach is based on an exotic technique (proton irradiation), the duplication into other substrate by mechanical stamping could be feasible.

Yet other techniques can be investigated. There are few doubts that the near future will be promising in new concepts for alignment procedures.

References

1. Gabor, D.: Light and Information, in Progress in Optics, vol.1, chapter 4, p 109 (1961).
2. Winthrop, J.T.: Propagation of structural information in optical wave fields, J. Opt. Soc. Am. 61, pp 15-30 (1971).
3. Streibl, N.: Beam shaping with optical array generators, J. Mod. Opt. 36, pp 1559-1573 (1989).
4. Psaltis, D., Gu X. and Brady D.: Fractal sampling grids for holographic interconnections, SPIE vol. 963 Optical Computing, pp 468-474 (1988).
5. Ozaktas, H.M. and Goodman, J.W.: Lower bound for the communication volume required for an optically interconnected array of points. J. Opt. Soc. Am. A, vol.7, pp 2100-06 (1990).
6. Feldman, M.R. and Guest, C.C.: Interconnect density capabilities of CGH for optical interconnections of VLSI circuits. Appl. Opt. 28, pp 3134-37, (1989). Schwider, J., Stork, W., Streibl N. and Völkel, R.: Possibilities and limitations of space-variant holographic optical elements for swithching networks and general interconnects. Appl. Opt., pp 7403-10 (1992).
7. Technical Digest, Topical Meeting on Optical Computing (OSA, Washington, DC, 1989, 1990 and 1991).
8. Iga, K., Oikawa, M., Misawa, S., Banno J. and Kokubun, Y.: Stacked planar optics; an application of the planar microlens. Appl. Opt. 21, pp 3456-60 (1982).

9. Agu, M., Akiba, A., Mochizuki T. and Kamemaru, S.: Multimatched filtering using a microlens array for an optical-neural pattern recognition system. Appl. Opt. 29, pp 4087-91 (1990).
10. Iga, K.: Active Parallel Microoptics. SPIE Vol. 1319, Optics in Complex Systems, pp 486-490 (1990).
11. Parker, J.W.: Progress in optical interconnection technologies and demonstrators under the ESPRIT II OLIVES programme. Optical Computing, OSA, Washington, DC, pp 304-307 (1991).
12. Jahns, J. and Huang, A.: Planar integration of free-space optical components. App. Opt., 28, pp 1602-05 (1989) and also Opt. Comm. 76, pp 313-317 (1990).
13. Huang, A.: Towards a digital optics platform. SPIE Vol. 1319 Optics in Complex Systems, pp 156-160 (1990).
14. Brenner, K.H., Frank, M., Kufner, M. and Kufner, S.: A new monolithic integration method for microprisms, microbeamsplitters and microlenses in PMMA. 8th Workshop on Optics in Computing, Paris 8-9 Sept. 1992, pp 51-54.

3 Reflective and refractive components

Ph. Lalanne and P. Chavel
Institut d'Optique, CNRS, Orsay

The previous chapter has described a few basic architecture usable for interconnects through free space. However, no details about the optical components were given. In this chapter, we study the classical elements encountered in optical setups, which permit a flexible shaping of waves thanks to refraction and reflection.

Section 3.1 introduces the concept of reflective and refractive components for optical interconnects. We make clear the distinction with diffractive components and we briefly discuss difficulties encountered during the fabrication of these components. Section 3.2 shows the importance of miniature lenses for optical interconnects; an aberration study in case of plane to plane information transportation systems argues in favour of the use of either macrolenses compensated from aberrations or very small lenses with diameter shorter than 1 mm. The last section is an overview of the different fabrication processes of microlenses.

3.1 Introduction and concept

3.1.1 Definition of reflective and refractive components

Light at an interface between two homogeneous media undergoes *reflection* and *refraction*; any limitation of a light beam, for example through a pupil, generates *diffraction*. Those are general phenomena and always coexist. It is nevertheless useful to distinguish between "refractive" optical components and "diffractive" ones. In this chapter, reflective components will be grouped with refractive ones because of the strong similarity between reflection and refraction.

It is the role of all optical components to modify the shape of light beams. If this modification is done by (reflection and) refraction while diffraction plays only the role of a limitation, the component is called refractive. This, of course, is the case of most common components: lenses, mirrors, prisms, plane parallel plates...If the modification is brought by the interference of the waves diffracted by several

clearly distinct zones of the component, it is called diffractive (even though refraction may play a role in each of the zones). This is the case for
- a diffraction grating, where the zones are the various grooves;
- a hologram, where the zones are the interference fringes recorded in the holographic material during fabrication;
- a zone plate (often called Fresnel zone plate), where each annular zone diffracts a fairly complicated beam and the various beams interfere constructively at the zone plate foci.

3.1.2. Reflection and refraction for interconnect components

Resulting directly from *Fermat's principle* which assigns a stationary value to the optical path connecting two points, the very simple laws of refraction and reflection are sufficient for the understanding of all components discussed in this chapter. Before giving the concept used in an arbitrary refractive component, let us briefly review these laws known as the *Snell-Descartes laws*. They can be stated as: when a plane wave propagating in a medium of refractive index n_1 reaches a medium of index n_2, it is generally partly reflected and partly transmitted. If we denote by i the angle of incidence (see figure 1), the reflected beam is backpropagated with an absolute deviation angle equal to 2i, while the refracted one is deviated by the angle $D=|i-r|$ with i and r obeying the following equation:

$$n_1\sin(i) - n_2\sin(r) = 0$$

Moreover, both the reflected and refracted wave normals belong to the plane of incidence of figure 2.1. This deflection is of prime importance with regard to interconnections. In addition, the state of polarisation and the angle of incidence influence the energy repartition between the two refracted and reflected beams. This dependency is given by the Fresnel coefficients [1].

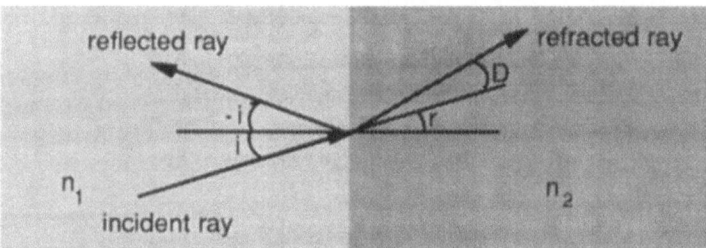

Fig. 3.1. Reflective and refractive deviations at a media interface.

Fig. 3.2 sketches a f-f-f-f configuration where an arbitrary interconnection is realized by a refractive component. This refractive component is a simple glass plate fabricated with a surface profile such that the incident ray   deviations implement the expected interconnection.

The plate consists in a set of prisms; each prism deflects a prescribed fraction of the incoming energy from object point A and brings it to a focus at point A_i in the back focal plane of the second lens (i=1,2,3,4 in the example of Fig. 3.1). Assuming that the object point B emits light with the same angle, its energy is split between points B_i with essentially the same relative efficiency if and only if the plate is the pupil of the system. This is the case if the plate is placed in the back focal plane of the first lens where each point corresponds to one direction of the incident illumination from the object. Whence the need of a f-f-f-f setup. For the sake of simplicity of the figure, points B_i have not been represented.

Diffraction comes into play if each individual prism is sufficiently small: a typical order of magnitude would be tens of micrometers, leading to an angular halfwidth of each beam of tenths of radians. In this chapter, we will consider the case of relatively large prisms, where diffraction can be neglected.

The problem with Fig. 3.2 is the fabrication of the individual prisms on the plate. Because of the small size of optical wavelengths, the surface definition and polishing must be accurate to typically a tenth of micrometer; an inaccuracy of the prism angles would lead to ill-placed points A_i; a departure from a plane surface would transform points A_i into large spots, and surface roughness would scatter light and create crosstalk.

Similarly, refractive components are sensitive to scratches, digs and coating deflects up to the submicron scale. Consequently, because of manufacturing difficulties encountered both at the etching and polishing level in the fabrication of the generic glass plate of Fig. 3.2, the latter is of no practical interest. That kind of arbitrary interconnects is in fact the realm of an arbitrary shape hologram rather than a refractive plate.

Generally speaking, spherical and flat shapes are the only surfaces quite easy to manufacture. The resulting optical components are also often quite expensive. This is probably a reason why only a small number of refractive component varieties realising a few simple functions are manufactured.

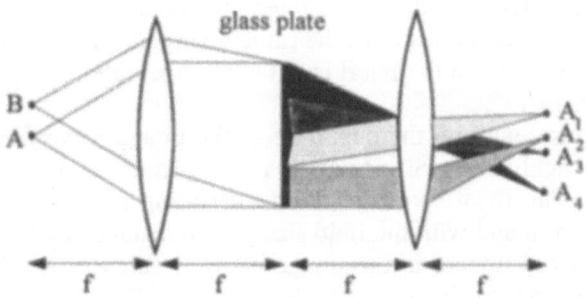

Fig. 3.2. Arbitrary refractive interconnects.

3.2 Interconnects with microlens arrays

3.2.1 Introduction

Lenses and now microlenses are well known for their ability to image one plane onto another. This concept has been discussed in the previous chapter. Another interesting property of lenses is that they provide a Fourier transform between their two focal planes [2]. This effect is related to diffraction and does not fall within the scope of this chapter; it will suffice here to say that a coherently illuminated f-f-f-f setup performs a spatial frequency filtering operation that can be monitored with a hologram in the pupil plane. More specific are the *aberration* properties of lenses. This section investigates in one simple ideal case, the influence of the focal length of lenses and microlenses on their ability to ensure an "accurate" (in a sense that will be defined) and dense imaging interconnection from an input to an output plane. This study will be made in the third order approximation for thin plano-convex lenses or microlenses used in a f-f-f-f setup with their spherical shape turned on the side of the pupil plane. We wish to illustrate with this example the influence on interconnect architecture of a general well-known law of geometrical optics, namely that downscaling brings a considerable reduction of aberrations. We do not claim that the configuration of figure 3.3 is optimal, it has just been selected for its simplicity.

3.2.2 Interconnection density provided by miniature lenses beyond the paraxial approximation

Let us discuss the influence of aberrations in the case where the optical interconnect scheme is a simple plane to plane information transportation. Fig. 3.3 sketches the interconnection which is quite similar to the arbitrary interconnect scheme shown in Fig. 3.2. The two planes to be interconnected are respectively the object and image focal planes of the two lenses. Let us suppose that the input plane contains laser diodes to be interconnected with photodiodes at the output plane. All the laser diodes have the same divergence angle θ and the lenses do not limit the beams so that the pupil is in the intermediate focal plane. For illustration, we consider $\theta=15°$.

Aberrations deal with the departure of the image wave from a spherical shape. A non spherical wave will not come to a rigorous focus, even in the absence of diffraction. The focal spot size of an aberrating imaging system increases with beam aperture θ and with the field size y. The distance δy between the ideal focus and the real ray impact in the image plane is called transversal aberration. It is usual to expand δy in the variables θ and y. For a system with a rotational symmetry, only odd powers starting with order 3 exist. Moreover, aberration calculations are more tractable in case of a symmetrical setup like that of Fig. 3.3; the aberrations named distortion and coma are then equal to zero for the third order terms. In this optical interconnect, the symmetry is realisable if both lenses are identical and if the distance between the two lenses is 2f (f being the image focal length of the lenses).

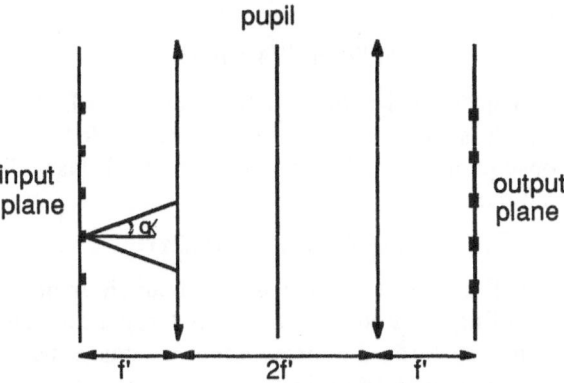

Fig.3.3. Plane to plane imaging setup.

In this particular case, often used for data communication, the third order transversal aberration δy takes the general form [3]:

$$\delta y = \alpha.tg^3\theta.cos\theta.f' + \beta.tg\theta.cos\theta.y^2/f' \qquad (3.1)$$

where α and β are two dimensionless coefficients that depend on the two lens shapes and refractive index, and where y is the object field. The first term of equation 3.1 is the spherical aberration while the second is the astigmatism term. Including the trigonometric factors in the coefficients α and β, we obtain a very simple equation:

$$\delta y = \alpha'.f' + \beta'.y^2/f' \qquad (3.2)$$

Now, let us introduce the "aberration pattern diameter" d, a new parameter such that, for $\delta y < d$, say all the energy (in the third order approximation) emitted by each laser diode is imaged onto each receiver. "d" is typically the size of a detector in the image plane. The maximal object field y_{max} is then given by:

$$y_{max} = [f'(d-\alpha'f')/\beta']^{1/2} \qquad (3.3)$$

Equation 3 indicates that d-α'f' must be positive. This gives the maximal focal length f'=d/α' that can be used so that the spherical aberration remains tolerable according to our criterion. If we denote by τ=f'/d and σ=y_{max} the focal length and maximal object field in normalized variables, we can rewrite the equation for y_{max} in the following form:

$$\sigma = [\tau(1-\alpha'\tau)/\beta']^{1/2} \qquad (3.4)$$

For focal lengths that are very small compared to d/α' i.e. in the domain where the spherical aberration is much smaller than astigmatism, it is easy to estimate the behaviour, as a function of f', of an important parameter for optical interconnects, the useful surface ratio S; S is defined for a given f' as the ratio between the surface of the object plane that can be interconnected according to our criterion and the surface needed to realise the interconnection:

$$S = [y_{max} / (y_{max} + f'tg\theta)]^2 \qquad (3.5)$$

For small f', S can be rewritten:

$$S = \tau/\text{ß'}/[(\tau/\text{ß'})^{1/2} + \tau.\text{tg}\theta]^2 \qquad (3.6)$$

For plano-convex lenses used with their flat face towards the input and output planes, the coefficient ß is equal to -12.65 (α=-2.33). So, for a divergence angle of 15°, ß'= -3.27. Consequently, for interconnections with laser diodes, the useful surface ratio is given by:

$$S = 0.31\tau/[(0.31\tau)^{1/2} + 0.27\tau]^2 \qquad (3.7)$$

As a function of τ=f'/d, figure 3.4 shows the behaviour of the useful surface ratio. Domains including very small and very large focal lengths have been omitted. The domain of large f' corresponds to cases where the spherical aberration cannot be neglected and where long distance interconnections can be achieved with special objectives partly compensated for aberrations. The curve indicates that short focal lengths realized with miniature optics offer increased density when simply interconnecting or imaging two planes. Let us notice that this curve is independent of diffraction since all results for the different focal lengths have been established at a given numerical aperture fixed by the beam divergence θ=15° of the laser diodes and therefore all diffraction patterns have the same size. Let us notice also that the benefit is even far larger on the volume required for the interconnection than on the surface given in figure 3.4.

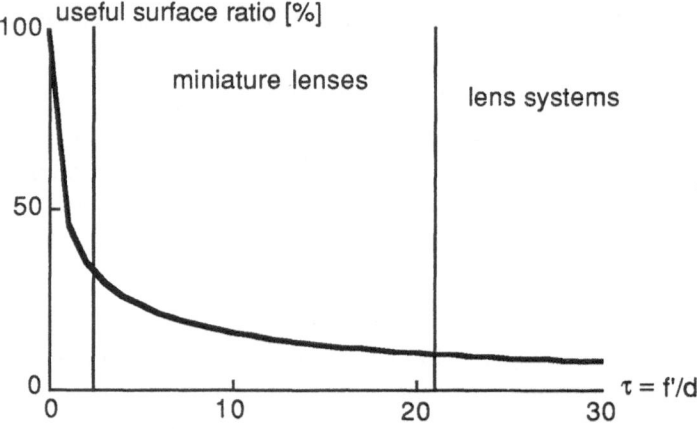

Fig. 3.4. Interconnect density ratio as a function of τ=f'/d in the symmetrical imaging setup.

Let us investigate a particular example: an array of 100x100 regularly spaced laser diodes whose distance is typically 100 μm (for instance, that corresponds to a typical case of parallel architecture with elementary processing cells containing some local intelligence). We assume that the array has to be imaged onto a similar array of photodiodes. This architecture may also be used as a backplane connector, see Chap. 16. As shown on figure 3.5, several optical configurations are available. Figure 3.5a shows the case where one microlens is assigned per channel. An intermediate case where several channels are grouped with one microlens is sketched in figure 3.5b. The last case corresponds to an interconnection realized with two macroscopic lenses (for instance, two 50 mm high-quality photographic lenses) conjugating two chips with object and image fields of 1 cm^2.

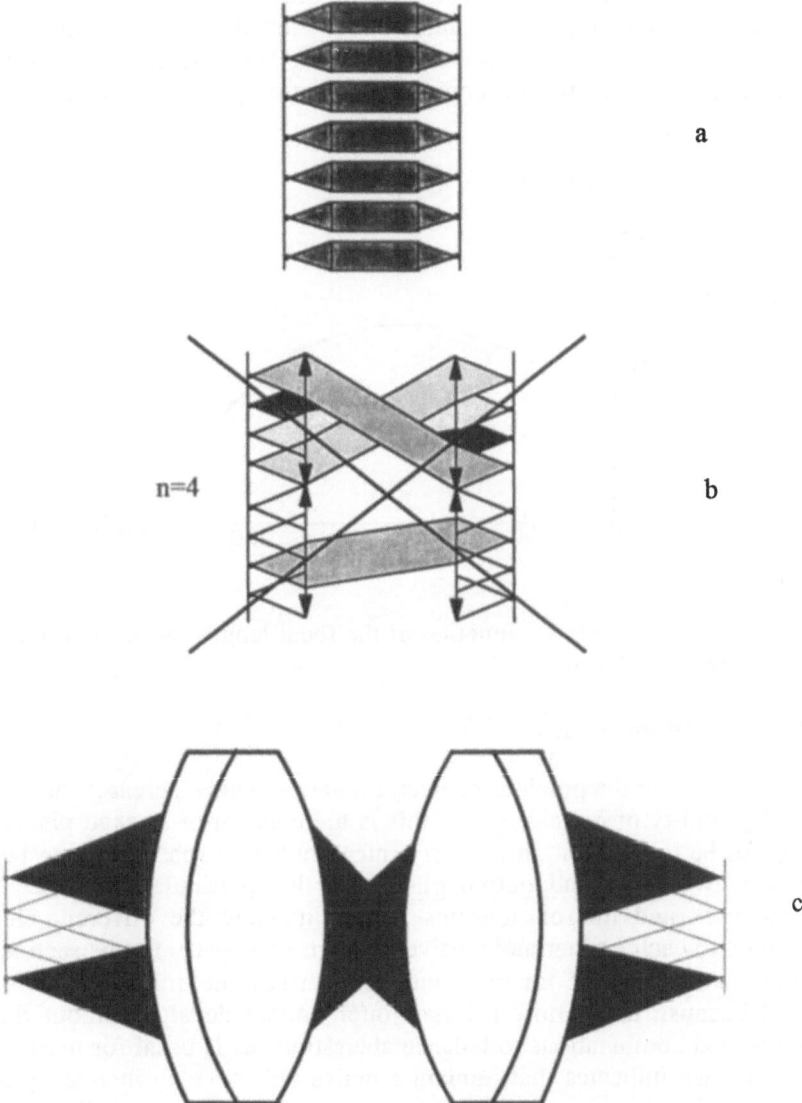

Fig. 3.5. Possible setups for 1 cm^2 chip to chip imaging.

Let us take a precise value for "d": suppose that the laser diode surfaces are 10x10 μm; for a photodiode area of 20x20 μm, an aberration pattern parameter d=10 μm seems a quite good value. As a function of the focal length, figure 3.6 shows the transversal dimension of the object (or image) field according to equation 3.3. Single lenses with large focal lengths are not tolerable because of the spherical aberration. The maximal field is obtained for f about 120 µm and is equal to 13 µm. An immediate consequence is that a configuration with n>1 (see figure 3.5b) is just never applicable to our problem. In conclusion, only two schemes compatible with our hypothesis are available. While the setup with the macroscopic lenses requires a volume of 500 cm^3, the microlens approach needs 1 cm^3. The volume reduction is then about 500. This relevant example indicates a supremacy of microoptics for compact and dense optical interconnects.

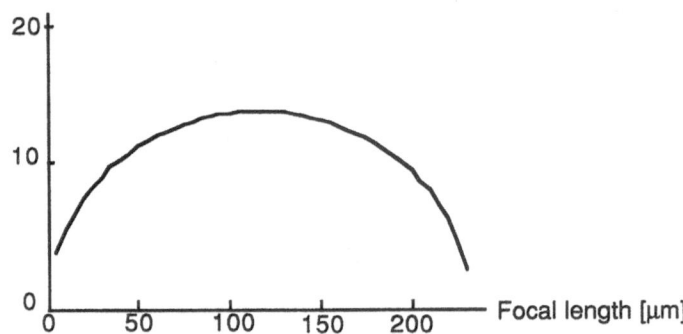

Fig. 3.6. Useful field as a function of the focal length for the setup of figure 3.3 with d=10 µm.

3.2.3 Concluding remarks

As was shown in the previous chapter, miniature optics increases the compactness and alignability of optical setups: this is the concept of stacked planar optics. It may also be used when an arbitrary interconnection scheme has to be achieved between two input and output planes. In this particular case (see Fig. 2.3), microlenses and microholograms locally manage the different channels by assigning to each emitter and receiver its own free space. In this section, we have shown that, in a space invariant interconnect scheme, miniature lenses are also useful because they allow a larger interconnect density without the need of complicated combinations to balance aberrations as is usual for macrolenses. All this evidence indicates that miniature optics will have an increasing role in the future of optical interconnects.

3.3 Refractive microlens arrays

Firstly, we review the few main applications which have triggered the mass production of refractive microlenses. Secondly, we introduce the different parameters useful for a description of microlenses in the paraxial approximation. And finally, we describe the main fabrication methods published and consider in each case the quality achieved. Cylindrical microlenses will not be covered here.

3.3.1 Fields of application

Since 1977, Glaser and coworkers [4] investigated the use of microlens arrays in the context of optical processing, firstly for imaging purposes and later for matrix-vector products. In the meantime, these arrayed components have reached a mature state of development because of applications that are not closely related to optical interconnects. Let us give a few examples:
- The need of compact and inexpensive components for coupling light into fibers has been felt ever since the birth of fiber communications. In this case, microlenses are rarely used in arrays, but mass production implies array fabrication of components that are discretized afterwards.
- Microlens arrays can be viewed as optical relays for use in colour imaging, photocopy or facsimile. Volume reduction of office machines then leads to replace a compound lens by a 1D microlens array whose focal length is in the millimetre range. Figure 3.7 shows a facsimile reading head with a special large size image sensor [5]. An image of the original document is cast onto the CCD sensor by the microlenses (Selfoc lens array in this case), and electrical signals aré produced. The distance between the object and the image is about 20 mm, so the head can be made very compact.

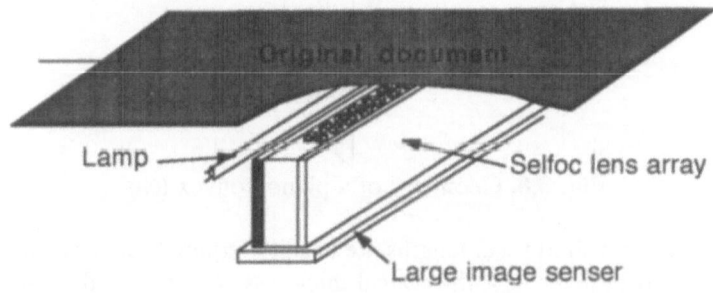

Fig. 3.7. Facsimile reading head.

- The method proposed by Hartmann to test the surface quality of large telescope lenses has known a recent progress with the birth of adaptive optics, correcting turbulent waves in real time.

Although they have been developed for such specific applications, microlens arrays may become valuable components for interconnects as they are now becoming more readily available. Several authors have already tested [6-7] microlens arrays as image multiplexers which duplicate many identical images of an input object. Multichannel correlators using this image duplication and Fourier transformation microlens arrays have also been proposed [7-8] (see also Chap. 12). Recent applications to interconnections may be found in references 28 and 29.

3.3.2. Geometry and parameters of microlenses

Figure 3.8 shows a plano-convex lens of refractive index n_1 immersed in a medium of index n_2. The spherical shape has a radius R, a centre thickness e and an edge thickness equal to zero. With the latter assumption, we neglect the substrate thickness, that can obviously be treated as a plane parallel plate covering the microlens. Among the set of parameters that caracterize the paraxial properties of the lens, we shall mainly use the focal length f and the *f-number* defined as $N_o = f/2r$ (see figure). The lens pupil radius r is a function of R and e: $r^2 = e(2R-e)$ and can be set equal to $(2Re)^{1/2}$ in the paraxial approximation. If we denote by δn the index difference, then it is easy to derive the expressions for:

- the focal length $f = R/\delta n = r^2/(2e\delta n)$ (3.8)

- the f-number $N_o = r/(4e\delta n)$ (3.9)

Extension to lens shapes other than plano-convex is straightforward.

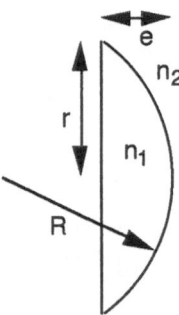

Fig. 3.8. Geometry of a plano-convex lens.

Small diameters and short focal lengths are usually required for microlenses. Given two of the four parameters r, f, the optical thickness $e\delta n$ and N_o, the two others are also fixed by equations 3.8 and 3.9. For most of the techniques used in microlens fabrication, the difficulty is to obtain large optical thickness. Figure 3.9 shows the domain of parameters corresponding to the different fabrication processes discussed in the next section; the black dots correspond to specific components described in the literature. The axes are the diameter 2r and the f-number N_o. The oblique lines show domains of equal focal length and equal optical thickness.

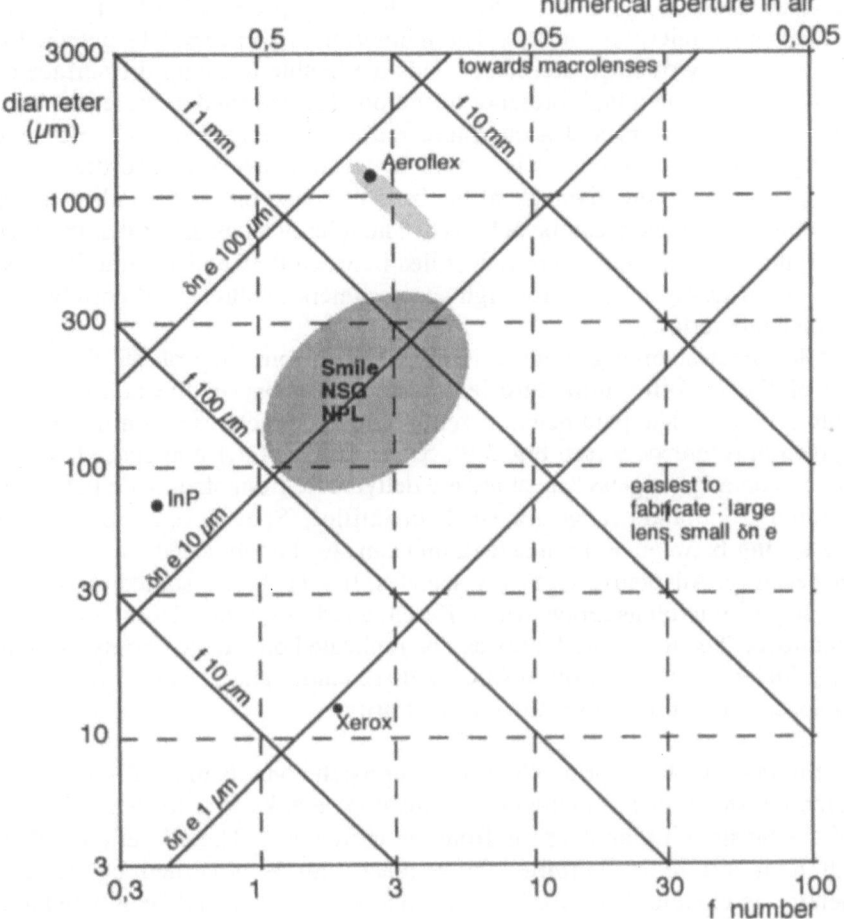

Fig.3.9. Parameter domains corresponding to techniques described in the next section.

The shape of the microlens (circular or square) and the distance between two neighbouring optical axes are two additional important parameters that have an effect upon the density of the useful area. In the case of a square pupil, r denotes the half of the square pupil and N_o is still defined as f/2r.

3.3.3 The different fabrication processes

This section discusses the different fabrication processes proposed in literature. Five different ways of structuring materials are described:
1. mechanical imprint
2. selective ion diffusion
3. chemical etching using wet or dry processes
4. photoresist melting
5. direct exposure of a suitable photosensitive material

Mechanical method. In 1981, a Soviet laboratory proposed [9] a mechanical method to realise microlens arrays. The method uses an accurately polished die containing an array of steel microballs. This die is able to stamp the surface of a plastic material during a high pressure operation. The reported microlenses have a diameter of about 1 mm and a curvature radius of about 5 mm. The distance separating two adjacent lenses is 1 mm at least. However, according to the working pressure and the diameter of the balls mounted on the die, a broad range of geometrical parameters can be achieved. The microlens diameter can vary from 0.05 to 2 mm while the curvature radius lies between 0.25 and 10 mm. This type of microlens arrays were made for high-speed camera, multichannel correlators or holographic memories.

A similar steel stamping method is also used in one step of the fabrication process of the Aeroflex miniature lens arrays [10], one of the commercially available products. The parameters given by the manufacturer are in the range of those given in reference 9 (see Fig. 3.9). 53x53 arrays of 1.1 mm spaced, squared microlenses on a "plexiglass" substrate are deliverable from stock with list price in the hundred US dollars range for small quantities. Specific components with diameters lying between 300 μm and 12 mm can also be fabricated.

More recently, Adaptative Optics Associates, Inc. [11] have started marketing a large variety of microlens arrays with different apertures, array sizes, focal lengths and f-numbers. The miniature lenses can be replicated on a large variety of optical materials including epoxies, fused silica, zinc selenides and plastics. The products are not specified on Fig. 3.9 for the sake of clarity.

Ion exchange. The basic principle of this approach consists in locally modifying the refractive index of an homogeneous material thanks to a local exchange of ions of the substrate for another ion from the molten salt. The migration obtained at high temperature is delimited by a mask and is governed by diffusion parameters and electric field. If the mask window is small enough, compared to the diffusion length, the window can be considered as a hemispherical diffusion source that generates a radial dependency of the index increment. First attempts made by Oikawa in 1981 used plastics [11]. The same year, this approach has been extended to glass substrates [12] where alkali boro-silicate ions are exchanged by heavier ions such as K^+ or Tl^+. The glass plate is immersed during tens of hours in a Tl_2SO_4 or K_2SO_4 melt; this technique rapidly allowed the fabrication of good quality microlenses with a diameter about 1.2 mm and a focal length of 25 mm. Thanks to a precise study of the heavy ions migration process, more recently higher numerical apertures have been demonstrated [13], in particular by the scientists at Nippon Sheet Glass ; commercial products are available [14]. For example, Oikawa and his co-workers at Nippon Sheet Glass (NSG) succeeded in the fabrication of diffraction limited microlenses of f-number equal to 0.52 and of 450 μm focal length; In 1990, thanks to a combination of the exchange effect and a slight inflation of the planar shape, this same group obtained a f-number about 0.83 with, however, a broadening of the point spread function by a factor 1.8 above the diffraction limit. A similar approach using diffusion and volume expansion in PMMA has also been reported recently [15].

InP chemical etching. For laser diode arrays, it is advantageous to fabricate microlens arrays in the same material, e.g. the indium phosphide constituting the exit face of the laser. Processes using ion exchange could be applied to this kind of materials but the refractive index increment would not be sufficient to achieve the numerical aperture needed for capturing all the light from the divergent beam of laser diodes (see Fig. 3.9 InP). However, because of the high refractive index of InP (n is about 3.5 for a wavelength of 1 μm), a small f-number can be achieved without requiring a large curvature radius. Refractive lenses have been formed in InP by wet-chemical etching and by angled ion-beam etching. The latter technique has potential difficulties in accurately controlling the miniature lens profile. The wet chemical etching technique is used in combination with high temperature mass transport. Firstly, repeated application of photolithography and wet etching are applied to an InP substrate so as to form cylindrical symmetric multilevel mesa structures. The result is shown in figure 3.10. The diameter of each successive cylinder is in the ten micrometer range. The smoothing mass transport that operates during several hours at a temperature about 800°C leads to improved quality components [16] with an accurately controlled profile for f-numbers between 0.7 and 1.25. Limitations of this technique come from the considerable challenge for controlling the mass transportation over large surfaces. Consequently, only small lens diameters are achievable. The authors estimate that a diameter of 130 μm is already a difficult task.

Fig. 3.10. Multilevel mesa structure formed by chemical etching. After 16.

Besides, although they are not properly speaking refractive elements, let us mention here the fabrication of binary diffractive lenses [17] on that kind of substrate with a quasi diffraction limited behaviour. This approach requires sophisticated lithographic and etching techniques.

Melting of photoresist materials. Photoresist materials are used at several levels of the integrated circuits technology. In a masking process, photoresist plates are created so as to etch the wafer with binary structures which are as spatially uniform as possible.

In a different way, during the microlens fabrication, the materials are used to provide continuous phase variations according to some specified shape. The Xerox (see figure 3.9) fabrication process described in references 18-19 by Popovic et al. requires several steps:

- step 1 consists in the formation of lens aperture holes. A thick aluminium film is then deposited onto a photoresist plate and patterned with an array of, for example, 15μm diameter holes.

- during step 2, a 1μm thick photoresist film is deposited to produce a pattern of 30 μm diameter circles centred above each hole. Each disk forms a pedestal which will support the next process.

- step 3 consists in a double deposition of an other kind of resist allowing the fabrication of 25 μm diameter and 15 μm height cylinders. The wafer is then exposed to near UV radiation. After development, the cylinders are 12 μm high with a flat top.

- finally, during the last step, the wafer is heated so that the resist melts. Due to surface tension forces, the melting and the resulting flow lead to a spherical microlens surface formed all above the pedestal. Figure 3.11 summarises the successive processes.

Popovic notes that the fabrication of concave microlenses through the design of cylindrical holes instead of cylindrical tumulus would be possible. The practical diameter of these lenses is 15 μm and the focal length is 36 μm ± 0.5 μm. They are diffraction limited components (Airy pattern size of 1.2 μm). Figure 3.12 represents a scanning electron micrograph of the microlens array fabricated by this process demonstrating the good quality of the smoothness and the sphericity achieved.

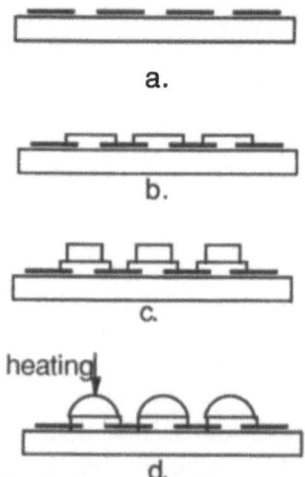

a.

b.

c.

heating

d.

Fig. 3.11. Processing steps for the fabrication of microlenses. After 18.

Fig. 3.12. After references 18 and 19. Microlens array. (photograph provided by the courtesy of Z.D. Popovic, Xerox Canada)

Hutley and his co-workers from the National Physical Laboratory (Teddington UK) have applied photoresist techniques [20] to the manufacture of microlenses of various shapes and sizes. Numerous quite similar, simple and effective methods were adopted in their different experiments; in each case, the crucial factor is that the resist is deeply developed down to the substrate in order to obtain spatially separated islands.

The first method is quite simple: a uniform resist is directly exposed through a contact printed mask marking the boundaries of the microcomponents. After exposure and development, they obtain a binary structure. A simple melting is then applied.

The second method is an improvement of the latter. Before melting, several masks are used to fabricate multilevel structures which can be understood as a pre-forming operation of the resist. The melting can then be considered as a smoothing operation. These two methods have been applied to generate 1D arrays of cylindrical lenses and 2D square microlenses.

The third method is an ingenious technique that allows the fabrication of close packed hexagonal arrays of microlenses based on three-wave interference recording. Extensive details may be found in reference 21.

Microlens diameters given by the researchers of the NPL lie between 60 μm (for f=110 μm) and 630 μm (for f=630 μm) in the case of square arrays, and between 200 and 500 μm for hexagonal microlenses. The latter have focal lengths varying from 400 to 1000 μm corresponding to numerical apertures between 0.3 and 0.5.

Similar results can be directly obtained with classical photolithography techniques. G. Artzner at CNRS in France has succeeded [22] in the tuning of a rigorous calibration of the photoresist sensitivity so as to obtain the expected profile after development. There is no need of melting in this multi-exposure method. A similar technique has also been developed by Eisenberg [23]. Reactive ion etching transfer of a photoresist microlens has been reported recently [30].

Other techniques. Several other techniques for fabricating microlens arrays have been published in the literature. Let us note the method proposed by Philips Research Laboratories [24] which make graded index half sphere microlenses integrated into a glass substrate. The lens material consists of Si_3N_4/SiO_2 which is deposited by a plasma-activated chemical vapour deposition technique.

Another process, developed by Corning and named "SMILE", produces 2D arrays of microlenses in a homogeneous photosensitive glass by a photothermal monolithic technique. The mechanism of the lens formation is based on the density change of the photonucleated microcrystalline phase developed relative to the unexposed glass. The regions surrounding the lens are rendered opaque without requiring any other treatment. Commercial products are available.

3.3.4 Concluding remarks

The large variety of techniques described above shows on one side the complexity of the purpose since no process seems preferable to cover the complete field of requirements, and on the other side the feeling of an increasing need. Moreover, the combination of refractive and diffractive components, not discussed above, may still extend the perspectives of miniature optical interconnects. However, the large number of requirements implies a large dispersion for the parameters. Consequently, the mass production of widely commercially available microlens arrays is still hardly in sight; among the set of components mentioned above, few are manufactured. Most of them have been designed by authors for their own use.

Microlens array is a fast moving area of research and development. Reference 31 is a recent conference digest containing indication about the latest progress on the fabrication, testing and applications of microlenses.

Some information about available products has been introduced in this section for the convenience of the reader. It is based on information available to the authors at the time of printing and may be incomplete. Manufacturers for which no indication of the fabrication process was known to the authors have not been mentionned. The authors thank L. Prod'homme for his contribution to the starting phase of this work and I. Glaser for his comments.

References

1. Born, M., Wolf, E.: Principles of Optics. Pergamon Press, Oxford, fifth edition 1975.
2. Goodmann, J.W.: Introduction to Fourier Optics. Wiley & Sons Editors 1976.
3. Chrétien, H.: Calcul des combinaisons optiques. Edition Masson, sections number 627 and 628, 1959, pp 588-590.
4. Glaser, I., Friesem, A.A.:in Applications of Holography and Optical Data Processing, E. Marom ed., Pergamon Press, 1977, p 467.
5. Toyama, M., Takami, M.: Luminous intensity of a gradient-index lens array. Appl. Opt. 21, 1982, pp 1013-1016.
6. Akiba, A., Iga, K.: Image multiplexer using a planar microlens array. Appl. Opt. 29, 1990, pp 4092-97.
7. Hamanaka, K., Nemoto, H., Oikawa, M., Okuda, E., Kishimoto, T.: Multiple imaging and multiple Fourier transformation using planar microlens arrays. Appl. Opt. 29, 1990, pp 4064-70.
8. Agu, M., Akiba, A., Mochizuki, T., Kamemaru, S.I.: Multimatched filtering using a microlens array for an optical-neural pattern recognition system. Appl. Opt. 29, 1990, pp 4087-91.
9. Ozerov, I.N., Petrov, V.M., Shishkina, V.A., Shor, V.M.: Shaping the contours of dies for manufacturing lens arrays having spherical elements. Sov. Journ. Opt. Technol. 48, 1981, pp 49-50.
10. Product Book: "MRP Miniature lens array, series MRP-110", Aeroflex Lab. Inc., Photo optical science division, 35 south service road, NY 11803.
11. Oikawa, M., Iga, K., Sanada, T.: A distributed index planar microlens made of plastics. Jap. J. Appl. Phys. 20, 1981, pp 51-54.
12. Oikawa, M., Iga, K.: Distributed index planar microlens. Appl. Opt. 21, 1982 pp 1052-56.
13. Oikawa, M., Okuda, E., Hamanaka, K., Nemoto, H.: Integrated planar microlens and its application. SPIE Vol. 898 Miniature Optics and Lasers, 1988, pp 3-11. Oikawa, M., Nemoto, H., Hamanaka, K., Okuda, E,: High numerical aperture planar microlens with swelled structure. Appl. Opt. 29, 1990, pp 4077-80.
14. Nippon Sheet Glass Co, Shimbashi Sumitomo Bldg. 11-3 Shimbashi 5 Chome, Minato-Ku, Tokyo, Japan.
15. Franck, M., Kufner, M., Kufner, S. and Testorf, M.: Microlenses in polymethyl methacrylate with high relative aperture. Appl. Opt. 30, 1991 pp 2666-67.
16. Liau, Z.L., Diadiuk, V., Walpole, J.N., Mull, D.E.: Large numerical aperture InP lenslets by mass transport. Appl. Phys. Lett. 52, 1988, pp1859-61.
17. Yap, D., Liau, Z.L., Walpole, J.N., Diadiuk, V.: Fabrication of miniature lenses and mirrors for InGaAsP/InP lasers. SPIE Vol. 898 Miniature Optics and Lasers, 1988, pp18-22.

18. Popovic, Z.D., Sprague, R.A., Connell, G.A.N.: A Process for Monolithic Fabrication of Microlenses on Integrated Circuits. SPIE Vol. 898 Miniature Optics and Lasers, 1988, pp23-25.
19. Popovic, Z.D., Sprague, R.A., Connell, G.A.N.: Technique for monolithic fabrication of microlens arrays. Appl. Opt. 27, 1988, pp 1281-86.
20. Daly, D., Stevens, R.F., Hutley M.C., Davies, N.: The manufacture of microlenses by melting photoresist. Meas. Sci. Technol. 1, 1990, pp 759-766.
21. Hutley, M.C.: Optical techniques for the generation of microlens arrays. Jour. of Mod. Opt. 37, 1990, pp 253-265.
22. Artzner, G.: Microlens arrays for Shack-Hartmann Wavefront Sensors. Opt. Eng.31, pp 1311-1322 (1992).
23. Eisenberg, N., Abitbol, M.: A New Process for Manufacturing Arrays of Microlenses. Sixth Meeting in Israel on Optical Engineering, proc. SPIE Vol. 1038, 1988.
24. Küppers, D., Schelhas, K.H., Biermann, U., Khoe G.D., Kock, H.G.: Microlenses prepared by the plasma-actived chemical vapor deposition technique. Advances in Low-Temperature Chemistry, Technology, Application, Ed. H.V. Boenig, vol.1, 1984, pp 92-107.
25. Borrelli, N.F., Morse, D.L., Bellman R.H., Morgan, W.L.: Photolythic technique for producing microlenses in photosensitive glass. Appl. Opt. 24, 1985, pp 2520-2525. Information on products can be obtained from Corning Incorporated Advanced Materials, MP 21-3-5, Corning, New York 14831.
26. Catalog: "Micro-optics", United Technology, Adaptative Associates, 54 CambridgePark Drive, Cambridge, MA 02140-1348.
27. Lohmann, A.W.: Image formation of dilute arrays for optical information processing. Opt. Communications 86, 1991, pp 365-370.
28. M.C. Hutley, P. Savander, M. Schrader : the use of microlenses for making spatially variant optical interconnections, J. Eur. Opt. Soc. A, 2, 1992, pp 337-340.
29. M.R. Taghizadeh, B. Robertson, P. Blair, K.J. Hughes, N. Ross: Applications of refractive lenslet arrays within an optical crossbar, Topical Meeting on Microlens Arrays, Teddington, May 1993, EOS Topical Meetings Digest Series, vol 2, p 92, 1993.
30. M. Eisner, S. Haselbeck, H. Schreiber, J. Schwider: Reactive ion etching of microlens arrays into fused silica, Topical Meeting on Microlens Arrays, Teddington, May 1993, EOS Topical Meetings Digest Series, vol 2, p 17-19, 1993.
31. Meeting Digest, Topical Meeting on Microlens Arrays, Teddington, May 1993, organised by the Institute of Physics and the European Optical Society, EOS Topical Meetings Digest Series, vol 2, p 17-19, 1993.

4 Diffractive components: holographic optical elements

Hans Peter Herzig and René Dändliker
Institute of Microtechnology, University of Neuchâtel, Neuchâtel, Switzerland

Besides refractive and reflective components, planar diffractive optical elements are promising for free space optical interconnects. The content of this chapter is focused on holographic optical elements (HOEs), which are diffractive elements fabricated by recording the interference pattern of an object and a reference wave in a photosensitive emulsion.

The first part introduces the basic interconnection components and the specific design and fabrication problems to be solved. In the second part, successful realizations of different types of interconnects, such as compact lenslet arrays and fan-out elements, will be presented.

4.1 Introduction

Holographic optical elements (HOEs) are diffractive structures which are fabricated by recording the interference pattern of an object beam and a reference beam in a photosensitive material. Compared with conventional refractive and reflective optics, they are thinner, lighter, and they can perform multiple functions such as beam splitting and focusing simultaneously. This chapter reports on the design and fabrication of HOEs for free-space optical interconnects.

The role of optical interconnects varies from one-to-one connections (e.g. routing and perfect shuffle) to more complex one-to-many connections (fan-out). Figure 4.1 shows three different types of interconnection elements. The multifacet element consists of an array of separated independent elements, where each elements connects one object point with one image point. The elements can be identical, as for lenslet arrays, or they can be different, as for perfect shuffle elements. The second type is a common facet fan-out element, which connects one or several object points with several image points. The imaging element acts as a lens, i.e. each object point is related to one image point by the same optical transfer function. These basic interconnection components can work in transmission or in reflection, and they can be combined in order to form more complex systems. In this chapter, we present the design and fabrication of HOEs for fixed optical interconnects.

Multifacet element   Fan-out element   Imaging element

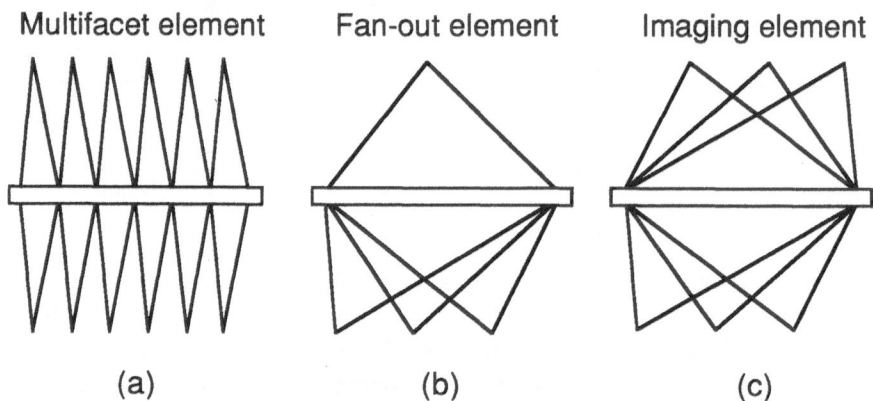

(a)   (b)   (c)

Fig. 4.1. Interconnection components: (a) Multifacet element, (b) fan-out element, (c) imaging element.

A basic element in holographic interconnects is a hologram that is recorded with two spherical waves, one diverging from a point source and the other converging. If the hologram is now replayed with the diverging wave, the converging wave will be reconstructed perfectly according to the holographic principle. The HOE maps the object point onto the image point, acting as deflecting and focusing element. This simple recording method fails if the holographic material is not sensitive at the readout wavelength, as it is the case for semiconductor laser sources. Aberrations also occur in the case of imaging HOEs, where different object points have to mapped by the same HOE function. Besides the recording of optimized single facet holograms, we are interested in systems of large HOE arrays. Each element may have a different optical function and has to be aligned precisely with respect to the others. Key elements for parallel interconnects are also the fan-out elements, which can be realized by recording a single HOE with multiple beams or by sequential superposition of HOEs. In both cases, unwanted interactions between the beams may significantly reduce the diffraction efficiency and the readout fidelity of the object points.

The review starts with a brief introduction to holographic optical elements (Sect. 4.2) followed by a summary of the standard holographic materials suitable for fixed optical interconnects (Sect. 4.3). In any hologram, the wave conversion process can be separated into two parts, one describing the change of wave propagation by diffraction, the other accounting for efficiency. Section 4.4 deals with the wavefront conversion. The design and analysis of HOEs will be discussed. The diffraction efficiency will then be subject of Sect. 4.5. Of special interest is the multiple beam recording of highly efficient fan-out elements. Finally, in Sect. 4.6, the different recording methods for optimized single-facet HOEs, arrays of HOEs, and fan-out HOEs are presented and supported with experimental results.

4.2 Types of holographic optical elements

In holography, we can distinguish between amplitude holograms, thin phase holograms, and volume holograms (thick phase holograms). Most attractive for HOEs are volume holograms because they achieve the highest diffraction efficiency (close to 100 %), whereas the efficiency of amplitude holograms is below 10 %. On the other hand, thin amplitude holograms, such as computer-generated holograms [1], are very helpful in testing of optical components (Chap. 6), or in the fabrication process as wavefront generators for aspheric waves (Sect. 4.6.1). Thin phase holograms generate several diffraction orders, thus they are interesting as fan-out elements. For example, a simple sinusoidal grating can generate a uniform fan-out of N = 3 beams with an efficiency close to 90 %. More complex on-axis structures could also be generated by other means than interferometrically, e.g., by copying gray level masks [2] or by laser beam scanning [3], see also Chap. 5.

In the following sections, we will present some of the principle characteristics of thin and thick holograms. For further reading we refer to the literature [4,5].

4.2.1 Thin holograms

A hologram is fabricated by recording the interference pattern of an object wave $A_O(x,y) \exp\{i\Phi_O(x,y)\}$ with a reference wave $A_R(x,y) \exp\{i\Phi_R(x,y)\}$ in a photosensitive material, as shown in Fig. 4.2. The exposure irradiance I in the hologram plane is given by

$$I(x,y) = |A_O e^{i\Phi_O} + A_R e^{i\Phi_R}|^2 = A_O{}^2 + A_R{}^2 + 2A_O A_R \cos(\Phi_O - \Phi_R). \qquad (4.1)$$

Depending on the holographic material and the processing, the exposure pattern is transformed into a modulation of the absorption (amplitude hologram), of the refractive index, or of the layer thickness (both phase holograms).

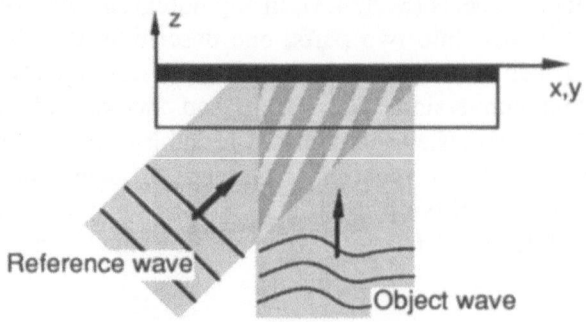

Fig. 4.2. Interferometric recording geometry.

If we assume linear recording, the transmission of the processed amplitude holograms becomes

$$t(x,y) = t_0 + t_1 \cos(\Phi_O - \Phi_R), \tag{4.2}$$

whereas phase holograms are described by

$$t(x,y) = \exp\left\{ i(\psi_0 + \psi_1 \cos(\Phi_O - \Phi_R)) \right\}. \tag{4.3}$$

In most cases of interest, object and reference wave are plane, spherical, or aspherical with slowly varying amplitudes $A_i(x,y)$. Therefore, the terms t_i and ψ_i in Eqs. (4.2) and (4.3), respectively, are roughly constant.

Let us consider the readout of a thin phase grating in more detail. According to the transparency theory [5], the electric field E_{out} at the hologram output is obtained by multiplying the incident field $E_{in} = A_{in}(x,y) \exp\{i\Phi_{in}(x,y)\}$ with the complex amplitude transmittance $t(x,y)$ of the film, given by Eq. (4.3):

$$E_{out} = A_{in}(x,y) \exp\{i\Phi_{in}(x,y)\} \exp\left\{ i\psi_1 \cos(\Phi_O - \Phi_R) \right\}. \tag{4.4}$$

The term $\exp\{i\psi_0\}$ has been neglected. If we develop this electric field in its space harmonics, we get

$$E_{out}(x,y) = A_{in}(x,y) \exp\left\{ i\Phi_{in}(x,y) \right\} \sum_{m=-\infty}^{\infty} i^m J_m(\psi_1) \exp\left\{ im(\Phi_O - \Phi_R) \right\}, \tag{4.5}$$

where m is the diffraction order, and J_m is the m-th order Bessel function. Thus, the amplitudes of the outgoing orders are determined by the Bessel functions, namely

$$A_{out}^m = A_{in} J_m(\psi_1), \tag{4.6}$$

whereas the phases are simply described by

$$\Phi_{out}^m = \Phi_{in} + m \, (\Phi_O - \Phi_R) \, . \tag{4.7}$$

If the hologram is illuminated by the reference wave, i.e. $\Phi_{in} = \Phi_R$, then the object wave is reconstructed in the first diffraction order ($m = 1$).

In general, several diffraction orders are generated. Equation (4.7) describes the relations between the wavefronts, but not the energy in the orders. If the hologram is not a pure phase element, or if the variation of the refractive index is periodic but not sinusoidal, e.g., if the response of the holographic material is not linear, then the light distribution in the different orders [Eq. (4.6)] may change considerably.

4.2.2 Thick holograms

If the thickness of the holographic medium is large compared with the fringe spacing, volume effects become important.

The distinction between thick and thin holograms (see also the section 13.2 of the chapter on interconnects with optically thick elements) is usually made with the aid of the parameter Q defined as

$$Q = \frac{2\pi\lambda d}{n\Lambda^2} \, , \tag{4.8}$$

where λ is the readout wavelength, n the refractive index of the recording material, d the thickness, and Λ the fringe spacing. When $Q > 10$, the hologram is considered to be thick. Otherwise, it is considered to be thin [6]. This criteria fails for strongly modulated holograms. In that case more accurate criteria are necessary [7].

The basic principle of holography are still the same as mentioned before in Sect. 4.2.1. The main difference is that volume holograms are selective and reconstruct, beside the zero-order, only one diffraction order. As a consequence, Eq. (4.6) is no longer valid. On the other hand, the transfer of the wavefront into the reconstructed order, typically $m = 1$, is still well described by Eq. (4.7).

4.3 Holographic materials for fixed interconnects

The ideal recording material for optical interconnects should have a spectral sensitivity well matched to available laser wavelengths, a linear transfer characteristics, high resolution and low noise. Candidates for highly efficient holographic materials are: bleached photographic emulsions with reasonable sensitivity at 400...700 nm, dichromated gelatin at 350...580 nm, photopolymers at UV...650 nm and photoresists at UV...500 nm [4].

Silver halide emulsions are widely used because of their high sensitivity and because they are commercially available. On the other hand, the efficiency is limited by scattering and absorption in the material.

Dichromated gelatin (DCG) is an ideal recording material for volume HOEs with the capability of large refractive index modulation, high resolution, low absorption and scattering. Efficiencies close to 100 % are possible. On the other hand, the wet process must be carefully controlled in order to get reproducible results.

Promising results for photopolymers show that they might become competitive with DCG [8], although the refractive index modulation is smaller than for DCG. Of great advantage is the dry processing.

High efficiencies (> 85 %) can also be achieved with surface relief gratings in photoresist. Relief structures are suitable for mass production. They can be replicated by embossing. Interferometrically recorded photoresist patterns can also be used as masks, which are then transformed into other materials by etching techniques (Chap. 5).

Unfortunately none of these materials is sensitive in the near IR for use with standard semiconductor lasers (700-1000 nm). Recording HOEs in the visible and reconstructing them in the IR requires careful control of all wavelength dependent parameters, such as Bragg angle, focal length and aberrations [9].

Photorefractive materials are used for active interconnect components. They will be described in Chap. 9.

4.4 Design of holographic optical elements

In any hologram, the wave conversion process can be separated into two parts, one describing the change of wave propagation by diffraction, the other accounting for the efficiency. This section deals with the wavefront conversion. If a HOE is recorded and replayed under the same conditions, then the generated wave has no aberrations. In many cases this is not possible. One reason for aberrations is the wavelength shift between the recording and readout due to the spectral sensitivity of the holographic material. In another case, different object points have to mapped by the same HOE function, e.g. imaging. The HOE design includes the calculation of the desired HOE phase structure and the analysis of its optical properties at the operating wavelength.

4.4.1 The hologram phase function

A hologram that is illuminated by an input wave $\Phi_{in}(x,y)$ generates an output wave $\Phi_{out}(x,y)$. According to Eq. (4.7), the wavefront conversion is described by

$$\Phi_{out}(x,y) - \Phi_{in}(x,y) = \Phi_O(x,y) - \Phi_R(x,y) \equiv \Phi_H(x,y). \tag{4.9}$$

The phase function $\Phi_H(x,y)$ of the hologram is defined by the wave fronts that are used to construct the HOE, namely $\Phi_O(x,y)$ the phase distribution of the object wave and $\Phi_R(x,y)$ the phase distribution of the reference wave in the hologram plane (x,y). Note, that we have assumed that only the first diffraction order is relevant.

In the case of a focusing HOE in the IR, e.g. at $\lambda_{in} = 780$ nm, for readout with a spherical wave Φ_{in}, the desired HOE phase function Φ_H is determined by Eq. (4.9), where both phase functions Φ_{out} and Φ_{in} are of the form

$$\Phi_i = \pm \frac{2\pi}{\lambda} \sqrt{(x-x_i)^2 + (y-y_i)^2 + (z_i)^2}. \tag{4.10}$$

The point (x_i, y_i, z_i) is the source or the focus of the spherical wave.

If only spherical waves are available for the recording at the wavelength λ_R different from λ_{in}, Eq. (4.9) cannot be fulfilled exactly. Thus, aberrations occur. For a fixed aperture the geometrical aberrations are proportional to the focal length. Compact systems require elements with short focal length. Consequently also the aberrations are small. In that case, an optimized recording with spherical waves is sufficient. A perfect phase function Φ_H can be realized by using aspheric waves (see Sect. 4.6.1).

4.4.2 Astigmatism and Bragg condition

Off-axis, focusing HOEs are the basic elements for one-to-one connections. Analytical approximations for beam deflection, focal length, and astigmatism of focusing HOEs are very helpful as guidelines for the design [10]. In the case of recording and reconstruction of point sources in a common plane of incidence (Fig. 4.3), the angles are determined by the grating equation

$$\sin\theta_{out} = \sin\theta_{in} + \mu \left(\sin\theta_O - \sin\theta_R \right). \tag{4.11}$$

For the radii of curvature ρ_i, we get

$$\frac{\cos^2\theta_{out}}{\rho_{out}^{\parallel}} = \frac{\cos^2\theta_{in}}{\rho_{in}} + \mu \left(\frac{\cos^2\theta_O}{\rho_O} - \frac{\cos^2\theta_R}{\rho_R} \right), \tag{4.12}$$

$$\frac{1}{\rho_{out}^{\perp}} = \frac{1}{\rho_{in}} + \mu \left(\frac{1}{\rho_O} - \frac{1}{\rho_R} \right). \tag{4.13}$$

θ_i are the angles of incidence in the air (refractive index n = 1) and $\mu = \lambda_{in}/\lambda_R$ is the wavelength ratio. The indices stand for the recording reference beam (R), the object beam (O), the readout beam (in), and the reconstructed astigmatic ray pencil (out) with the principle radii of curvature $\rho_{out}^{\parallel}$ parallel and $\rho_{out}^{\perp}$ perpendicular to the plane of incidence. Wavefronts without astigmatism require that $\rho_{out}^{\parallel} = \rho_{out}^{\perp}$.

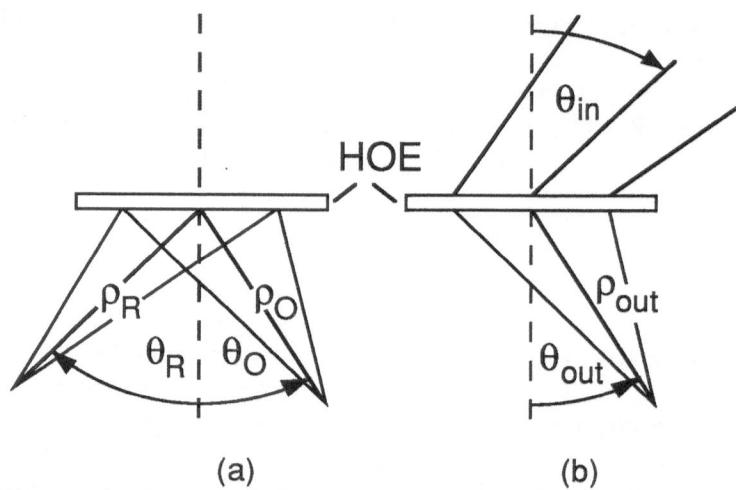

(a)   (b)

Fig. 4.3. (a) HOE recording geometry, (b) HOE readout geometry.

Up to now, we have considered the aberrations but not the efficiency. To get high diffraction efficiency from thick volume holograms into the first diffraction order, the Bragg condition has to be satisfied. In a holographic emulsion with a refractive index n, the wavevectors are related by

$$\mathbf{k}_{out} - \mathbf{k}_{in} = \mathbf{k}_O - \mathbf{k}_R, \tag{4.14}$$

with $|\mathbf{k}_{in}| = 2\pi n/\lambda_i$

If now a readout geometry for a focusing HOE is determined by the angles θ_i and radii of curvature ρ_i, we can find from Eqs. (4.11-14) the recording parameters for a readout without astigmatism ($\rho_{out}^{\parallel} = \rho_{out}^{\perp}$) and fulfilling the Bragg condition simultaneously.

For larger apertures and focal lengths, second order aberrations do not describe the HOE sufficiently well. Higher order aberrations become dominant. They can be analyzed by ray tracing or by using scalar diffraction theory (Sect. 4.4.4).

4.4.3 Optimum design for imaging elements

In the case of imaging HOEs and holographic Fourier lenses, a set of different input waves have to be transformed into another set of output waves by the same optical transfer function [Fig. 4.1(c)]. There is no interactions between the different waves as in the case of fan-out elements. The optimum design of such elements has been reported by different authors [9,11,12]. A summary of the design methods will be presented in Chap. 5.

4.4.4 Analysis of holographic optical elements

The performance of HOEs can be analyzed by ray tracing or by scalar diffraction theory. Ray tracing yields the geometrical aberrations but not the exact light distribution in the image or focal plane. Diffraction by a finite aperture or a locally varying illumination of the HOE is not taken into account. A more accurate, but also more time consuming method, uses the scalar diffraction theory.

Ray tracing. The aberration properties of HOEs can be analyzed by geometrical optics using ray tracing [13], which is very common in conventional lens design. In ray tracing through lens systems the path of the light is determined with the help of elementary geometry by successive application of the law of refraction (or reflection). In holography, the law of refraction has to be replaced by the law of grating diffraction.

The holographic recording and reconstruction process is essentially governed by the condition of phase matching in the hologram plane (x,y), which is given by Eq. (4.7). The phase matching condition in the hologram plane yields relations for the normal projections k_{iH} of the wavevectors k_i (i = out, in, O, R) onto that plane, namely

$$k_{outH} = k_{inH} + m [k_{OH} - k_{RH}], \tag{4.15}$$

where the phase functions Φ_i and the vectors k_{iH} are related by $k_{iH}(x,y) = \text{grad}[\Phi_i(x,y)]$.

The length of the wavevectors at the reconstructing wavelength is given by $|k| = |k_{in}| = 2\pi/\lambda_{in}$. For the component k_{out}^z of the outgoing wave normal to the hologram plane follows for a transmission HOE

$$k_{out}^z = \text{sign}(k_{in}^z) [|k_{in}|^2 - |k_{outH}|^2]^{1/2}, \tag{4.16}$$

where $\text{sign}(k_{in}^z)$ denotes the sign of k_{in}^z. For reflection holograms, $\text{sign}(k_{in}^z)$ in Eq. (4.16) has to be replaced by $-\text{sign}(k_{in}^z)$.

Equations (4.15) and (4.16) describe the grating diffraction, which allows the tracing of a bundle of finite rays through the holographic component. The results are usually presented as spot diagrams, which are the points of intersection of the calculated rays with the image plane.

Wave propagation calculated by scalar diffraction theory. The scalar diffraction theory is based on wave optics [14]. According to Rayleigh-Sommerfeld the complex amplitude $U(x',y',z_0)$ in the "image" plane is given by

$$U(x',y',z_0) = \frac{1}{i\lambda} \int_S U(x,y) \frac{e^{ikr}}{r} \cos(\mathbf{n},\mathbf{r}) \, dxdy, \qquad (4.17)$$

where $U(x,y)$ is the complex amplitude in the hologram plane, $\mathbf{r}$ is the vector between the points (x,y) and (x',y',z_0), $r = |\mathbf{r}|$ and $(\mathbf{n},\mathbf{r})$ is the angle between the normal $\mathbf{n}$ of the HOE plane and the vector $\mathbf{r}$.

The complex amplitude $U(x,y)$ in the HOE plane is given by

$$U(x,y) = A(x,y) \exp[i\Phi_{out}(x,y)], \qquad (4.18)$$

where $A(x,y)$ is the amplitude of the illumination within the pupil and $\Phi_{out}(x,y)$ is obtained through Eq. (4.9) from the HOE phase function Φ_H and the readout wave Φ_r. In the case of homogeneous illumination, $A(x,y)$ is equal to 1 inside the pupil and 0 outside. $A(x,y)$ can also describe inhomogeneous illumination, such as readout with a gaussian beam.

Equation (4.17) allows an accurate calculation of the diffraction pattern. For specific problems, the mathematical manipulations can be simplified by using Fresnel and Fraunhofer approximations [14]. For the computing they take profit of the powerful Fast Fourier Transform (FFT) algorithms.

4.5 Diffraction efficiency of volume holograms recorded with one or multiple object beams

The diffraction efficiency of volume holograms is mostly analyzed by using coupled wave theory. This method was first applied to holography by Kogelnik [6]. His two wave model leads to simple analytic expressions for the efficiency, which are still a valid guideline for the understanding of volume holograms. In the meantime many papers have been published to improve the accuracy and to widen the scope of its application. The actual state of the art is well described in Syms' book [5].

Interesting for realizing fan-out elements [Fig. 4.1(b)] is the recording of HOEs with N object beams. Different models are reported in the literature [5,15,16], but they do not describe sufficiently well regular fan-out elements. We will show that the relative phases of the object beams are important parameters at recording, at least if the thickness of the hologram is less than about 50 wavelengths. The recording conditions are optimum if the N object beams generate a uniform irradiance in the hologram plane. We distinguish between simultaneous recording (Sect. 4.5.2) and sequential recording (Sect. 4.5.3).

4.5.1 Diffraction efficiency of single volume gratings

In most cases, Kogelnik's coupled wave model describes the diffraction efficiency of volume gratings sufficiently well. This theory assumes that only the zero and the first diffraction order are significant. The diffraction efficiency η of a transmission HOE is then given by

$$\eta = \frac{\sin^2\left(\nu d\sqrt{1 + (\Delta k_z/2\nu)^2}\right)}{1 + (\Delta k_z/2\nu)^2},$$ (4.19)

with

$$\nu_s = \frac{\pi\,\Delta n}{\lambda\,\sqrt{\cos\theta_{out}\,\cos\theta_{in}}}, \qquad \nu_p = \nu_s\cos(\theta_{out} - \theta_{in})\,,$$ (4.20)

where ν_s and ν_p hold for the s- and p-polarization, respectively. The angles with respect to the surface normal are θ_{in} for the zero and θ_{out} for the first diffraction order. Δk_z is the mismatch of the Bragg condition in the z-direction. If the Bragg condition [Eq. (4.14)] is fulfilled, i.e. $\Delta k_z = 0$, the diffraction efficiency is highest. The maximum, ideally $\eta = 1$, is achieved for $\nu d = \pi/2$. For $\Delta k_z \neq 0$ the diffraction efficiency is reduced. For thinner emulsions larger Δk_z can be accepted, but this requires larger Δn for the same maximum efficiency.

The corresponding equations for reflection HOEs have also been derived by Kogelnik [6]. The analytic expressions for the diffraction efficiency can easily be incorporated into the HOE design. For focusing HOE, the phase mismatch Δk_z can be calculated by ray tracing from Eqs. (4.14-16), which leads to an estimation of the overall efficiency.

4.5.2 Simultaneous recording of N object waves

In the case of simultaneous recording of a fan-out element, N object waves $E_i(x) = A_i \exp\{-jk_i x + j\phi_i\}$ and one reference wave $E_0(x) = A_0 \exp\{-jk_0 x + j\phi_0\}$ are present at the same time, as shown in Fig. 4.4. The waves are characterized by their wavevectors $\mathbf{k}_i = (k_i{}^y, k_i{}^z)$, amplitudes A_i, and phases ϕ_i, where $\mathbf{x} = (y,z)$ are the coordinates. Only the s polarization is considered.

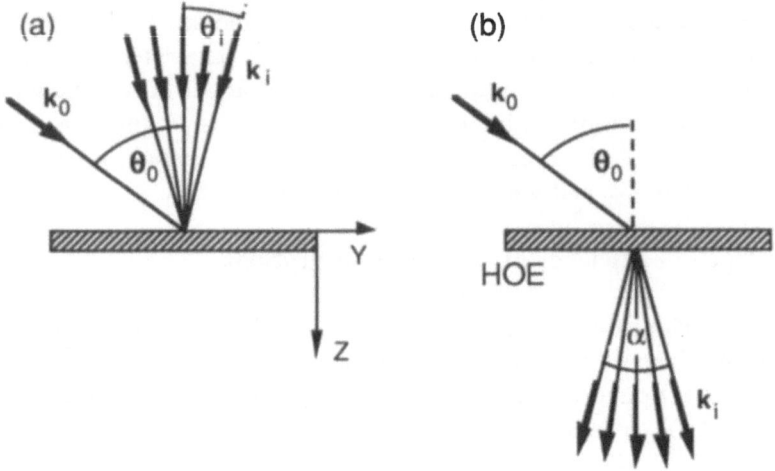

Fig. 4.4. (a) Recording and (b) readout of fan-out elements. The angles are defined inside the recording medium with refractive index $n = \sqrt{\varepsilon_a}$, α is the full fan-out angle.

We assume, that the recording material responds linearly to the accumulated energy during exposure, i.e.

$$\varepsilon(y,z) = \varepsilon_a + \delta \left| \sum_{i=1}^{N} E_i + E_0 \right|^2 . \tag{4.21}$$

The dielectric permittivity $\varepsilon(y,z)$ after exposure becomes then

$$\varepsilon(y,z) = \varepsilon_a' + \sum_{i=1}^{N} \Delta\varepsilon_{0i} \cos(\mathbf{K}_{0i}\,\mathbf{x} - \Phi_{0i}) + \sum_{q>p=1}^{N} \Delta\varepsilon_{pq} \cos(\mathbf{K}_{pq}\,\mathbf{x} - \Phi_{pq}), \tag{4.22}$$

where $\Phi_{pq} = \phi_p - \phi_q$ and $\Delta\varepsilon_{pq} = 2\delta A_p A_q$. Besides the desired N primary gratings $\mathbf{K}_{0i} = \mathbf{k}_0 - \mathbf{k}_i$, $N(N-1)/2$ unwanted intermodulation gratings $\mathbf{K}_{pq} = \mathbf{k}_p - \mathbf{k}_q$ are recorded. At readout, they generate intermodulation waves, which are coupled with the desired reconstructed beams. As a consequence, efficiency and uniformity of the fan-out suffer.

Minimized intermodulations. The intermodulations are reduced with respect to the primary gratings, if we increase the reference-to-object beam ratio B [16], defined by

$$B = \frac{A_0^2}{\sum\limits_{i=1}^{N} A_i^2}.$$

(4.23)

Then, the modulations of the desired primary gratings become dominant ($\Delta\varepsilon_{0i} \gg \Delta\varepsilon_{pq}$). Unfortunately this method requires a high dynamic range of the recording material.

Better results are achieved by optimizing the phases Φ_{pq} of the recorded gratings. For regular fan-out elements the intermodulation gratings can be eliminated nearly perfectly [17]. In the case of regular elements, the projections K_{pq}^{y} of the grating vector $\mathbf{K}_{pq}$ in the hologram plane (y-axis) are all integer multiples of the lowest frequency $2\pi v$ formed by the interference between two neighboring object beams. Thus, we can write the intermodulation term in Eq. (4.22) as

$$\sum\limits_{q>p}^{N} \Delta\varepsilon_{pq} \cos(\mathbf{K}_{pq}\mathbf{x} - \Phi_{pq}) = \sum\limits_{q>p}^{N} \Delta\varepsilon_{pq} \cos(2\pi m v y + K_{pq}^{z} z - \Phi_{pq}),$$

(4.24)

where m = p − q. By adding the gratings with the same frequency, but optimized phase shifts Φ_{pq}, one can minimize the intermodulations.

Depth of the optimum plane. The intermodulations vary in the z-direction with K_{pq}^{z} [Eq. (4.24)]. As a consequence, they can be minimized only for one specific plane parallel to the hologram plane (z = const). For on-axis object beams, shown in Fig. 4.4, the z-variation is rather slow, i.e. K_{pq}^{z} is small. The largest component K_{pq}^{z} is formed by the interference between the central beam, propagating in the direction of the z-axis, and the marginal beams of the fan-out. We obtain

$$(K_{pq}^{z})_{max} = \frac{2\pi}{\lambda} n(1 - \cos\frac{\alpha}{2}) = \frac{2\pi}{\Lambda},$$

(4.25)

where λ is the wavelength, α the full angle (k_1, k_N) of the fan-out, n the refractive index, and Λ the periodicity of the grating $(K_{pq}^{z})_{max}$. The intermodulations remain small within a depth t_1, i.e.

$$t_1 = \frac{\Lambda}{5} = \frac{\lambda}{5n(1 - \cos(\alpha/2))}.$$

(4.26)

We can conclude, that the hologram plane must be normal to the z-axis, within the tolerances given by Eq. (4.26). This restricts the optimum recording geometry.

Special aspects of wave coupling in fan-out elements. We will not repeat the derivation of the coupled wave equations here, but we will discuss some important aspects of fan-out elements. For further reading we refer to Ref. 18. In the coupled wave models the electrical field is given as a sum of plane waves E_m, which interact through the recorded gratings K_{pq}. The art consists in a good choice of the relevant waves and their interactions.

In the case of multiple beam recording, we have to consider that each beam E_m is diffracted by all gratings. Furthermore, all gratings with equal periodicity, i.e. equal $K_{pq}{}^y$, diffract light in the same direction. Their coupling coefficients have to be added coherently.

Figure 4.5 shows the spectrum of a fan-out of $N = 3$. The reference reconstructs 3 object waves (O-waves) through the primary gratings K_{01}, K_{02}, K_{03}. These gratings are thick and the Bragg condition is fulfilled. The reference wave reconstructs also the intermodulation waves (I-waves) through the thin intermodulation gratings K_{12}, K_{23}, K_{13}. Because the gratings are thin, also higher diffraction orders can be generated. Additionally second order diffraction occurs. Once the O-waves and I-waves exist, they are diffracted again by all gratings and generate the secondary waves (S-waves).

Light can now be coupled from one beam to another through the primary gratings K_{0i} and intermodulation gratings K_{pq}. Figure 4.5 shows possible interactions for O-wave no. 2. As different waves contribute coherently to the same wave, this process is sensitive to the relative phases of the beams. Note, that a high reference-to-object beam ratio B reduces the influence of the intermodulation gratings K_{pq}, but not the cross-coupling through the primary gratings K_{0i}, see Eq. (4.23).

For small interbeam angles $\Delta\alpha$, the secondary diffraction at the thick primary gratings are rather strong, because they are only slightly off-Bragg. The coupling remains efficient, if the phase mismatch is less than 2π. For the arrangement shown in Fig. 4.4, these off-Bragg interactions are important for a hologram thickness smaller than

$$t_2 = \lambda/(n \tan\theta_0 \Delta\alpha), \tag{4.27}$$

where $\Delta\alpha = \alpha/(N-1)$ is the interbeam angle between two neighboring beams, θ_0 the reference beam angle, λ the wavelength, and n the refractive index. As a consequence, for larger interbeam angles $\Delta\alpha$ and thicker holograms, the fan-out properties become less sensitive to the phases.

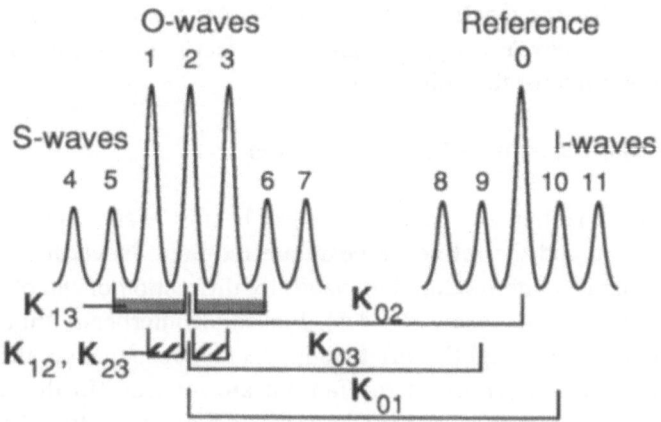

Fig. 4.5. Spectrum of the waves included in our model for a fan-out of N = 3 and possible interactions for O-wave no. 2 through primary gratings K_{0i} and intermodulation gratings K_{pq}.

Results. A numerical model that includes N object waves (O-waves), 2(N − 1) secondary waves (S-waves) and 2(N − 1) intermodulation waves (I-waves), has been presented by Herzig et al. [18]. The best results are obtained with optimized phases in order to eliminate the intermodulation terms.

Figure 4.6 shows the total useful diffraction efficiency η_T versus the modulation amplitude $\Delta\varepsilon$, for a fan-out of N = 9 and the following recording parameters: $\lambda = 0.488$ μm, n = 1.5, B = 1, $\theta_0 = 30°$, $\Delta\alpha = 0.1°$, d = 15 μm. For a reference-to-object beam ratio equal to unity, the fan-out element shows already good properties. For example, for N = 9 and $\Delta\alpha = 0.1°$ the maximum diffraction efficiency becomes $\eta_{Tmax} \sim 94\%$ with a uniformity error smaller than ±10 %. In the worst case, if all gratings are in phase ($\Phi_{pq} = 0$) the maximum diffraction efficiency drops to 45 %.

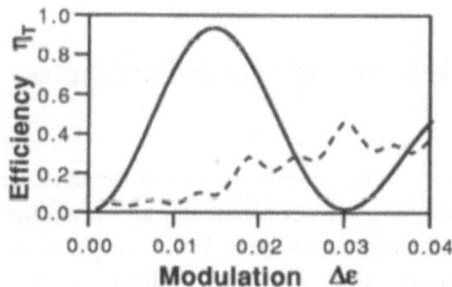

Fig. 4.6. Results for a fan-out of N = 9. Diffraction efficiency η_T versus the modulation amplitude $\Delta\varepsilon$, for the corrected phases (solid line), and the worst case (dashed line).

It turns out, that for small interbeam angles $\Delta\alpha$ the corrected phases give a much better efficiency. For larger angles the effect is less significant. A similar behavior is observed for the uniformity error e.

4.5.3 Sequential recording of N object waves

In the case of sequential recording of N object beams, N subsequent exposures to each object beam and the reference beam are required. Nevertheless, secondary diffraction generates intermodulation waves by diffraction of the object waves at the primary gratings $\mathbf{K}_{0i}$ (see Fig. 4.5). For small interbeam angles $\Delta\alpha$, these interactions are only slightly off-Bragg. As a consequence the diffraction efficiency and the uniformity of the fan-out are reduced. In that case a phase control at recording is also necessary for achieving good results. These off-Bragg interactions can be neglected for a phase mismatch greater than 2π, i.e. for a thickness $> t_2$, see Eq. (4.27).

High efficiency can be reached for sequential recording as well as for simultaneous recording. A major drawback of sequential recording is that the exposure energy for the same modulation of the refractive index is higher. The maximum number of fan-out beams is then limited by the dynamic range of the holographic material. Therefore, we recommend sequential recording for very thick holograms, thickness $> t_2$, where all off-Bragg interactions are inefficient. As the required modulation of the refractive index is small for large thicknesses, the dynamic range is no longer a limiting factor.

In order to give an example, we obtain $t_2 = 323 \, \mu m$ for the following typical parameters: wavelength $\lambda = 488$ nm, refractive index n = 1.5, reference beam angle $\theta_0 = 30°$, full fan-out angle $\alpha = 10°$, and interbeam angle $\Delta\alpha = 0.1°$. This shows that sequential recording is rather suitable for materials like photorefractive crystals, which are used for reprogrammable components (see Chap. 9).

4.6 Fabrication methods and selected examples

In the introduction three basic interconnect components have been identified, namely multifacet elements, fan-out elements, and imaging elements (Fig. 4.1). This chapter deals with their realization. First, the fabrication of an optimized HOE will be presented. This can be a single facet focusing HOE for semiconductor wavelengths or a specially designed imaging element requiring aspheric recording waves. Then, we will show how to realize arrays of HOEs (multifacet elements). Finally, the recording of fan-out holograms will be summarized.

4.6.1 Optimized holographic optical elements

Holographic optical elements for use with semiconductor lasers or for imaging need an optimized design, as shown in Sect. 4.4. The resulting HOE phase function $\Phi_H(x,y)$ is fabricated by recording the interference pattern of an object and a reference wave in a holographic emulsion. There are many possibilities to record one and the same phase function. Practically, however, the choice is restricted to easily available waves, mainly spherical waves. Furthermore, for recording efficient HOEs, the Bragg condition [Eq. (4.14)] has to be respected. Aspheric waves can be produced by computer-generated holograms (CGHs) enabling complex hologram structures. This is not always necessary. In many cases spherical waves are sufficient to approximate the desired function.

Recording with spherical waves. The recording geometry has already been shown in Fig. 4.2. Object and reference waves are plane or spherical waves generated by a laser and standard optics, such as pinholes and lenses. For a readout in the near IR the recording parameters, i.e. the position of the sources, have to be optimized. Second order theory, Eqs. (4.11-13), leads to initial solutions, which can be improved by numerical optimum design [19]. This may include the minimization of the spot diameter obtained by ray tracing (Sect. 4.4.4) and the deviation Δk_z from the Bragg condition (Sect. 4.5.1). Different numerical optimization methods are known from the literature, such as damped least-squares, downhill simplex, and simulated annealing [20].

Recording with computer-generated holograms. Perfect HOE structures can be realized with the aid of computer-generated holograms (CGHs) [1,21]. CGHs are binary structures, which allow the creation of optical waves from numerical data by using well known encoding techniques (Lohmann, Lee, Burch, Arnold etc.). Today, a CGH is usually fabricated by laser beam writing or by e-beam lithography, permitting a large space band-width product.

A typical recording set-up is shown in Fig. 4.7. The laser beam is split into a plane wave branch for the reference and a spherical wave branch for the object. The CGH is inserted into the plane wave branch of the recording set-up. A telescopic lens system creates a 1:1 image of the CGH at the HOE. The carrier frequency of the CGH, which is a binary hologram, separates the diffraction orders. The desired aspherical wave is obtained by inserting a spatial filter for the first order in the Fourier plane. Systems using two CGHs have been proposed recently [22].

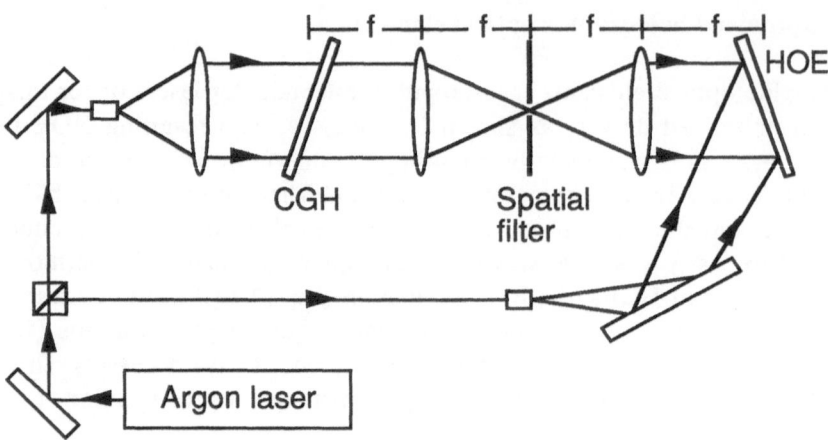

Fig. 4.7. Recording set-up to realize a HOE with the aid of a computer-generated hologram.

Copying. Reasons for copying HOEs may be their mass production or for changing the three dimensional structure (Bragg angle) in order to get higher diffraction efficiency. Surface relief HOEs can be copied by embossing or casting. These methods will be discussed in Chap. 5. Volume phase holograms can also be copied [5]. One possibility is to replay the HOE and to add a new reference wave. This technique has the advantage of great flexibility. Contact copying is another possibility. Then, the transmitted and the diffracted light of the master hologram are the reference and object beam for the copy. This method is insensitive to vibrations or to variations of the illuminating beam, as long as the master and the copy do not move relative to each other.

4.6.2 Sequentially recorded multifacet elements

A simple lenslet array is fabricated by sequentially recording of single holographic lenses as shown in Fig. 4.8. The apertures are imaged onto the HOE plane. This allows individual lenslets to be closely packed. Depending on the requirements, a CGH can be inserted into one of the beams, as discussed in the previous section. The critical points are the high positioning accuracy necessary for the step and repeat process and the high and uniform efficiency over all elements of the array.

As an example, we quote the results achieved by researchers at the Heriot-Watt University [23]. They realized a 56x56 holographic lenslet array for the readout wavelength $\lambda = 850$ nm. Each lenslet has a diameter of s =180 µm, a focal length of f = 900 µm, and an average diffraction efficiency of η = 92 % with a maximum variation of ±5 %. The HOE array generates spot sizes of 9.5 µm.

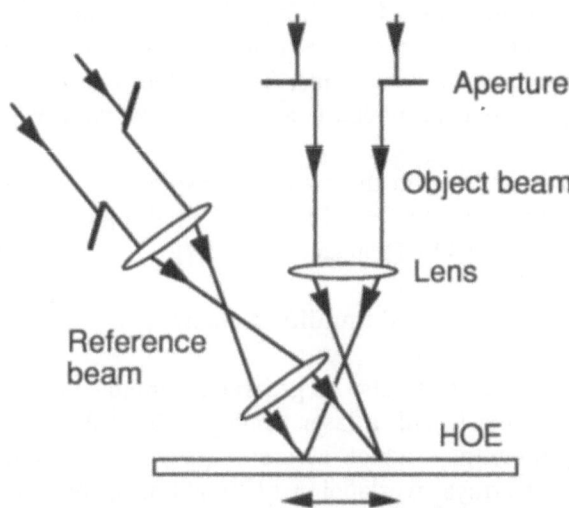

Fig. 4.8. Recording of a holographic lenslet array with a plane reference wave and a spherical object wave. The apertures are imaged onto the HOE plane.

The step and repeat process shown in Fig. 4.8. generates only identical HOE lenses. By using additional elements, however, the deflection angle can be varied, which enables the recording of more complex systems [24].

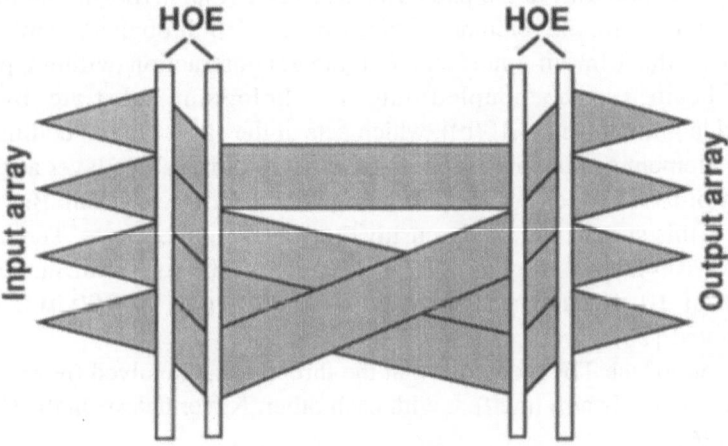

Fig. 4.9. On-axis interconnect system with doublet HOE.

To record a general interconnection pattern, which permits any optical wiring with optimized recording conditions, a programmable positioning of the sources in 3-D space is required. This could be realized with the aid of fiber optics. However, high position accuracy is required. Uncontrolled variations in intensity or polarization are not permitted. Thus, actual propositions prefer a doublet HOE arrangement as shown in Fig. 4.9 [24,25]. The first element is an array of equal focusing HOE, whereas the second element is an array of varying planar gratings to change the direction of the beams. The doublet can be cemented together yielding a compact element, which is also less sensitive to wavelength changes [26].

As was briefly discussed in the previous chapter, the transmission geometry shown in Fig. 4.9 can also be folded in order to build a compact system. Similar arrangements are reported by Haumann et al. [27].

4.6.3 Simultaneously recorded multifacet elements

Arrays of spherical waves can also be generated simultaneously by illuminating a pinhole array on a mask with a laser beam. Such masks can be fabricated by electron beam lithography, which assures high position accuracy of the focal points. For lenslet arrays overlapping of neighboring beams in the hologram plane has to be avoided.

Transmission holograms are recorded by interfering an object and a reference wave, where both sources have to be on the same side of the hologram. For recording holographic lenses with short focal lengths, a problem occurs by reasons of geometry. This is overcome by using total internal reflection (TIR) near-field holography [28]. A pinhole array mask is recorded holographically by passing a collimated beam through the mask, which is placed in close proximity (typically 50...500 μm) to the photosensitive layer [Fig. 4.10(a)]. The transmitted wave interferes with the reference wave, which is fed through a prism and totally reflected at the film-air interface. For the reconstruction without prism, the readout beam can be coupled into the hologram substrate by another holographic grating [Fig. 4.10(b)], which acts at the same time as collimator. This coupling element can be fabricated in the same photosensitive layer as the lenslet array. Collimator and lenslet array form together a compact system. Because of its symmetry, this system is quite insensitive to wavelength changes. The fabrication of a 100×100 lenslet array with a focal length of 400 μm, a diffraction limited spot size of 10 μm and a spacing between the spots of 100 μm, has been demonstrated [26].

A problem of the TIR method is that the three beams involved (reference beam, TIR beam, object beam) interfere with each other. Nevertheless, high efficiency is possible [36].

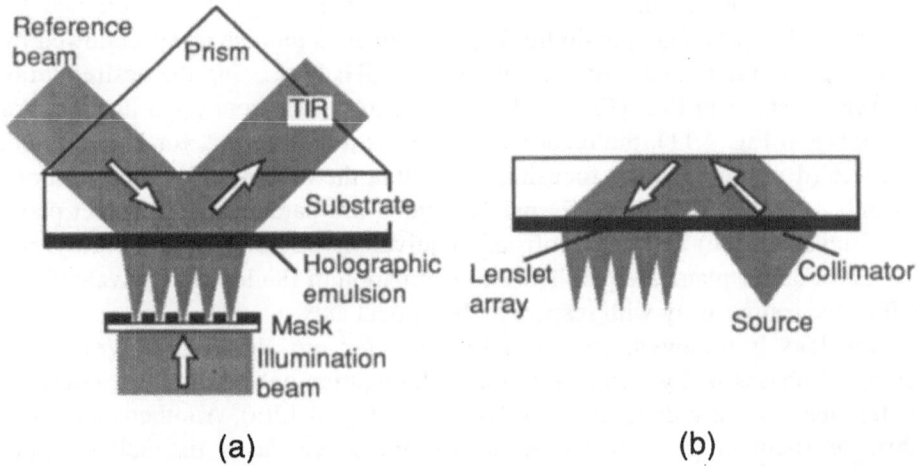

(a)   (b)

Fig. 4.10. (a) Recording and (b) readout geometry for holographic lenslet arrays fabricated by TIR-holography.

4.6.4 Simultaneous recording of fan-out elements

Holographic fan-out elements can be realized by recording the interference pattern of N object waves and one reference wave. It is known from Sect. 4.5.2 that high efficiency and uniformity can be achieved by optimized phases of the object beams. Here we will discuss successful recording methods.

Figure 4.11(a) shows the recording set-up for on-axis fan-out elements. The object is an array of coherent sources. The recording conditions are optimum if the irradiance of the object beam is uniform in the hologram plane. This is not the case if all light beams are in phase. For many object points, random phases lead to a reduction of the intermodulations [29]. However, random is not the best case.

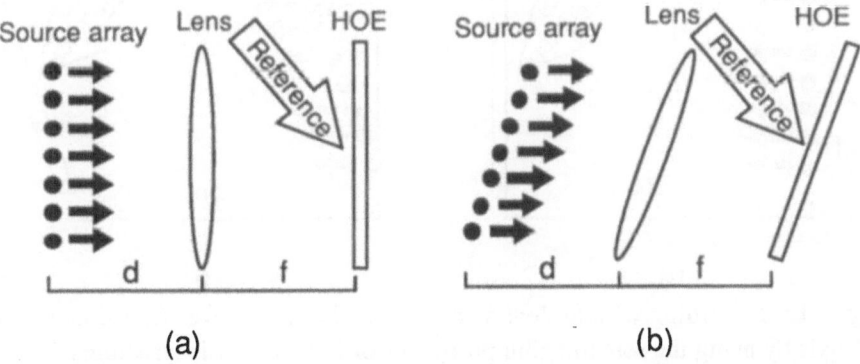

(a)   (b)

Fig. 4.11. Object waves (a) on-axis and (b) off-axis; the HOE becomes focusing for d > f and non-focusing for d = f.

An array of object beams with optimized phases ϕ_i can be obtained by different techniques. One possibility is to illuminate a pinhole array, followed by an appropriate phase plate. Another is to use CGHs generating the desired array (see Figs. 6 and 7 in Ref. [17]). If the sources are in the front focal plane of the lens (d = f in Fig. 4.11), the recorded element is non-focusing, for larger object distances (d > f) it becomes focusing. Note, that the minimum intermodulations can only be achieved in specific planes, which are parallel to the object plane [18]. Figure 4.11(b) shows the off-axis equivalent where the source array, the lens, and Fourier plane are parallel. The inclination of the lens is equivalent to a shift of the source array with respect to the optical axis.

If the lens is removed, we get a focusing fan-out element. However, the optimized phases will generate a uniform illumination only if the hologram is in the far-field, i.e. at a distance $d > (Ns)^2/\lambda$, see Fig. 4.12(a). Another method to fabricate focusing fan-out elements without a lens uses the self-imaging properties of large periodic structures [30]. In this case the object is a regular array of coherent sources with identical phases $\phi_i = 0$. Considering the beam propagation in free space, planes of reduced intermodulations are found, which are suitable for recording efficient holograms. These planes are parallel to the object plane, as in the case of optimized phases. In Fig. 4.12(b) we incline the object, thus, we have also to incline the hologram. The self-imaging (Talbot) distance is proportional to s^2 and for inclined objects proportional to $(s \cos\theta)^2$. If we incline a regularly spaced 2-D array (same spacing s in x and y direction) with respect to the y-axis, we get two different distances for the optimum planes, depending on s^2 and $(s \cos\theta)^2$ respectively. However, a common minimum plane can be determined. This problem can be avoided if the initial array has two different periods Λ in the x and the y direction, namely $\Lambda_x = s$ and $\Lambda_y = s/\cos\theta$.

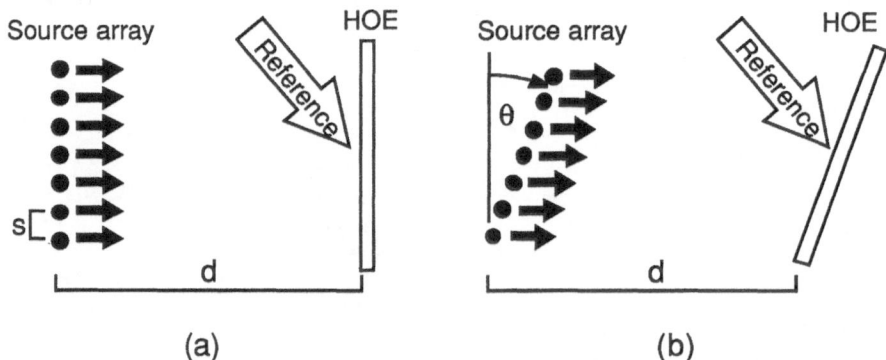

(a)   (b)

Fig. 4.12. Recording without lens with spherical object waves (a) on-axis and (b) off-axis by using the self-imaging properties of large periodic structures.

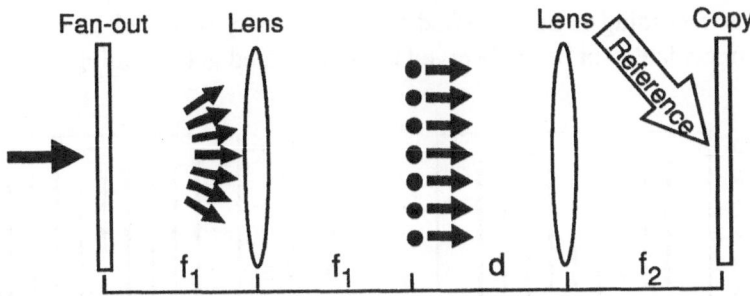

Fig. 4.13. 4f-system for copying fan-out elements with magnification $m = f_2/f_1$, where $d = f_2$. Focussing power can be included if $d > f_2$.

Fan-out elements in volume holograms can also be fabricated by copying the phase structure of already existing fan-out elements as e.g. Dammann gratings [31]. A simple image formation with a single lens would destroy the phase structure and therefore the properties of the fan-out. This can be avoided by using a 4-f imaging system as shown in Fig. 4.13, which applies twice a Fourier transform thereby conserving the phase distribution. Analog to Figs. 4.11 and 4.12, there exists also an off-axis arrangement of Fig. 4.13.

Robertson et al. [32] have fabricated a linear fan-out of N = 8 beams in dichromated gelatin by copying an amplitude mask. The period of the fan-out was reduced from 2.5 mm (mask) to 1 mm (copy) by choosing $f_1 = 2.5\,f_2$ in Fig. 4.13. The diffraction efficiency in the central eight orders was measured to be 96 % and the uniformity error was found to be ±3.4 %. Comparable results have also been obtained by Herzig et al. [33] for a fan-out of N = 9 (see Fig. 4.14).

4.6.5 Sequential recording of fan-out elements

In sequential recording of N object beams, N exposures to the reference and each object beam are required. The advantage of this method is the absence of intermodulation gratings formed by the interference between the object beams.

For small interbeam angles $\Delta\alpha$, a control of the relative phases between the exposed N gratings is required (Sect. 4.5.3). First experimental results for fan-out elements (N = 9) recorded with optimized phases have shown an overall efficiency of 75 % while the uniformity error was ±11 % [33]. This corresponds to a diffraction efficiency of 85 % if antireflective coatings are used. The results for the efficiency are about 10 % lower than those measured for simultaneously recorded HOEs. Theoretically, the results for sequential recording should be as high as for simultaneous recording. Fig. 4.14 shows a comparison between simultaneously and sequentially recorded fan-out elements.

Beside the control of the object phases, material properties may be responsible for the lower diffraction efficiency. For silver halide materials the linearity assumed at recording is not satisfied for sequentially recorded holograms [34]. Similar effects have not been observed in dichromated gelatin holograms [35].

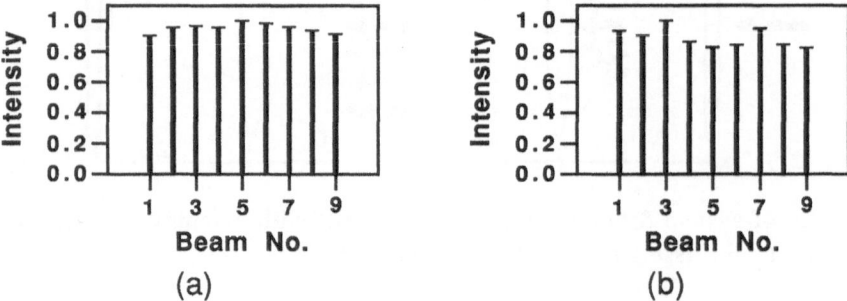

Fig. 4.14. (a) Simultaneously and (b) sequentially recorded fan-out elements: number of object waves N = 9, reference beam angle $\theta_0 = 30°$, interbeam angle $\Delta\alpha = 0.057°$.

4.7 Conclusions

Holographic optical elements are well suited to fulfill the needs of optical interconnects. They are efficient and serve as focusing elements, imaging elements, beam deflector, and fan-out elements. Almost any structure shape can be realized. Their high flexibility enables complex and compact interconnection system. However, all this advantageous are also available from computer-generated components that will be presented in Chap. 5. An advantage of HOEs is the high resolution of the holographic materials (> 3000 lines/mm) which is important for the fabrication of off-axis elements in compact systems. Also it needs only relatively simple equipment (laboratory, laser, etc.) to start with. On the other hand, it is not easy to control the recording process sufficiently well to get reproducible results.

Actual efforts are rather concentrated on computer-generated components, which are fabricated by modern microfabrication technologies (see Chapt. 5). These technologies are well established and their feasibility has been proven under industrial conditions.

References

1. W. H. Lee: Computer-generated holograms: techniques and applications, in *Progress in Optics XVI*, E. Wolf, ed. (North-Holland, 1978), pp. 119-232
2. H. Andersson, M. Ekberg, S. Hard, S. Jacobsson, M. Larsson, T. Nilsson: Single photomask, multilevel kinoforms in quartz and photoresist: manufacture and evaluation, Appl. Opt. **29**, 4259-4267 (1990)
3. B. Robertson, M. R. Taghizadeh, J. Turunen, A. Vasara: Techniques for fabricating periodic Fourier-transform kinoforms, Proc. SPIE **1136**, 20-26 (1989)
4. P. Hariharan: *Optical Holography*, (Cambridge University Press, Cambridge, 1984)
5. R. R. A. Syms: *Practical Volume Holography* (Clarendon Press, Oxford, 1990)
6. H. Kogelnik: Coupled wave theory for thick hologram gratings, Bell. Syst. Tech. J. **48**, 2909-2947 (1969)
7. M. G. Moharam, T. K. Gaylord, R. Magnusson: Criteria for Bragg regime diffraction by phase gratings, Opt. Commun. **32**, 14-18 (1980)
8. W. K. Smothers, T. J. Trout, A. M. Weber, D. J. Mickish: Hologram recording in Dupont's new photopolymer materials, *Holographic Systems, Components and Applications*, Bath, UK, Conference Publication No. 311 (Institution of Electrical Engineers, London, 1989), pp. 184-189
9. H. P. Herzig, R. Dändliker: Holographic optical elements for use with semiconductor lasers, in *International Trends in Optics*, J. W. Goodman, ed., (Academic Press, New York, 1991), pp. 57-75
10. H. P. Herzig: Holographic optical elements (HOE) for semiconductor lasers, Opt. Commun. **58**, 144-148 (1986)
11. J. Kedmi, A. A. Friesem: Optimized holographic optical elements, J. Opt. Soc. Am. A **3**, 2011-2018 (1986)
12. J. N. Cederquist, J. R. Fienup: Analytic design of optimum holographic optical elements, J. Opt. Soc. Am. A **4**, 699-705 (1987)
13. W. T. Welford: *Aberrations of optical systems*, (Adam Hilger Ltd, Bristol, 1986)
14. J. Goodman: *Introduction to Fourier Optics* (McGraw-Hill, San Francisco, 1968)
15. L. Solymar, D. J. Cooke: *Volume holography and volume gratings* (Academic Press, London, 1981)
16. R. K. Kostuk: Comparison of models for multiplexed holograms, Appl. Opt. **28**, 771-777 (1989)
17. H. P. Herzig, D. Prongué, R. Dändliker: Design and fabrication of highly efficient fan-out elements, Jpn. J. Appl. Phys. **29**, L1307-L1309 (1990)

18. H. P. Herzig, P. Ehbets, D. Prongué, R. Dändliker: Fan-out elements recorded as volume holograms: optimized recording conditions, Appl. Opt. **31**, 5716-5723 (1992)

19. Y. Ono, N. Nishida: Holographic optical elements with optimized phase-transfer functions, J. Opt. Soc. Am. A **3**, 139-142 (1986)

20. W. H. Press, B. P. Flannery, S. A. Teukolsky, W. T. Vetterling: *Numerical Recipes* (Cambridge University Press, Cambridge, 1987)

21. O. Bryngdahl, F. Wyrowski: Digital holography - computer-generated holograms, in *Progress in Optics XXVIII*, E. Wolf, ed., (Amsterdam, North Holland, 1990), pp. 1-86

22. N. Lindlein, J. Schwider: Practical notes on designing and analysing thick holographic optical elements, Pure Appl. Opt. **1**, 111-125 (1992)

23. I. R. Redmond, E. J. Restall, A. C. Walker: High performance holographic optics for the visible and near Infra-Red, *Holographic Systems, Components and Applications*, Bath, UK, Conference Publication No. 311 (Institution of Electrical Engineers, London, 1989), pp. 190-194

24. B. Robertson, E. J. Restall, M. R. Taghizadeh, A. C. Walker: Space-variant holographic optical elements in dichromated gelatin, Appl. Opt. **30**, 2368-2375 (1991)

25. J. Schwider, W. Stork, N. Streibl, R. Völkel: Possibilities and limitations of space-variant holographic optical elements for switching networks and general interconnects, Appl. Opt. **31**, 7403-7410 (1992)

26. D. Prongué, H. P. Herzig: HOE for clock distribution in integrated circuits: experimental results, Proc. SPIE **1281**, 113-122 (1990)

27. H.-J. Haumann, H. Kobolla, F. Sauer, W. Stork, N. Streibl, R. Völkel: Optical bus based on light-guiding-plates, in *Conference Record of Topical Meeting on Optical Computing*, Kobe, Japan 1990 (Japan Society of Applied Physics, Tokyo, 1990), pp. 162-163

28. K. Stetson: Holography with total internally reflected light, Appl. Phys. Lett. **11**, 225-226 (1967)

29. R. J. Collier, C. B. Burckhardt, L. H. Lin: *Optical Holography* (Academic Press, New York, 1971), p. 486

30. I. Seyd-Darwish, P. Chavel, J. Taboury, Y. Malet: Array illuminator hologram based on the Talbot effect, in *Conference Record of Topical Meeting on Optical Computing*, Kobe, Japan 1990 (Japan Society of Applied Physics, Tokyo, 1990), pp. 294-296

31. B. Robertson, M. R. Taghizadeh, J. Turunen, A. Vasara: High-efficiency, wide-bandwidth optical fanout elements in dichromated gelatin, Opt. Lett. **15**, 694-696 (1990)

32. B. Robertson, J. Turunen, H. Ichikawa, J. M. Miller, M. R. Taghizadeh, A. Vasara: Hybrid kinoform fan-out holograms in dichromated gelatin, Appl. Opt. **30**, 3711-3720 (1991)

33. H. P. Herzig, D. Prongué, P. Ehbets, R. Dändliker: Holographic fan-out elements: simultaneous versus sequential recording, *8th Workshop on Optics in Computing*, Paris, Topical Meetings Digest Series (European Optical Society, 1992), Vol. 1, pp. 59-62

34. K. M. Johnson, L. Hesselink, J. W. Goodman: Holographic reciprocity law failure, Appl. Opt. **23**, 218-227 (1984)

35. I. R. Redmond: Holographic optical elements in dichromated gelatin, Ph.D. Thesis, Heriot-Watt University, U.K. (1989)

36. P. Ehbets, H. P. Herzig, R. Dändliker: TIR holography analyzed with coupled wave theory, Opt. Commun. **89**, 5-11 (1992)

5 Diffractive components: computer-generated elements

H. P. Herzig
Institute of Microtechnology, University of Neuchâtel, Neuchâtel, Switzerland
M. T. Gale, H. W. Lehmann, R. Morf
Paul Scherrer Institute, Zürich, Switzerland

Highly efficient diffractive optical elements (DOEs) can also be realized using modern microfabrication technologies. Computer-generated data for arbitrary phase profiles can be transformed into optical elements. These elements offer optimum design freedom and established fabrication technology.

In this chapter, we will discuss the design and fabrication of binary, multilevel and continuous microrelief elements for optical interconnects. First, an overview of the different DOEs will be given. In Sect. 5.2, we consider the optical function of DOEs, which can be described by the transparency theory, as well as the optimum design methods. For fine surface relief gratings (feature size $\approx \lambda$), an approach using scalar diffraction theory is not sufficient. What is required is a rigorous solution of Maxwell's equations, which is usually only available in numerical form. Section 5.3 will be concerned with recent developments in rigorous diffraction theory. Section 5.4 summarizes modern microfabrication technologies for microrelief elements. In the last part (Sect. 5.5), selected examples of computer-generated DOEs for optical interconnects are presented.

5.1 Diffractive optical elements

Diffractive optical elements (DOEs) are components which make full use of the wave nature of light. By relying on diffraction and interference rather than on reflection and refraction, unique and novel properties can be realized complementing and exceeding the possibilities of traditional lenses, prisms and mirrors. Interesting aspects for parallel optical interconnects are the full design freedom, the possibility of realizing small elements that can be aligned precisely to form large arrays, and the capability of multiple beam splitting (fan-out).

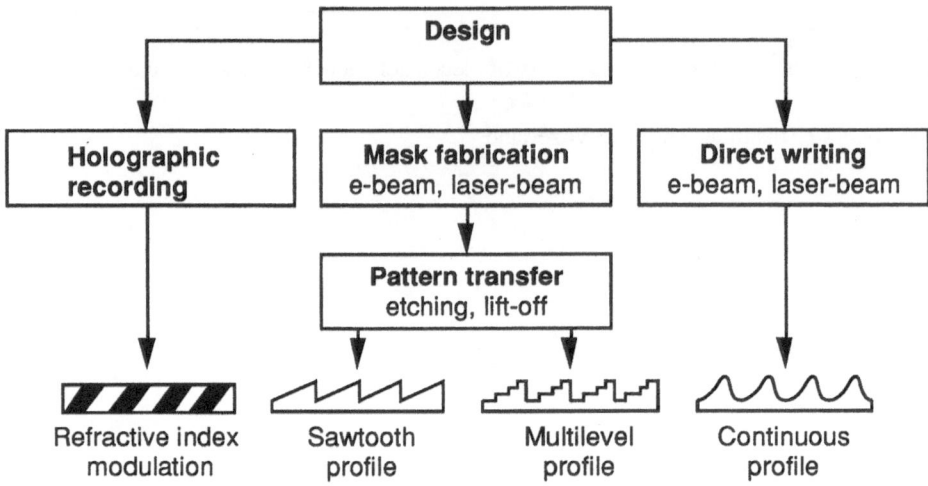

Fig. 5.1. Fabrication of diffractive optical elements (DOEs).

We begin with a short overview of the different types of DOEs. Since efficiency is an important parameter, the discussion is concentrated on phase elements. In Fig. 5.1, they are classified with respect to their volume or relief aspect, which is closely related to the fabrication technology. We can distinguish between a modulation of the refractive index and a modulation of the surface-relief (sawtooth, multilevel, continuous). Holographic optical elements (HOEs) are diffractive structures that are fabricated by recording the interference pattern of a reference and an object wave in a photosensitive emulsion. In materials like silver halide, dichromated gelatin and photopolymer, it results a modulation of the refractive index. In other materials, such as photoresist, the interference pattern is converted into a relief structure. Diffractive optical elements can also be realized by techniques which have become standard in VLSI-technology. Masks are generated by e-beam lithography or by laser beam writing lithography. Then, to achieve high efficiency, they are transformed into surface-relief structures by etching into materials like silicon nitride, quartz, silicon or optical glass, or by thin film deposition, e.g. SiO. Typical examples are two-level phase gratings, also called Dammann gratings [1]. Using several masks, multilevel profiles can be generated to improve the efficiency (blazing) [2]. Anisotropic etching is another method to introduce blaze effects resulting in sawtooth gratings. An interesting technique is the direct writing of the phase structure by laser beam or e-beam lithography in photoresist [3]. The developed photoresist relief can be converted into a metal master relief by electroplating to emboss or cast low-cost replicas. This method allows one to fabricate continuous relief structures, such as lenslet arrays [4] or highly efficient fan-out elements [5].

Holographic optical elements (HOEs) have been discussed in Chapt. 4. Here, the emphasize will be placed upon high efficiency computer-generated elements, which have surface-relief structures. The elements needed for optical interconnects are multifacet elements, fan-out elements and imaging elements as shown in the last chapter (Fig. 4.1). The design of these elements will be discussed in the following section. For the fabrication, the distinction between the different elements is less important. The fabrication of an array of 100x100 Fresnel microlenses is not more difficult than the fabrication of a single Fresnel microlens.

5.2 Design of diffractive optical elements

In this section, we will present the design of DOEs. For the example of a diffractive lens, the different structures, i.e. binary, multilevel, and continuous microrelief, are introduced before discussing more general structures. Section 5.2.2 summarizes the optimum design methods for imaging and beam shaping (e.g. fan-out elements).

5.2.1 The phase function of diffractive components

Diffractive optics were proposed already in the last century. Probably the earliest and best known diffractive element is the Fresnel zone plate [6]. The element consists of circular zones which are alternately transparent and opaque, as shown in Fig. 5.2. The radii R_m of the concentric rings are approximately given by

$$R_m = \sqrt{mf\lambda}. \tag{5.1}$$

Such an element acts as a lens when illuminated with a plane wave at the wavelength λ. The light emitted from the transparent zones interferes and forms focussed light spots on the z-axis at the distances f/p, where $p = 1,3,5,...$. The z-axis is perpendicular to the drawing plane in Fig. 5.2. The light intensity is maximum when the waves interfere constructively, i.e. the phase difference is a multiple of 2π. Amplitude zone plates have low diffraction efficiency. Only a few percent of the light is found in the first order focus ($p = 1$).

The efficiency can be increased up to 40 % with a phase-reversal zone plate. Rather than blocking out every other zone, the thickness of alternate zones are increased, thereby retarding their phases by π.

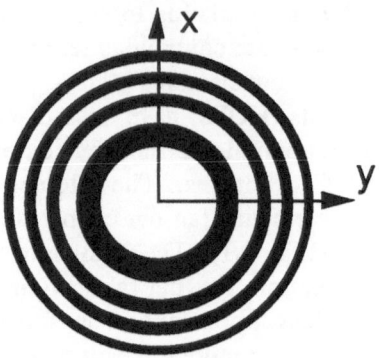

Fig. 5.2. Fresnel zone plate.

The zone plate described above is a binary amplitude or phase grating with a local grating period of $R_{m+2} - R_m$. Ideally, the retardation is made gradually over a grating period as shown in Fig. 5.3(a). The phase profile modulo 2π is described by

$$\Phi(x,y) = \frac{2\pi}{\lambda} \sqrt{x^2 + y^2 + f^2}, \tag{5.2}$$

Now, all the incoming light is diffracted into the focus spot at the distance f. This element is called a phase Fresnel zone plate [7], or also micro-Fresnel lens [8]. For elements with a high numerical aperture, this approach leads to very fine outer zones and is limited in practice by fabrication considerations.

The phase profile $\Phi(x,y)$ can be generated directly as smooth, continuous surface-relief element by e-beam [9] or by laser beam writing lithography [11]. More common is the quantization of the continuous phase profile into discrete phase levels, as shown in Fig. 5.3(b). Such a structure can be made by repeated etching steps. For 8 phase levels the diffraction efficiency is already 95 %. The diffraction efficiency as a function of the number M of phase levels is given by (e.g. [7])

$$\eta = \left[\frac{\sin(\pi/M)}{(\pi/M)}\right]^2. \tag{5.3}$$

An advantage of DOEs is that almost any structure shape, including non-rotationally symmetric aspherics, can be generated. The complex amplitude transmittance of a DOE that converts an input wave Φ_{in} into an output wave Φ_{out} is simply given by

$$t(x,y) = \exp[i\Phi(x,y)] = \exp[i(\Phi_{out}(x,y) - \Phi_{in}(x,y))]. \tag{5.4}$$

The phase function $\Phi(x,y)$ modulo 2π is then realized as thickness variation, as shown in Fig. 5.3. The operation of modulo 2π, is equivalent to an encoding of the phase function Φ as a lateral position of the local grating, like a hologram. The shape of each grating period determines the energy distribution into the different diffraction orders. For strongly off-axis elements or small feature size, the simple operation of modulo 2π may need some modifications, which have to be determined for the specific case [10].

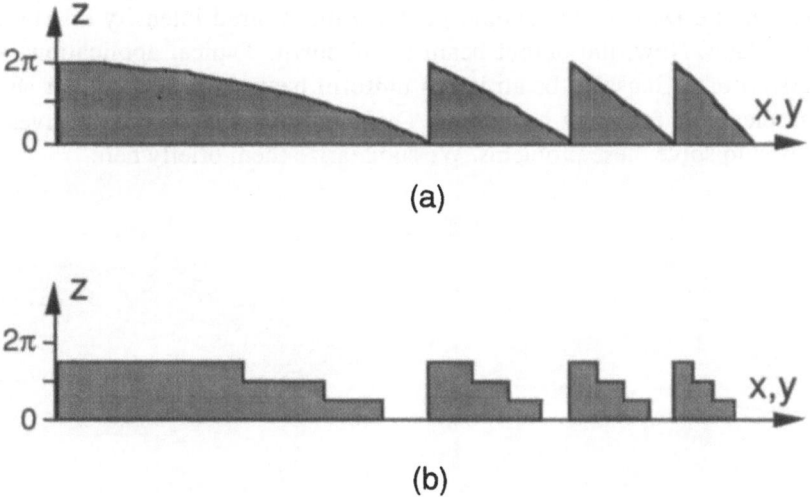

Fig. 5.3. Phase profiles for a phase Fresnel zone plate: (a) continuous phase profile, (b) 4-level phase profile.

Such diffractive elements can be analyzed by using ray tracing or scalar diffraction theory as described in Sect. 4.4.4 for HOEs (see also [12]). There also exist numerous commercially available lens design programs, which allow the computation of diffractive phase profiles. Ray tracing describes the geometrical properties, such as focal lengths and aberrations. They are determined by the lateral position of the diffractive zones. If the energy distribution in the image plane is important, then scalar diffraction theory is required to design the relief structure of each zone.

Equation (5.4) handles well the case where only two waves (Φ_{out}, Φ_{in}) are involved and where both waves are known. For imaging many different waves, or for generating a desired intensity pattern in the far-field, the determination of the phase function $\Phi(x,y)$ becomes an optimization problem, which will be discussed in the following Sect. 5.2.2.

5.2.2 Optimum design for imaging and beam shaping

Figure 5.4 shows (a) an imaging and (b) a beam shaping element. In the case of imaging a set of continuous waves emitted by the object has to be converted into another continuous set of output wavefronts forming the image. The ideal input and output waves are known, but not the optical element that images well all input waves. The term "imaging" stands here for mathematically similar problems, such as the design of Fourier lenses. In the case of beam shaping, one input wave illuminates the DOE which should generate the desired intensity distribution in another plane. Now, the output beam is unknown. Typical applications are the conversion of a Gaussian beam into a uniform beam with rectangular shape, or fan-out elements for array generation. Different methods have been investigated in the past to solve these problems. We summarize them briefly here.

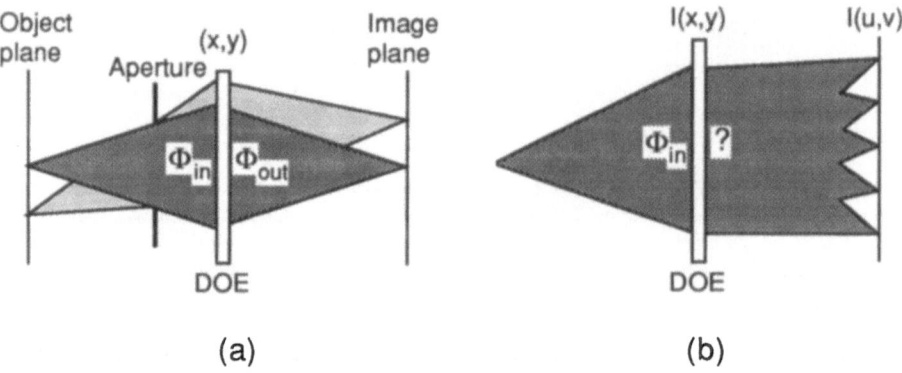

Fig. 5.4. (a) Imaging, (b) beam shaping (fan-out).

Optimum design for imaging. As already mentioned above a continuous a set input waves $\Phi_{in}(x,y,t)$ has to be converted into another continuous set of output waves $\Phi_{out}(x,y,t)$ [13,14]. The ideal waves are spherical waves emitted by points in the object plane which are focussed onto the image plane. The parameter t describes for example the direction of the wave or the position of the focus. Thus, for each pair of input and output waves we know the desired phase function $\Phi_d(x,y,t)$ of the optical element as the difference between the phases of the input and the output waves [Eq. (5.4)], i.e.,

$$\Phi_d(x,y,t) = \Phi_{out}(x,y,t) - \Phi_{in}(x,y,t). \tag{5.5}$$

The phase $\Phi_d(x,y,t)$ varies with the parameter t. We are now looking for a continuous DOE phase function $\Phi(x,y)$, which is as close as possible to $\Phi_d(x,y,t)$ for all t. The performance of the optical element is considered to be optimum when the value of the mean-squared difference between the desired local functions $\Phi_d(x,y,t)$ and the real DOE phase $\Phi(x,y)$ is minimum:

$$\int W(t)\, P(x,y,t)\, [\Phi_d(x,y,t) - \Phi(x,y) + \phi(t)]^2\, dtdxdy \rightarrow \min. \qquad (5.6)$$

Note, that if a discrete number of waves are involved, the integral will become a sum. The absolute phase of the desired function $\Phi_d(x,y,t)$ is not significant, therefore an arbitrary phase $\phi(t)$ can be added. $W(t)$ is a weighting function and $P(x,y,t)$ is the pupil function. The t-dependence of the pupil function indicates that the readout wave may illuminate different parts of the DOE for different values of the parameter t.

Another approach is based on analytic ray tracing and relies on propagation vectors and grating vectors rather than on the phase functions [15]. This means that the phases Φ_i in Eqs. (5.5) and (5.6) are replaced by the projection of the wave-vectors onto to the x,y-plane, $k_{iH}(x,y) = \text{grad}[\Phi_i(x,y)]$. The function $\phi(t)$ disappears.

For either approach variational methods are applied to find the optimum phase function $\Phi(x,y)$ for the given functions $W(t)$, $P(x,y,t)$ and $\Phi_d(x,y,t)$. In many cases of practical interest the desired phase function [$\Phi_d(x,y,t)$ in Eq. (5.6)] is not precisely known, for example, if distortions or field curvature can be tolerated. Or, the optimum position of the aperture stop has first to be found in order to determine the pupil function $P(x,y,t)$. A good insight into these problems is obtained by an alternative method [16], which compares the local derivatives of the desired hologram phase function $\Phi_d(x,y,t)$ at the position of the pupil $P(x,y,t)$, and the real phase $\Phi(x,y)$. The method has been developed initially for holographic scanners, but also works well for the design of imaging elements.

These methods yield analytical solutions for the problem of imaging. Alternatively, numerical methods such as damped least-squares, downhill simplex, simulated annealing and others, have also been applied [17].

Optimum design for beam shaping. In general, the beam shaping problem has no analytical solution. Many different numerical methods have been investigated. Some representative methods are described here. After this, we will report theoretical results obtained for fan-out elements which convert an input wave into a one- or two-dimensional array of light spots. For experimental results we refer to Sect. 5.5.2.

In the case of beam shaping, a prescribed intensity distribution has to be generated in the far-field. The phase distribution is free, and can therefore be used for the optimization process. This phase retrieval problem can be solved well by the Gerchberg-Saxton algorithm [18]. A number of schemes have since evolved from this basic iterative method [19,20,21]. The process starts with the desired intensity distribution $I(u,v)$ and a random phase distribution in the far-field of the desired DOE. The corresponding complex amplitude $G(x,y)$ in the DOE plane is obtained by a Fourier transform and the result is projected onto the constraints of the DOE. In the case of a phase only element this means $|G(x,y)| = 1$. The inverse

Fourier transform is then taken and projected onto the design constraints, and the process is then repeated until it converges. This method is fast, but sensitive to the start parameters. It is difficult to avoid that the process stagnates in a local minimum. The advantage of the method is that large amounts of data can be handled.

Downhill simplex is a different method [22]. Here we vary the parameter set which describes the shaping element and observe the result in the image plane. The result is defined by a merit function. From different parameter sets the direction to the minimum is determined. The method converges rapidly to the next minimum, which is not always the global one. This method is well suited for few parameters.

Good results have been obtained by simulated annealing [23]. This method is able to climb out of local minima. A merit function $E(p)$ has to be defined, which describes the error as a function of the parameter set p. During the process the parameter set is altered, which yields a change of ΔE in the merit function. If ΔE is reduced, the change is accepted. If ΔE increases, the change is accepted with probability $\exp(-\Delta E/T)$. T is a control parameter analog of temperature. After a number of cycles, the temperature T is lowered slowly and the process continues. As a result, the probability of a change being accepted, with $\Delta E > 0$, is reduced. The annealing schedule is very important for the success of the method. The method works well, but is rather slow.

Simulated annealing has been applied successfully to design fan-out elements with different surface-relief profiles, such as Dammann gratings [24], multilevel phase gratings [25], and continuous phase structures [26]. As the method is slow, it has been combined with other algorithms, such as damped least-squares or the greedy downhill algorithm. Besides fan-out elements which generate uniform arrays of light spots, weighted fan-out elements have also successfully been designed [27].

Fabrication considerations favour the use of Dammann gratings which have a binary profile, i.e. two phase levels. Elements generating arrays up to $N = 1024$ have been calculated with typical efficiencies of 80 % [28]. Two-dimensional fan-out elements generating equally intense diffraction orders can be constructed from-one dimensional solutions. As a result the diffraction efficiency of the one-dimensional solution is squared. Thus, we obtain 64 % for Dammann gratings. This can be avoided by looking for non-separable solutions. The drawback is an increase of the number of optimization parameters. Such a design, based on simulated annealing, has been reported recently by Dames et al. [27]. For two-level phase gratings they have obtained diffraction efficiencies between 75 % and 80 %, depending on the spot patterns to be generated.

The success of nonlinear optimization depends not only on the algorithm but also on the choice of the parameters and the merit function. A very fast method to calculate smooth surface-relief structures for fan-out elements has been presented by Herzig et al. [5]. The relief structure is described by the amplitude and phases of virtual object sources. The optimization criterion is the variance of the object irradiance $<I^2> - <I>^2$ in the plane of the DOE. This procedure converges rapidly by using the downhill simplex method. The highest efficiency of 99.3 % and perfect uniformity has been obtained for a fan-out of N = 9. For larger fan-out numbers the efficiency remains higher than 97 % [29]. We have also determined non-separable solutions for continuous relief gratings. A strong improvement for non-separable design is found for low fan-out numbers N < 7. For example, the efficiency of 2x2 fan-out increases from 64 % to 92 %, and for a 6x6 fan-out the efficiency increases from 79 % to 93 %. For larger fan-out numbers, the difference is only a few percent. In some cases, e.g. efficiency of 98.6 % for a 9x9 fan-out, no better solution has been found. Note, that the values given above for the continuous relief gratings are obtained for a uniformity error < 0.1 %.

Multilevel and continuous relief gratings offer higher diffraction efficiencies than Dammann gratings, on the other hand the fabrication tolerances are more critical. Therefore, they are preferable where the highest benefit is achieved compared with Dammann gratings. This is the case for low fan-out numbers and for weighted fan-out elements.

5.3 Diffraction theory

Computational methods for calculating the diffraction efficiency of diffractive optical elements depend on the dimensions of the microstructure features relative to the wavelength of the read-out illumination (typically $\lambda = 500 - 900$ nm). For coarse structures (features $>> \lambda$), scalar diffraction theory can be used to accurately calculate the diffraction efficiency of the microstructures [30]. For finer structures (features $\approx \lambda$), the approximate methods are no longer valid, although they may serve as a guideline for practical design. However, for a detailed analysis a rigorous diffraction theory is required.

In this section, we will introduce approximate diffraction theories and then present recent developments in rigorous diffraction theory for the treatment of fine (features $\approx \lambda$) structures.

5.3.1 Approximate diffraction theory

A number of approximate methods have been devised for the computation of the diffraction amplitudes, in particular for coarse structures (features $>> \lambda$). They are typically based on incomplete expansions of the wavefield. A very important simplification results if the vector character of the electromagnetic field is neglected. The most widely used method for the design of surface-relief gratings is the scalar diffraction theory [30]. For a periodic structure, the amplitudes of the various diffraction orders can be determined by a simple Fourier transform of the grating transmittance function. The result is independent of the grating period and the state of polarization of the incident light. These predictions are valid for gratings whose geometry is coarse on the scale of the wavelength. The region of validity can be extended down to a few wavelengths, if the optical path through the grating is taken into account [31]. Descriptions of a number of approximate methods can be found in Ref. [32]. The relationships between various grating diffraction theories in terms of fundamental approximations have been discussed in Ref. [33].

5.3.2 Rigorous diffraction theory

Several different methods have been developed over the years for the accurate solution of Maxwell's equations. These have been reviewed extensively in the literature. We refer the interested reader in particular to the book edited by R. Petit [34]. Rigorous diffraction theory has also been applied to analyze Dammann gratings [35].

Gratings with arbitrary profiles, which are not very deep compared to the wavelength, can be calculated by integrating the partial differential equations, as well as by Green's function methods [36]. Deep gratings pose severe problems to schemes based on direct numerical integration of the differential equations. While differential methods are straightforward to implement [37]. Differential methods include the rigorous coupled wave theory and the rigorous modal theory, which are the most common methods for analyzing diffraction gratings. Both of these approaches can produce exact formulations without approximations. In their full rigorous forms these formulations are completely equivalent [38]. They differ mainly in the way the electromagnetic field inside the grating is represented. In the case of the coupled wave theory the field inside the grating is expanded in space harmonics [39], whereas in the case of the modal theory the field is expanded in terms of allowed modes of the periodic structure [40].

Computer-generated DOEs often have binary or multilevel profiles, which can be described by stacks of lamellar gratings. In the next section, we will discuss an efficient method to compute these gratings.

5.3.3 Rigorous modal theory for one-dimensional lamellar gratings

A class of diffraction gratings, which permits a particularly efficient numerical treatment, are one-dimensional lamellar gratings or stacks of lamellar gratings. Examples are binary Dammann gratings or multilevel phase gratings. Even more general grating structures can be approximated by stacks of lamellar gratings, as shown in Fig. 5.5.

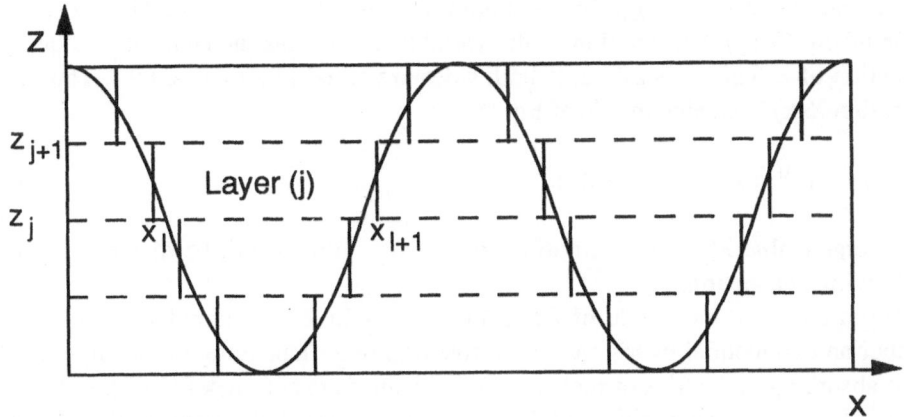

Fig. 5.5. Approximation of sinusoidal profile by stack of lamellar gratings.

For this type of grating, the dielectric constant $\varepsilon(x,z)$, which is periodic in the x-direction, does not depend on the variable z within a layer $z_j < z < z_{j+1}$. It results a natural separation of the variables. We will illustrate this for the example of s-polarization, where the electric field vector **E** is parallel to the y-direction, i.e. parallel to the grooves in Fig. 5.5.

The Helmholtz equation reads for this case:

$$\frac{\partial^2 E(x,z)}{\partial x^2} + \frac{\partial^2 E(x,z)}{\partial z^2} + k_0^2\, \varepsilon(x,z)E(x,z) = 0, \tag{5.7}$$

where $k_0 = 2\pi/\lambda_0$ is the wavenumber and λ_0 the vacuum wavelength. For each layer (j), in the interval $z_j < z < z_{j+1}$, we may write

$$\varepsilon(x,z) = \varepsilon^{(j)}(x). \tag{5.8}$$

As mentioned above, we are looking for separable solutions of the following type:

$$E(x,z) = X(x)Z(z). \tag{5.9}$$

Inserting $E(x,z)$ described by Eq. (5.9) into the Helmholtz equation (5.7) yields

$$X''(x)Z(z) + X(x)Z''(z) + k_0^2\, \varepsilon^{(j)}(x)X(x)Z(z) = 0, \tag{5.10}$$

or, as usual, after dividing by $X(x)Z(z)$

$$\frac{X''(x)}{X(x)} + k_0^2\, \varepsilon^{(j)}(x) = \frac{Z''(z)}{Z(z)}. \tag{5.11}$$

Here and in the following, differentiation is denoted by a prime. The left-hand-side of Eq. (5.11) is a function of the variable x, whereas the right-hand-side is a function of z. This is possible, if both sides are equal to a constant λ^2. Thus, the function X(x) is an eigenmode of Eq. (5.12):

$$X_n''(x) + k_0^2 \varepsilon^{(j)}(x)X_n(x) = \lambda_n^2 X_n(x), \tag{5.12}$$

with eigenvalue λ_n^2. The calculations of the eigenmodes will be discussed in the following subsection.

From the right-hand-side of Eq. 5.11, we obtain $Z(z) = \exp(\pm i\lambda_n z)$, which is either an exponential or trigonometric function of z, depending on the sign of λ_n^2. For absorbing dielectrics or metals the eigenvalues are complex [41,42]. In layer j, this allows a decomposition of E(x,z) in terms of eigenmodes $X_n^{(j)}(x)$ of the Helmholtz- equation, as

$$E(x,z) = \sum_{n=1}^{N} X_n^{(j)}(x)Z_n^{(j)}(z), \tag{5.13}$$

where

$$Z_n^{(j)}(z) = a_n^{(j)}\exp(-i\lambda_n z) + d_n^{(j)}\exp(+i\lambda_n z). \tag{5.14}$$

The coefficients $a_n^{(j)}$ and $d_n^{(j)}$ have to be determined from the boundary conditions (continuity of E(x,z) and its derivative at y_j) and the radiation condition at $\pm\infty$, as usual. The calculation of the diffraction properties thus proceeds in two steps: (1) For each grating layer (j), the eigenmodes $X_n^{(j)}(x)$ of Eq. (5.13) are calculated - since these do not depend on the thickness of a particular layer, calculation of diffraction properties for varying thicknesses in an optimization involves this step only once; (2) the boundary conditions lead to a system of coupled linear equations for the coefficients $a_n^{(j)}$ and $d_n^{(j)}$, which can be solved by standard techniques. This system of linear equations, though, suffers from numerical instability due to fast exponential growth of $Z_n^{(j)}(z)$ for high-order modes. However, there are methods to overcome this numerical instability without the need for high-precision arithmetic.

Calculation of eigenmodes. For the calculation of the eigenmodes $X_n(x)$ of Eq. 5.12, different procedures have been developed. Fourier decomposition of X(x) and the dielectric constant $\varepsilon(x)$ leads to an algebraic eigenvalue problem [40]. Alternatively, Eq. 5.12 together with the requirements of periodicity and continuity of E(x,z) and $\partial E(x,z)/\partial x$, can be used to derive a transcendental equation for the eigenvalue λ^2 [41,42]. The eigenmodes $X_n(x)$ are then given analytically in terms of λ_n. The Fourier method suffers from unnecessarily poor convergence, since the discontinuity of the refractive index at the various interfaces leads to a Gibbs phenomenon. This poor convergence is avoided, if one determines the exact eigenvalue λ_n by the solution of the transcendental equation. However, its solution poses numerical problems, in particular for complex dielectric constants [41,42].

For this reason, we have used an alternative method, which eliminates both disadvantages. The electromagnetic field is infinitely many times differentiable, except at the interfaces between different materials. There, either the first (p-polarization) or the second derivative of the field is discontinuous. So, the poor convergence of the Fourier expansion results from the fact, that these discontinuities are not correctly treated. This is avoided by representing the field as a superposition of analytic functions in each region of constant refractive index, separately and by imposing the correct boundary conditions at the interfaces. As basis functions, we choose the Legendre polynomials $P_i(x)$. The vertical positions of the interfaces in layer (j) are denoted by $x_1^{(j)}, 1 = 1,..,L^{(j)}$, see Fig. 5.5. Then, the eigenmode $X_n^{(j)}(x)$ is represented as

$$X_n(x) = \sum p_i^{(l)} P_i(\xi), \qquad x_1^{(j)} \le x \le x_{l+1}^{(j)}. \tag{5.15}$$

For simplicity, we have dropped the superscript j, which denotes grating layer j. The reduced variable

$$\xi = \frac{2x - x_1^{(j)} - x_{l+1}^{(j)}}{x_{l+1}^{(j)} - x_1^{(j)}} \tag{5.16}$$

results from the mapping of the interval $[x_1^{(j)}, x_{l+1}^{(j)}]$ to the unit interval $[-1,1]$, on which Legendre polynomials $P_n(x)$ are orthogonal and complete. Expressing the derivatives of the Legendre polynomials again in terms of sums of Legendre polynomials, and using the appropriate boundary conditions for field and derivative at the vertical interfaces $x_1^{(j)}$, Eq. (5.12) is transformed into an algebraic eigenvalue problem, which can be solved by standard methods.

Examples. In this section, we present two examples of grating structures optimized for a particular application.

The first example is a grating for beam fan-out or beam splitting. The goal is the design of a diffraction grating for a wavelength of $\lambda = 900$ nm, with the following properties: s-polarization should be split into three beams (zero-, ±first-order) of equal intensity, p-polarization should be transmitted with minimal diffraction.

The structure found is shown in Fig. 5.6. In the case of s-polarization, the computed values for transmission in zero- and ±first-order are all 26.3 %. In reflection, the values are 10.4 % for zero-order and 5.4 % for ±first order.

84

For the p-polarization, the value for transmission in zero-order is 75.9 % and in ± first-order it is 6.0 %. In reflection, we have calculated 8.6 % in zero-order and 1.8 % in ±first order.

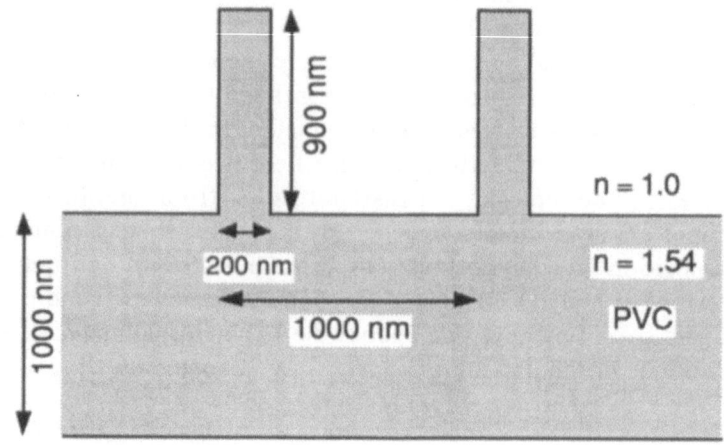

Fig. 5.6. Optimized grating for beam fan-out.

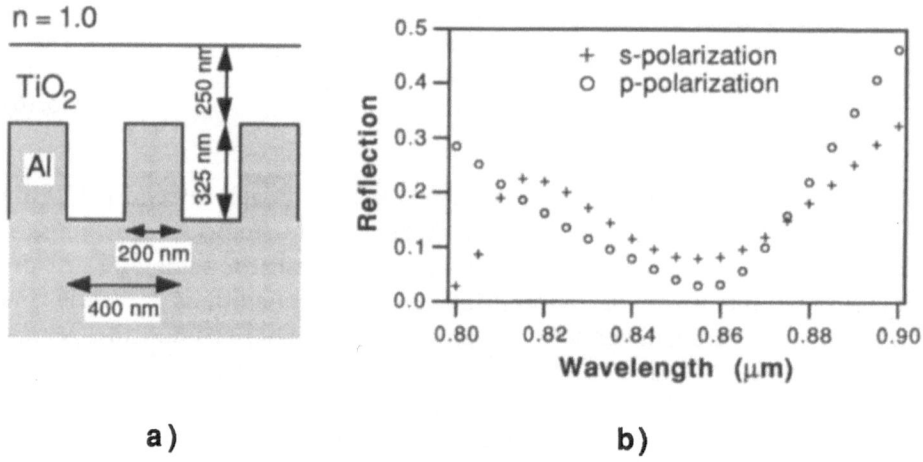

a)   b)

Fig. 5.7. Highly absorbing grating structure ("black aluminum"): **a)** microstructure and **b)** computed reflectivity.

The second example is the "Black aluminum". Here, the goal is the design of a grating which makes an aluminum layer essentially black for the wavelength range 800 - 900 nm. Such a grating would have applications in thermally activated optical (reflection) switching elements for optical computing and interconnects. A solution for the grating microstructure is shown in Fig. 5.7(a), together with the computed reflectivity versus the wavelength [Fig. 5.7(b)]. Note, that neither the grating, nor the TiO_2 overcoat alone, can achieve such low reflectivity. It is the optimized combination which leads to this result.

5.4 Fabrication technology for diffractive optical elements

Fabrication techniques for realizing the microstructures resulting from the design of diffractive optical elements (DOEs) are based on a variety of high resolution lithographic and optical processes. This section describes techniques suitable for the fabrication of surface-relief (phase) microstructures, such as binary and multilevel grating structures, and continuous microreliefs. The typical procedure is to generate a mask by e-beam lithography or by laser beam writing lithography. Then, to get high efficiency, they are transformed into surface-relief structures by dry or wet etching. Using several masks, multilevel profiles can be generated to improve the efficiency. Another technique is the direct writing of the DOE phase relief in photoresist by e-beam or laser beam. The developed photoresist relief can be converted into a metallized master relief by electroplating to emboss or cast low-cost replicas.

The major fabrication processes can be grouped into the following categories:
   - Primary pattern fabrication (Sect. 5.4.1)
   - Pattern transfer by etching and material deposition (Sect. 5.4.2)
   - Direct writing of continuous microreliefs (Sect. 5.4.3)
   - Replication technologies (Sect. 5.4.4)

A detailed description of these techniques is beyond the scope of this chapter, which will summarize the main features and limitations. An exhaustive account of equipment and processing for modern, high-resolution semiconductor IC fabrication has been published by Moreau [43]. The state of the art in microlithography is summarized in the proceedings of *Microcircuit Engineering 90* [44]. A review of dry etching technology and other thin film processes can be found in Ref. [45]. Numerous references to planar fabrication technologies for optical interconnects are given in Ref. [46].

5.4.1 Primary pattern fabrication

The primary pattern defines the geometrical (fringe) pattern for a DOE. Typical examples include simple linear gratings, curved grating patterns, and complex grating patterns, such as those required for Dammann gratings. The primary pattern is typically a chrome-on-glass mask, which can be used to expose a photoresist-coated substrate, or it is written directly in photoresist. The pattern is then processed to the ultimate microrelief structure by a series of lithographic and etching or deposition steps. In some cases, the mask or the patterned resist film can be used directly as DOE. Table 5.1 summarizes the main approaches and gives references to more detailed descriptions.

For multilevel structures, a number of masks may be required. The mask approach has the advantage that the complex and/or high resolution pattern must be fabricated only once and can then be used for fabricating many DOEs. It follows the general fabrication route of semiconductor processing.

Table 5.1. Primary pattern fabrication techniques

Technology	Resolution [a]	Comments
e-beam writing [47,49]	100 - 200 nm	Commercial mask fabrication services available. Data must be prepared in standard format.
Focused ion beam writing	100 - 200 nm	Allows maskless processing. Used for mask repair and IC failure analysis
Optical writing [50]	500 - 1000 nm	Commercial mask fabrication services available. Data must be prepared in standard format.
Laser interference [51]	150 - 250 nm	Resolution depends on laser wavelength [b]. Curved gratings by shaping interfering wavefronts.

[a] Finest features - approximate limit for corresponding technology; higher resolution obtainable using special techniques (e.g. shadow evaporation [52])

[b] Typical examples range from the HeCd laser (λ = 442, 325 nm) to the frequency doubled Ar-laser (λ = 256 nm).

Electron-beam writing. The most straightforward technique to fabricate chrome masks is to use a commercial e-beam writing system. Since such systems are very expensive ($2-4 million), most users prefer to buy the masks in a commercial mask shop. Data must typically be prepared in an accepted, standard format. Mask resolution of 500 nm (smallest line dimensions) is routinely available. Typical cost for a mask is in the range $2000 - $4000, depending upon complexity, size etc. Higher resolution is offered by certain shops undertaking advanced jobs for research and development applications. Current capabilities and limits of e-beam writing technology are summarized in [47]. An alternative technique which is of growing importance, focused ion beam writing, is summarized in [48]. The technique has the potential advantage of allowing in-situ, maskless processing; writing parameters such as resolution, accuracy and writing field are similar to those for e-beam writing.

Laser beam writing. Many mask shops also offer binary masks written by commercial laser writing systems (not to be confused with the laboratory systems for writing continuous microrelief structures as described in Sect. 5.4.3). The mask cost is somewhat lower than that of e-beam written masks, but maximum resolution is limited to about 500 nm. Data must again be prepared in a standard format.

Mask printing into photoresist. The transfer of a chrome mask pattern into photoresist can be achieved by contact printing, or by projection printing. Contact printing is a low-cost process suitable for R&D work. The mask is pressed onto a photoresist-coated substrate, exposed by UV radiation from a high pressure mercury lamp. Resolution of better than 500 nm is readily achieved in the laboratory. Commercial mask aligners cost in the range $200,000 - $500,000. They can be used for advanced (i.e. reproducible, large area) work.

Projection printing has the advantage that an image reduction can be used, but it requires a high performance micrographic lens. Such lenses are commercially available, as are complete projection printing mask aligners (typical cost $1 - 2 million). The resolution is limited by the lens. For a high quality lens, the resolution is typically about 500 nm over imaging fields of 1-2 cm diameter.

The developed photoresist structure can form the final phase DOE if fabricated on a transparent substrate (e.g. glass or quartz). It can also form a master microrelief for the fabrication of a metal replication shim (Sect. 5.4.4). Alternatively, it can function as a mask for subsequent subtractive (etching) or additive (plating or lift-off) pattern transfer techniques (Sect. 5.4.2).

An alternative to the chrome mask approach is to fabricate the primary pattern directly in resist. In principle, this can be achieved by direct e-beam writing (which requires an expensive e-beam writing system) or by laser beam writing (limited resolution). A more common approach used in many R&D laboratories is the patterning of photoresist by laser interference ('holographic exposure').

Laser interference techniques. Submicron grating structures can conveniently be fabricated by laser interference exposure of a photoresist film, as shown in Fig. 5.8. A review of various recording configurations can be found in Ref. [53]. Curved and chirped grating geometries can be recorded by shaping the interfering wavefronts [51].

If the substrate is a chrome-on-glass blank, subsequent etching of the chrome produces a high resolution mask suitable for further lithography (see above). Practical limitations in the chrome etching limit the resolution to about 500 nm for standard (commercial) chrome blanks (~100 nm chrome thickness), or about 100 nm using special thin chrome layers. For fine structures (linewidths < 500 nm) the best approach is generally to use the resist microrelief directly for processing the substrate, e.g. dry-etching into quartz, or lift-off techniques (see Sect. 5.4.2).

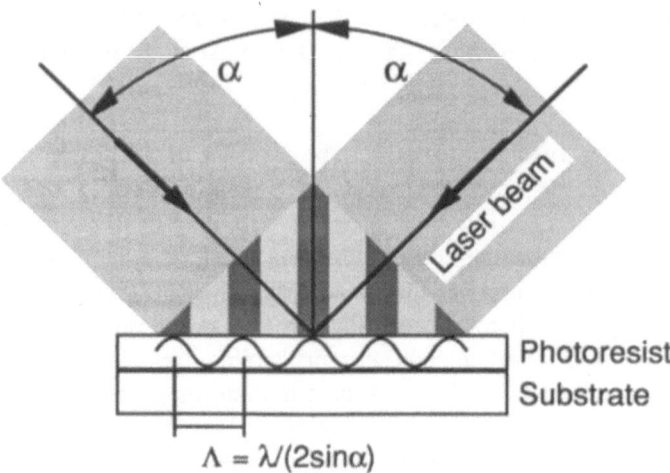

$$\Lambda = \lambda/(2\sin\alpha)$$

Fig. 5.8. Laser interference for the fabrication of fine grating structures in photoresist.

5.4.2 Pattern transfer

The primary pattern can be transferred into a more durable material by either subtractive (e.g. wet chemical etching, plasma etching) or additive (e.g. lift-off) techniques. Fig. 5.9 schematically shows the pattern transfer by lift-off and etching. In both cases the resist pattern acts as a mask for the pattern transfer step. It is obvious that the exact shape of the mask profile can have a large influence on the final profile. These two techniques will be described in more detail. Typical materials that have to be patterned include metals (Au, Al), semiconductors (Si, GaAs), or dielectrics (polymers, glass, SiO_2, $LiNbO_3$, Ta_2O_5 or TiO_2), either as bulk material or thin films. Although it is possible to establish some general rules for microfabrication of gratings, each material poses its own set of problems which have to be addressed individually. In addition, the exact shape of the final profile not only depends on the material and the technique, but in many cases also on the chosen masking material and its profile.

If the pattern transfer step is done by *subtractive techniques*, the primary mask pattern protects the substrate surface in desired areas and the unwanted material is removed by *wet or dry etching*. In order to faithfully transfer the primary pattern into the substrate, this etching step usually has to be highly directional or anisotropic. *Additive techniques* typically involve one of two basic processes: *lift-off* of material deposited by a physical process such as evaporation, or masked growth of material deposited by a chemical process such as *electroplating*. In the following, we will discuss the different techniques.

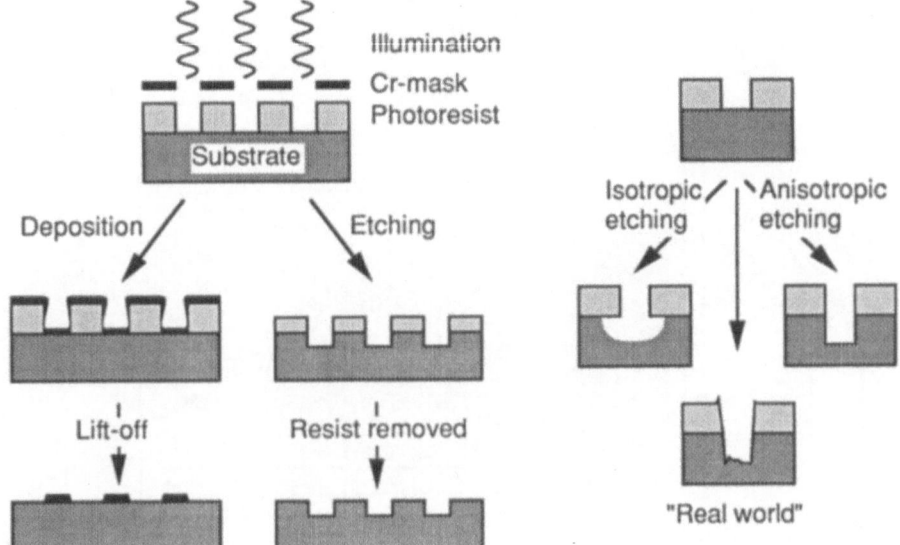

Fig. 5.9. Pattern transfer by additive (left side of figure) or subtractive (right side) technique. The lift-off technique adds material to be patterned to the substrate, etching leads to a structured surface of a given substrate.

Fig. 5.10. Isotropic and anisotropic etching processes. A "real world" dry etching profile is shown in the center.

Wet etching. Wet chemical etching mostly leads to isotropic etching, i.e. etching proceeds at the same rate in all directions. However, in single crystal materials such as Si, it is possible to make use of the difference in wet chemical etch rates along different crystallographic directions. Highly anisotropic structures with aspect ratios (height to width of lines) in excess of 650:1 have been achieved [54]. Figure 5.10 shows idealized isotropic and anisotropic etch profiles, as well as a typical "real" profile to illustrate some of the problems which are encountered in plasma-assisted dry etching. Basically, the anisotropic nature of transferred patterns is due to the highly directional ion impact, also other effects contribute.

Dry etching. Dry etching is a term used to describe unassisted (or direct) as well as assisted (e.g. by ions) gas-solid reactions. It is an important process for the fabrication of binary or multilevel microstructures in substrate materials such as quartz. Since fabrication of DOEs generally requires very specific profiles over relatively large areas, the discussion of dry etch processes will be limited here to a family of methods which are capable of producing highly anisotropic profiles, namely plasma-assisted etching.

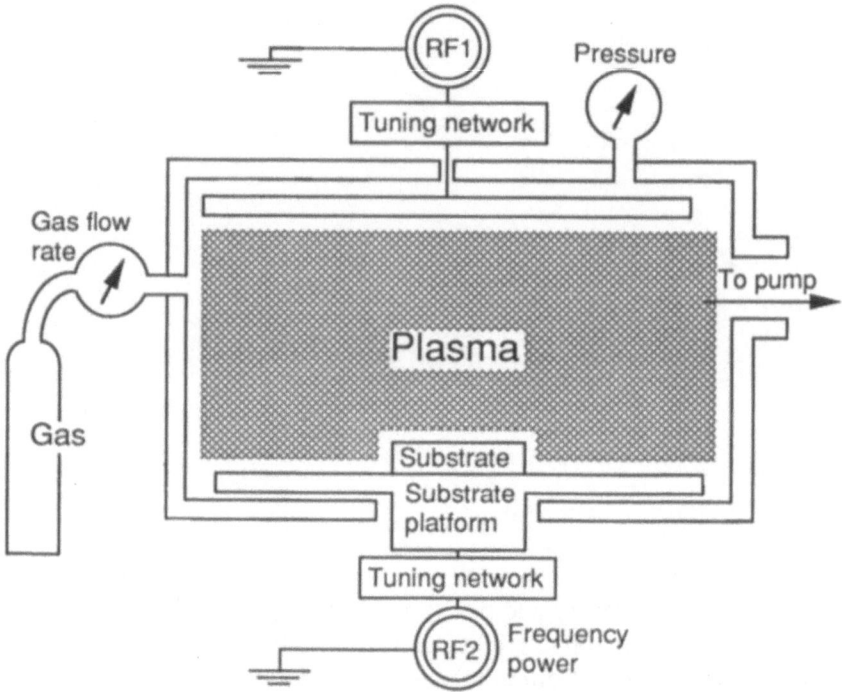

Fig. 5.11. Typical parallel plate reactor used for *reactive ion etching (RIE)* or *plasma etching*. Parameters which can be independently controlled include frequency and power of applied rf-voltage, gas pressure and flow rate, pumping speed, electrode material and substrate temperature. The area ratio of powered to grounded electrode determines the energy of the bombarding ions through the plasma potential. Polymer deposits, which form in C-containing plasmas (e.g. CF_4), can influence the plasma chemistry.

Figure 5.11 serves to illustrate some of the fundamental concepts of reactive ion etching; for a more complete discussion of dry etching see Ref. [55]). The figure schematically shows a typical reactive ion etching chamber with some of the instrumentation required for controlling the glow discharge in the chamber. The figure caption describes the various important parameters. The system consists of a metal vacuum chamber. Etching is typically done at pressures in the range between 1 and 100 mTorr, typical flow rates being in the range between 5 and 200 sccm. The application of a high RF-voltage of the order of a few hundred volts between the electrode supporting the substrate and the metal chamber leads to the formation of a plasma above the substrate electrode.

Basically, the dry etching process can be described as follows: A gas discharge is ignited by applying dc- or rf-power to the electrodes. A specific molecular gas is flown through the chamber. Activation of the gas occurs in the glow discharge forming ions and/or radicals through excitation, dissociation or ionization. A relatively complex combination of physical and chemical processes between these plasma-produced species and the solid surface will result in etching. It is very important that the majority of the etched solid material is chemically converted into volatile products which can be removed by the vacuum system. If the material is only sputtered, as e.g. in purely physical sputter etching or ion milling, it will be redeposited on the sidewalls of steep features (e.g. masking pattern) leading to an undesirable deformation of the etched profile.

Depending on the specific application, there are a number of considerations to be made in order to optimize the pattern transfer step by dry etching. In addition to the requirement of a high fidelity pattern transfer from the mask into the substrate, uniformity and selectivity have to be closely controlled and the etch process should lead to minimum contamination.

A number of factors can influence the *etch uniformity* across the whole area of the substrate: Gas flow (relative position of inlet and pumping port, total gas flow), gas pressure and - closely related to flow rate and pressure - residence time of the feed gas. Also the relative reactivity of the substrate surface with respect to that of the electrode material can also greatly influence uniformity.

An ideal etch process should result in a fast removal of the material of the substrate and in a minimal etch rate for the mask, i.e. the process should be *highly selective*. Plasma-assisted etching processes have a considerably lower selectivity than corresponding wet-chemical etch processes. Relative dry etch rates for two materials rarely differ by more than a factor of 20 and often much less. The transfer of grating patterns mostly involves the use of photoresist as the masking material. Its dry etch rate can be reduced to a certain degree by some special techniques [56]. The selectivity can also be greatly affected by the composition of the plasma.

Contamination and damage are two undesired side-effects in plasma-assisted etching which have to be minimized. High energy particles, in particular ions, have large enough momentum to remove material from any surface which comes in contact with the plasma. Since this material is usually not volatile, it will be redeposited the next time it strikes a surface, which might be on the substrate surface leading to contamination. Furthermore, the substrate to be etched is constantly exposed to high energy radiation (UV and X-rays) emanating from the plasma. Problems due to deposition of contaminants and radiation damage are intimately related.

Due to the large parameter space available, plasma-assisted etching offers many different ways to control the profile of a transferred pattern. In many applications, highly anisotropic pattern transfer is required, but in other cases the final grating profile in the substrate should have walls with a specific angle or it should have rounded bottoms etc. In the following, some of the basic ion-surface effects affecting profile control will be discussed.

Ion-surface interactions play a dominant role in pattern transfer by plasma-assisted etching. The high degree of anisotropy, usually required in the transfer of grating patterns, can be obtained by highly directional bombardment of the surface by ions. Since the ions follow the electric field lines which are normally perpendicular to the substrate, etch profiles show sidewalls arranged perpendicularly to the substrate surface. But the pattern transfer process is not only dependent on these physical effects due to ion-surface interactions, but also influenced by purely chemical reactions between reactive neutral species in the plasma and vertical surfaces (in particular sidewalls of etched profiles which receive practically no ion bombardment).

Chemical reactions can lead to undercutting of masking patterns (see Fig. 5.10) or to the deposition of so-called sidewall blocking layers. Undercutting can be undesirable for one kind of process (i.e. square-wave grating patterns) but highly desirable for other processes (i.e. generation of lift-off profiles). Deposition of sidewall blocking layers can be very important in obtaining anisotropic etch profiles. A thorough understanding of the various physical and chemical effects is essential to optimize the profiles of etched gratings.

One of the main effects leading to a deformation of the etched profile by dry etching is *redeposition* of non-volatile material sputtered either from the substrate or from chamber fixtures and walls in contact with the processing plasma. Backscattering can also have significant effects in reactive ion etching. If the plasma comes in contact with materials which do not form volatile products with the etch gas (e.g SiO_2-masking pattern on Si in Cl_2-plasma or Si-based material resting on a substrate holder made from aluminium and etched in F-containing plasmas), backscattering can cause problems in reactive ion etching which are far more serious than in normal glow discharge sputter etching. When a non-volatile species arrives at and condenses on a surface which is rapidly etching forming volatile compounds, the non-volatile species acts like a micromask [57].

Faceting is yet another problem directly related to physical ion bombardment. It is due to the angle-dependent sputter yield [58] and causes sharp edges to develop facets at an angle to the incoming ions which maximizes sputter yield. Normally it only affects the masking pattern and its influence on the fidelity of the pattern transfer process can be minimized by making the mask sufficiently thick.

As a result of redeposition, faceting, and generally rather poor selectivity, pure physical sputtering using glow discharges of Ar or other noble gases is only used for pattern transfer in materials which do not readily form volatile reaction products in reactive gas environments. For such applications the preferred pattern transfer method is *ion milling* [59]. In ion milling, an ion beam is able to glow discharge sputter etching for the following reasons: 1) less parameters, 2) no backscattering (due to the lower pressure and consequent larger mean free path) and 3) the possibility of varying the angle of incidence of the sputtering ions and thus being able to etch structures such as blazed gratings [60].

Special techniques - squarewave profile fabrication. It is sometimes very difficult or even impossible to use holographically produced submicron grating patterns as etch masks for dry etching, because these patterns usually only have a thickness of about half the periodicity. One elegant method to convert a primary pattern with a sinusoidal profile into a squarewave profile by slope evaporation of a periodic etch mask, has been described by Anderson et al. [61] and is illustrated in Fig. 5.12.

An oxygen-plasma resistant material (e.g. Cr or SiO) is evaporated at an angle of 10 - 15° with respect to the substrate surface onto a holographically recorded grating. The grating lines act as individual shadow masks for the evaporated material. The total photoresist thickness can actually be much greater than the desired grating depth, because the final definition of the pattern is done by RIE in an O_2-plasma removing the photoresist between the grating lines. The ensuing photoresist grating pattern (with evaporated material covering its top) has now a profile approaching a square-wave profile and its aspect ratio (height to width) can be made easily as high as 5:1. This pattern can again be used as an etch- or lift-off mask for further processing.

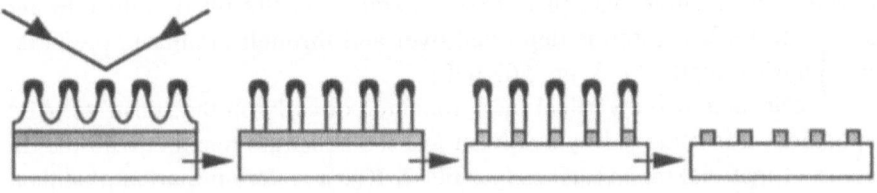

(a) Slope evaporation   (b) Resist RIE in   (c) Etch substrate   (d) Strip resist
of hard mask O_2 plasma   e.g. SiO_2 in CHF_3

Fig. 5.12. Fabrication of etched squarewave profile from sinewave resist mask by slope evaporation of an etch mask.

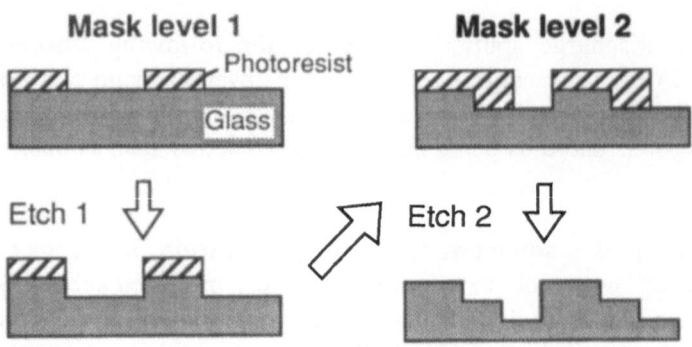

Fig. 5.13. Fabrication of multi-level microrelief structures by repeated mask and etching steps.

Multi-level microstructures. Multilevel microstructures with more than 2 relief levels are required for efficient diffractive optical elements [7,62]. They can be fabricated by repeated etching steps, shown schematically in Fig. 5.13. If the substrate is resist-masked and etched a total of n times, a microrelief structure with 2^n relief levels is generated. Each mask exposure must be carefully aligned with the existing pattern, a requirement which can best be fulfilled using a commercial mask aligner [63].

Lift-off. If the surface of a material (on a specific substrate) cannot easily be structured by dry etching, or if such methods are undesirable for some other reasons (e.g. radiation damage), lift-off techniques can be employed. An example is the fabrication of diffractive microlenses by thin film deposition [64].. In these processes, a negative pattern is first generated on the blank substrate and the material to be structured is deposited over and through the mask, preferably by some physical method [43, pp. 567-630].

The resist acts like a vertical mask mounted directly on the substrate. After the deposition, it is removed by a solvent and the substrate remains with the desired pattern of material on its surface (see Fig. 5.10). It is very important that there is a "break" (e.g. a small crack) in the deposited metal film where it meets the vertical resist wall to enable access for the resist solvent. Therefore, the mask geometry is of prime importance for the lift-off process. Many process variations exist for the preparation of improved masks in which the resist profile is undercut to provide an enhanced shadowing effect [52]. For example, Shipley resist AZ-1350J can be soaked in chlorobenzene before exposure and development, to create an undercut profile due to differential solubilities of the upper and lower parts of the resist in the developer after exposure [65].

For certain applications, a lift off mask that will withstand temperatures up to 900°C is required. An elegant method to form a T-shaped mask utilizes the lower solubility in buffered HF of Si_3N_4, prepared by low frequency (50 kHz) plasma enhanced deposition (PECVD), as compared to Si_3N_4, prepared by high frequency (13.56 MHz) PECVD [66]. Another high-temperature lift-off technique allowing processing temperatures up to 420°C, but more compatible with existing photoresist techniques, involves the use of photosensitive polyimide [67].

The deposition of the material to be structured by lift-off should only be carried out in a directional fashion, since in an ideal case the sidewalls of the resist pattern should receive a minimum amount of deposited material to provide ready access for the solvent. This requires an evaporation source (resistance or e-beam heated) with an emitting angle of no more than ±15° [68]. Furthermore, the evaporated beam should meet the substrate surface as vertically as possible. Sputtering does not fulfill the above requirements due to the large diameter of the target and its proximity to the substrate in most commercial systems.

Electrodeposition. An alternative additive technique, which can be used on conducting surfaces, is pattern generation by electrodeposition. In this approach, the photoresist pattern or any non-conductive masking pattern (e.g. an organic polymer structured by O_2-plasma) applied to a conductive surface acts as a direct template for the metal to be deposited. The technique is often used for producing Au-absorber patterns on X-ray masks [69].

Equipment costs. A wide variety of commercial equipment for dry etching has been developed for the semiconductor industry. The cost of an etching system depends upon its complexity and is strongly affected by 'optional' features such as automatic wafer handling. Table 5.2 shows a comparison of the approximate cost of equipment required for the etching processes described above - the costs given are for basic systems suitable for fabrication of DOE microstructures in the laboratory.

Table 5.2. Etching equipment costs

Technology	Typical cost of equipment	Comments
Wet etching	$100	Requires only beaker and etching chemicals Anisotropic etching only for crystalline substrates
Dry etching		
Reactive ion etching (RIE)	$300,000	e.g. etching in plasma of SF_6, $CHClF_2$, CHF_3, ... (material specific)
Ion beam milling	$300,000	Ion beam etching with Ar^+ of practically all materials
Reactive ion beam etching (RIBE)	$500,000	Reactive ion beam, e.g. with Cl_2 for etching GaAs

5.4.3 Direct writing of continuous microreliefs

Diffractive optical elements constituting continuous microoptical relief can be fabricated by direct writing in photoresist [4,11]. Fig. 5.14 shows the laser beam writing system at the Paul Scherrer Institute Zürich [4]. A photoresist-coated substrate is raster scanned by a focused laser beam, synchronously controlled in intensity, to write a fully programmable, 2-dimensional exposure pattern. Development of the resist then results in a microrelief of the desired structure. The exposure date have to be computed from the desired final microrelief and the resist development characteristic.

The ability to program complex xyz microrelief profiles enables a wide range of structures to be fabricated, ranging from arrays of 1- and 2-dimensional lenslets, fresnel lenslets to highly complex phase structures for new types of DOEs. Typical periods of 10-100 µm and a maximum relief amplitudes of about 5 µm can conveniently be realized. Fig. 5.15 shows examples of recorded microlens and fan-out structures. The microreliefs can be used directly as phase structures (resist on glass) or for fabricating replication shims.

Continuous microrelief structures can also be written directly by e-beam lithography [70], although the maximum achievable profile depth and the writing speed are in general lower than for laser beam writing.

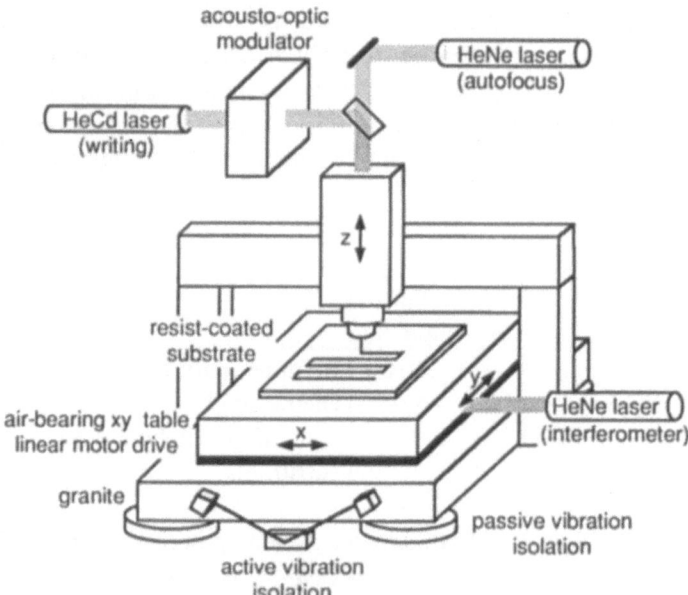

Fig. 5.14. Laser beam writing system for the fabrication of continuous profile microoptical components.

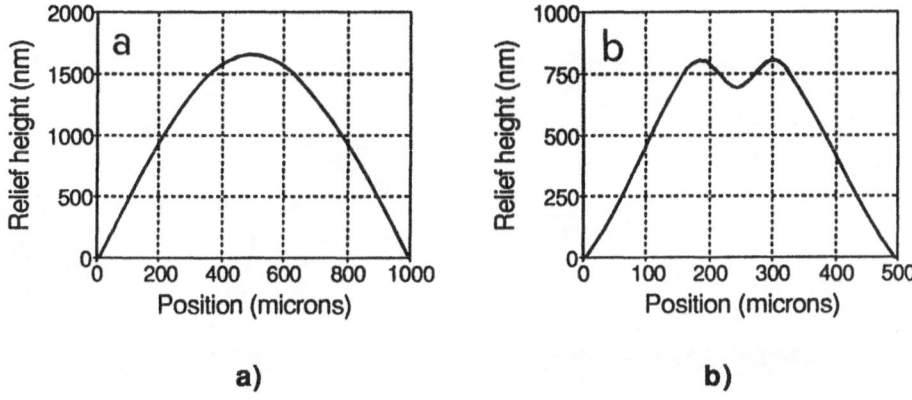

a)   b)

Fig. 5.15. Measured profiles of **a)** microlens and **b)** kinoform microstructures
fabricated by laser beam writing.

5.4.4 Replication techniques

An important feature of DOE structures recorded as microreliefs is the ability to
reproduce the structures by replication techniques such as embossing and
casting. In recent years embossing technology has become industrially
widespread for the low-cost mass-production of grating and hologram structures
with resolution down to submicron dimensions. The technology has been driven
by the commercial success of diffractive foil for display and packaging
applications, as well as for security features for credit cards, postage stamps, etc.
Embossing or casting is also a convenient process for the replication of high
resolution diffractive structures in the laboratory.

The first step in all replication techniques is the generation of a metal copy of
the required microrelief, generally known as the replication shim. A common
procedure is shown in Fig. 5.16. The required (master) relief is first fabricated in
any material using one or a combination of the techniques described above. A
faithful copy is then produced in nickel by electroplating. The original microrelief
must first be made conductive, either by evaporating a thin gold film onto the
surface or by electroless Ni plating (sensitizing the surface with adsorbed Pd by
successive treatments in an acidic $SnCl_2$ solution, followed by acidic $PdCl_2$ [71]).
Ni is then electroplated to a thickness of about 100 μm and subsequently
separated to produce the required shim. In the laboratory, shims of about
$10x10$ cm^2 area are readily fabricated. Commercial embossing shims are typically
greater than $20x30$ cm^2 in size. Further information and references on modern
replication technology, together with a list of companies in this field can be
found in [72].

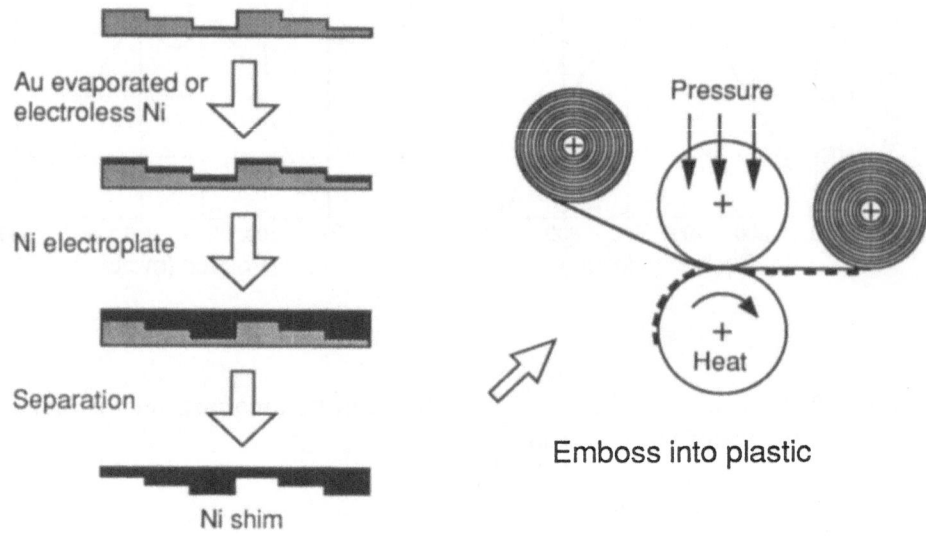

Fig. 5.16. Replication of DOE microrelief structures.

Replication can follow by a number of techniques:

Embossing. Hot embossing into thermoplastic film produces excellent microrelief copies with faithful replication of the submicron relief features. For high quality optical elements, plastics, such as polycarbonate, offer the best optical quality. Reflection DOEs can be produced by metallizing the embossed structures with a thin gold or aluminium layer. High pressure embossing presses can also emboss directly into metallized film. Commercial embossing machinery can emboss plastic film of width about 200 cm at speeds in excess of 5 m/sec.

Casting. Replicas can be fabricated by casting epoxy or UV-curable materials on the replication shim. This technique is long used for the fabrication of high quality spectroscopic gratings. Separation of the cured cast from the shim can be a major difficulty and suitable release coatings are well-guarded industrial secrets. A simple approach for the laboratory is to sputter a thin layer of teflon or teflon-like material onto the shim prior to casting.

Moulding. Compact (as well as audio and video) disks are produced by injection moulding. The required machinery is relatively expensive, but can produce high quality replicas, for example in polycarbonate.

5.5 Selected experimental results for DOEs

The basic diffractive elements for optical interconnects are on- and off-axis diffractive lenses and specially designed phase gratings. They are useful for efficient one-to-one connections, as well as for array generators [73]. With these basic elements more complex interconnect systems can also be implemented, such as perfect shuffle and Banyan network. They will be described in Part 2.1 of this book.

In this section, we shall present some experimental results for lenslet arrays and fan-out gratings in order to show the actual feasibility. In the case of diffractive lenses, the optical function, i.e. the generation of diffraction limited spot sizes, as well as the diffraction efficiency (first order) are of main importance. Fan-out elements generate more than one diffraction order. Thus, also the balancing between the orders is important (uniformity error).

5.5.1 Lenslet arrays

Researchers at the Lincoln Laboratory have fabricated an array of phase Fresnel zone plates to collimate a linear laser diode array [74]. The lens spacing was chosen to be 50 μm and the focal length to be 100 μm (f/2) in the lateral direction. To compensate for the astigmatism of the lasers, the transversal focal length was chosen to be 69 μm, which yields a high f-number of f/1 (vertical aperture 69 μm). The lenses were fabricated by reactive-ion-etching to produce four discrete levels. Measured efficiencies ranged from 70 % across an f/2 region of the lenslet to 80 % in the central f/4 region (theoretically 81 %)

A 10x10 array of diffractive lenslets was fabricated by Jahns and Walker [64]. Each lens has a rectangular aperture and a size of 1.5x1.5 mm. The focal length is 46.5 mm. This corresponds to a lens with f/31. The array was produced by depositing thin films of silicon monoxide on a quartz glass substrate and photolithographic techniques (lift-off). The diffraction efficiency of the phase structure with eight discrete levels was measured to be 91 %.

A higher efficiency is achieved with more phase levels [Eq. (5.3)]. On the other hand the feature size becomes smaller and the alignment errors increase. The relation between the minimum feature size w and the number of phase levels L for a diffractive lens is given by

$$w = \frac{2\lambda f/no.}{L}. \tag{5.17}$$

As an example, to make an f/8 with eight phase levels for a wavelength of $\lambda = 0.8$ μm, we need a lithography with a resolution of 1.6 μm. This is achievable with standard technology and corresponds well to the results obtained by Leger et al. [74], see above.

The fabrication methods of two- or multilevel structures are most advanced, because they are based on well established technologies known from microelectronics. However, there is also a progress in fabricating smooth relief or sawtooth profiles for blazed gratings.

Electron-beam lithography is a method to obtain excellent blazed gratings. The profile is drawn directly into a resist film by varying the electron-dose. As example, we refer to the results obtained by a Japanese group [75]. They have made micro-Fresnel lenses with a focal length of 0.88 mm at 633 nm wavelength and a size of 150x100 μm^2. The efficiency was measured to be 74 %. Also higher apertures up to NA = 0.5 have been reported [8].

An interesting alternative to e-beam lithography is laser beam lithography. The potential of fabricating smooth surface relief profiles into photoresist by scanning with an intensity-modulated focused laser beam has been demonstrated 1983 by Gale and Knop [4]. They have formed refractive lenslet arrays, with a diameter of 45 μm and a focal length of 260 μm. Haruna et al. [11] have obtained a sawtooth shaped relief grating in photoresist above the 3 μm period with a depth of 1 μm. Diffractive microlenses are possible with a NA $\leq$ 0.2, i.e. f/2.5. Two diffractive microlenses have been fabricated for the readout wavelength λ = 633 nm. A diffraction efficiency of 65 % was obtained for the first lens, with a focused spot size (FWHM) of 6.6 μm, which was the diffraction limit. The lens has a diameter of 9.6 mm and a numerical aperture of NA = 0.05. A lower diffraction efficiency of 50 % was measured for the second lens, having an aperture of NA = 0.21 and a diameter of 212 μm.

5.5.2 Fan-out elements

Fan-out elements split an incoming laser beam into one- or two-dimensional arrays of output beams. Such elements are Dammann gratings [1], multilevel phase gratings [76], or smooth surface relief structures [5]. Regular and also irregular patterns can be generated. Unfortunately, the uniformity error increases with the number of desired diffraction orders and with the angular separation between the orders [62]. Thus, the fabrication tolerances limit the maximum number of fan-out beams.

Different groups have successfully fabricated Dammann gratings [77] and four-level phase gratings [76] of varying sizes up to 128x128 [28]. The measured diffraction efficiencies were typically 65 %, which is close to the calculated values, and uniformity errors were about ±5 %. The uniformity error depends on the grating period of the fan-out element. Large fan-outs, such as a 128x128 fan-out with a uniformity error of ±10 % are feasible with grating periods in the millimeter-range [28].

Researchers at the British Telecom Research Laboratories have reported on the design and fabrication of fully 2-D surface relief DOE that can split a single collimated beam into many beams in an arbitrary intensity distribution [27]. These DOEs have been made by electron beam lithography and subsequent reactive ion etching of the pattern with two or four phase levels into silica glass substrates. They have fabricated two phase level elements to produce large splits of 100 and 400 beams with a measured efficiency of ~74 % (relative to the transmitted light) and they can achieve accuracies in the beam intensity ratios of ~1 % (peak to peak). Furthermore, they have realized four phase level elements to produce weighted beam splitting.

Binary Dammann gratings become highly efficient if the fan-out angles are large, so that most of the undesired grating orders become evanescent. They are difficult to fabricate, because the grating period is in the order of a few wavelengths. These gratings are, at present, feasible for infrared light ($\lambda = 10.6$ μm). A reflection-type 5-beam array generator has been demonstrated [28] with a grating period of 2.9 λ.

A challenge is the fabrication of smooth surface relief structures, which enable maximum efficiencies. Promising results have been obtained by laser-beam writing into photoresist [5,29]. For a fan-out of 9 beams, an efficiency of 92 % and intensity variations of ± 7 % have been measured. Recently, also two-dimensional elements, as shown in Fig. 5.17, have been realized by the same technique [78]. For a 9x9 array, the experimental results show an efficiency of 94 % (relative to the transmitted light) and a uniformity error within ±8 % of the average diffracted beam power. Within one line or one row the uniformity is better than ±5 %.

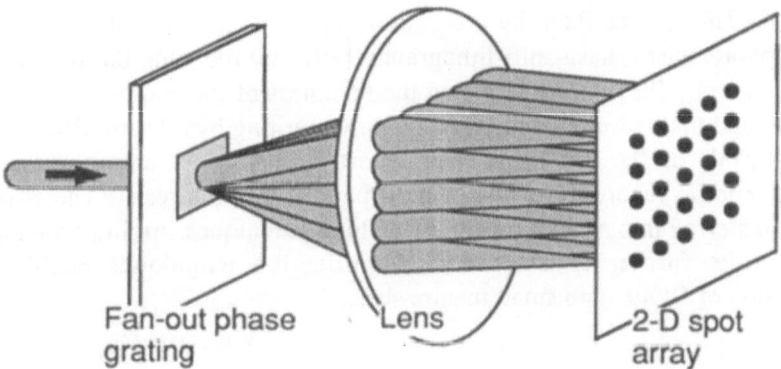

Fig. 5.17. Read-out of fan-out element.

5.6 Future perspectives for diffractive optical elements

Today, many of the proposed interconnect systems use holographically recorded optical elements (HOEs). Highly efficient lenslet arrays and fan-out elements have been realized, permitting complex optical wiring. An advantage of the interferometric recording is the high resolution offered by holographic materials (> 3000 lines/mm). On the other hand, it is difficult to control the recording process in order to get reproducible results. The recording of 100x100 lenslet arrays by step and repeat, where each HOE has a different Bragg angle, is very time consuming. A copy of such an element in a single shot is possible, but needs a very sophisticated illumination element in order to adjust the Bragg angles correctly.

Microfabrication techniques used for synthetic (computer-generated) elements, are well established and their feasibility has been proven under industrial conditions. However, it is still a problem to make properly blazed structures for high spatial frequencies (> 1000 lines/mm), if the structure is more complex than a simple grating.

Most common is the multiple-mask fabrication process for realizing multilevel phase DOEs. The mask approach has the advantage that complex pattern must be fabricated only once and can then be used for fabricating many DOEs. This process has two critical limitations. One is the mask alignment (actually 0.1 μm) and the second is the diffraction of the UV exposure light in the photoresist when the mask is contact printed. In practice, feature sizes for multilevel profiles are larger than 1 μm already enabling high aperture Fresnel microlenses. However, submicron feature sizes have bean achieved for simple patterns, e.g. gratings. This can be done by controlling the propagation of the exposure light in the photoresist (phase-shift lithography [79]), by reducing the wavelength (x-ray lithography [80]), or by reducing the thickness of the photoresist.

A method that is well advanced is the direct writing by e-beam. Blazed gratings with periods of 0.6 μm have been reported [70]. Very common is also the interferometric recording of the desired pattern in photoresist. The structure is then transferred into a relief profile by etching techniques. An important question will be the further development of replication techniques enabling mass production of DOEs with small feature sizes.

References

1.   H. Dammann, K. Görtler: High-efficiency in-line multiple imaging by means of multiple phase holograms, Opt. Commun. **3**, 312-315 (1971)

2. G. J. Swanson: Binary optics technology: the theory and design of multi-level diffractive optical elements, MIT Lincoln Lab, Technical Report 854 (1989)

3. M. T. Gale, G. K. Lang, J. M. Raynor, H. Schütz, D. Prongué: Fabrication of kinoform structures for optical computing, Appl. Opt. **31**, 5712-5715 (1992)

4. M. T. Gale, K. Knop: The fabrication of fine lens arrays by laser beam writing, Proc. SPIE **398**, 347-353 (1983)

5. H. P. Herzig, D. Prongué, R. Dändliker: Design and fabrication of highly efficient fan-out elements, Jpn. J. Appl. Phys. **29**, L1307-L1309 (1990)

6. E. Hecht: *Optics* (Addison-Wesley, Reading, 1987).

7. G. J. Swanson, W. B. Veldkamp: Diffractive optical elements for use in infrared systems, Opt. Eng. **28**, 605-608 (1989)

8. S. Aoyama, S. Ogata, T. Inoue, T. Yamashita: Laser diode source integrating a high-diffraction-efficiency micro-Fresnel lens with 0.5 N.A. fabricated by electron-beam lithography, in Technical Digest of Conference on Lasers and Electro-Optics (Optical Society of America, Washington, DC, 1988), paper THM49

9. H. Hosokawa, T. Yamashita: ZnS micro-Fresnel lens and its uses, Appl. Opt. **29**, 5106-5110 (1990)

10. M. Rossi, R. E. Kunz, M. T. Gale: Phase-matched Fresnel reflectors, Proc. Applied Optics and Opto-Electronics, Leeds, England, 145-147 (1992)

11. M. Haruna, M. Takahashi, K. Wakahayashi, H. Nishihara: Laser beam lithographed micro-Fresnel lenses, Appl. Opt. **29**, 5120-5126 (1990)

12. H. P. Herzig, R. Dändliker: Holographic optical elements for use with semiconductor lasers, in *International Trends in Optics*, J. W. Goodman, ed. (Academic Press, New York, 1991), pp. 57-75

13. J. Kedmi, A. A. Friesem: Optimized holographic optical elements, J. Opt. Soc. Am. A **3**, 2011-2018 (1986)

14. J. N. Cederquist, J. R. Fienup: Analytic design of optimum holographic optical elements, J. Opt. Soc. Am. A **4**, 699-705 (1987)

15. E. Hasman, A. A. Friesem: Analytic optimization for holographic optical elements, J. Opt. Soc. Am. A **6**, 62-72 (1989)

16. H. P. Herzig, R. Dändliker: Holographic optical scanning elements: Analytical method for determining the phase function, J. Opt. Soc. Am. A **4**, 1063-1070 (1987)

17. J. Fagerholm, J. Turunen, E. Byckling: Optimization of holographic optical systems by damped least squares and wavefront matching techniques, Proc. SPIE **883**, 20-27 (1988)

18. R. W. Gerchberg, W. O. Saxton: A practical algorithm for the determination of phase from image and diffraction plane pictures, Optik **35**, 237-246 (1972)

19. M. T. Eismann, A. M. Tai, J. N. Cederquist: Iterative design of a holographic beamformer, Appl. Opt. **28**, 2641-2650 (1989)

20. F. Wyrowski, O. Bryngdahl: Iterative Fourier-transform algorithm applied to computer holography, J. Opt. Soc. Am. A **5**, 1058-1065 (1988)
21. F. Wyrowski: Diffractive optical elements: iterative calculation of quantized, blazed phase structures, J. Opt. Soc. Am. A **7**, 961-969 (1990)
22. W. H. Press, B. P. Flannery, S. A. Teukolsky, W. T. Vetterling: *Numerical Recipes* (Cambridge University Press, Cambridge, 1986)
23. S. Kirkpatrick, C. D. Gelatt, M. P. Vecchi: Optimization by simulated annealing, Science **220**, 671-680 (1983)
24. J. Turunen, A. Vasara, J. Westerholm, G. Jin, A. Salin: Optimization and fabrication of grating beamsplitters, J. Phys. D: Appl. Phys. **21**, S102-S105 (1988)
25. J. N. Mait: Design of binary-phase and multiphase Fourier gratings for array generation, J. Opt. Soc. Am. A **7**, 1514-1528 (1990)
26. J. Turunen, A. Vasara, J. Westerholm: Kinoform phase relief synthesis: a stochastic method, Opt. Eng. **28**, 1162-1167 (1989)
27. M. P. Dames, R. J. Dowling, P. McKee, D. Wood: Efficient optical elements to generate intensity weighted spot arrays: design and fabrication, Appl. Opt. **30**, 2685-2691 (1991)
28. A. Vasara et al.: Binary surface-relief gratings for array illumination in digital optics, Appl. Opt. **31**, 3320-3336 (1992)
29. D. Prongué, H. P. Herzig, R. Dändliker, M. T. Gale: Optimized kinoform structures for highly efficient fan-out elements, Appl. Opt. **31**, 5706-5711 (1992)
30. J. W. Goodman: *Introduction to Fourier Optics* (McGraw-Hill, San Francisco, 1968)
31. G. J. Swanson: Binary optics technology: theoretical limits on the diffraction efficiency of multilevel diffractive optical elements, MIT Lincoln Lab, Technical Report 914 (1991)
32. R. R. A. Syms: *Optical Volume Holography* (Oxford University Press, Oxford, 1990)
33. T. K. Gaylord, M. G. Moharam: Analysis and applications of optical diffraction by gratings, Proc. IEEE **73**, 894-937 (1985)
34. R. Petit, ed.: *Electromagnetic Theory of Gratings* (Springer-Verlag, Berlin, 1980)
35. A. Vasara, E. Noponen, J. Turunen, J. M. Miller, M. R. Taghizadeh: Rigorous diffraction analysis of Dammann gratings, Opt. Commun. **81**, 337-342 (1991)
36. D. Maystre: Integral methods, in 34., pp. 63-100
37. P. Vincent: Differential methods, in 34., pp. 101-121
38. R. Magnusson, T. K. Gaylord: Equivalence of multiwave coupled wave theory and modal theory for periodic-media diffraction, J. Opt. Soc. Am. **68**, 1777-1779 (1978)
39. M. G. Moharam, T. K. Gaylord: Diffraction analysis of dielectric surface-relief gratings, J. Opt. Soc. Am. **72**, 1385-1392 (1982)

40. K. Knop: Rigorous diffraction theory for transmission phase gratings with deep rectangular grooves, J. Opt. Soc. Am. **68**, 1206-1210 (1978)
41. L. C. Botten, M. S. Craig, R. C. McPhedran, J. L. Adams, J. R. Andrewartha: The finitely conducting lamellar diffraction grating, Optica Acta **28**, 1087-1102 (1981)
42. L. C. Botten, M. S. Craig, R. C. McPhedran: Highly conducting lamellar diffraction gratings, Optica Acta **28**, 1103-1106 (1981)
43. W. Moreau: *Semiconductor Lithography: Principles, Practices and Materials* (Plenum, New York, 1988)
44. Proc. ME'90, Microelectronic Engineering **13** (1991)
45. J. L. Vossen, W. Kern, eds.: *Thin Film Processes II* (Academic Press Inc., San Diego, 1991)
46. J. Jahns, A. Huang: Planar integration of free-space optical components, Appl. Opt. **28**, 1602-1605 (1989)
47. H. C. Pfeiffer, T. R. Groves: Progress in e-beam masks making for optical and X-ray lithography, ME'90, Microelectronic Engineering **13**, 141-149 (1991)
48. J. Melngailis: Focused ion beam technology and applications, J. Vac. Sci. Technol. B **5**, 469-495 (1987)
49. J. M. Stauffer, Y. Oppliger, F. Vasey: Fabrication of optoelectronic devices on AlGaAs using electron beam lithography, Proc. ME'90, Microelectronic Engineering **13**, 193-196 (1991)
50. For example, Laser Pattern Generator Systems CORE-2000 (ATEQ Corp., Oregon, USA) or Micronic LRS-10 (Micronic Laser Systems AB, Täby, Sweden)
51. H. Nishihara, M. Haruna, T. Suhara: *Optical Integrated Circuits* (McGraw-Hill, New York 1987)
52. D. C. Flanders, A. E. White: Application of ~100A linewidth structures fabricated by shadowing techniques, J. Vac. Sci. Technol. **19**, 892-896 (1981)
53. M. C. Hutley: *Diffraction gratings* (Academic Press, London, 1982)
54. K. E. Bean: Anisotropic etching of Silicon, IEEE Trans. Electron Dev. **ED-25**, 1185 (1978)
55. H.W. Lehmann: Plasma-assisted etching, in 45., pp. 673-748
56. J. C. Matthews, M. G. Ury, A. D. Birch, M. A. Lashmann: Microlithography techniques using a microwave powered deep UV source, Proc. SPIE **394**, 172-183 (1983)
57. J. L. Vossen, J. Appl. Phys. **47**, 544-546 (1976)
58. R. E. Lee: Inhibition of chemical sputtering of organics and C by trace amounts of Cu-surface contamination, in *Plasma Processing for VLSI, VLSI Electronics Microstructure Science 8*, N. G. Einspruch, D. M. Brown, eds. (Academic Press, Orlando, 1984) p. 34
59. P. R. Puckett, S. L. Michel, W. E. Hughes: Ion beam etching, in 45., p. 749

60. S. Matsui, T. Yamato, H. Aritome, S. Namba: Fabrication of SiO_2 blazed holographic gratings by reactive ion-etching, Jap. J. Appl. Phys. **19**, L126-L128 (1980)

61. E. H. Anderson, C. M. Horowitz, H. I. Smith: Holographic lithography with thick photoresist, Appl. Phys. Lett. **43**, 874-875 (1983)

62. J. Turunen, J. Fagerholm, A. Vasara, M. R. Taghizadeh: Detour-phase kinoform interconnects: the concept and fabrication considerations, J. Opt. Soc. Am. A. **7**, 1202-1208 (1990)

63. M. B. Stern, M. Holz, S. S. Medeiros, R. E. Knowlden: Fabricating binary optics: Process variables critical to optical efficiency, J. Vac. Sci. Technol. B **9**, 3117-3121 (1992)

64. J. Jahns, S. J. Walker: Two-dimensional array of diffractive microlenses fabricated by thin film deposition, Appl. Opt. **29**, 931-936 (1990)

65. M. Hatzakis, B. Canavello, J. Shaw: Single-step optical lift-off process, IBM J. Res. Dev. **24**, 452 (1980)

66. P. Buchmann, V. Graf, Th. O. Mohr, P. Vettiger: High-temperature-stable Si_3N_4 dumy T-gate and lift-off mask, ME'86, Proc. of the Internat. Conference on Microlithography, Interlaken, 395 (1986)

67. M. T. Gale, R. E. Kunz, B. J. Curtis, O. Parriaux, G. Voirin: Waveguide grating fabrication and optimization for integrated optic polarization interferometry, Proc. IOOC'89, Kobe, Japan, **1**, 54-55 (1989)

68. H. W. Lehmann, R. Widmer: Limitations of a single level lift-off process, Proc. ME'84, Berlin (North Holland, Amsterdam, 1985), pp. 493-500

69. W. Windbracke, H. Betz, H.-L. Huber, W. Pilz, S. Pongratz: Critical dimension control in X-ray masks with electroplated gold absorbers, Proc. ME'86, Interlaken, (North Holland, Amsterdam, 1986) p. 73

70. J. M. Stauffer, Y. Oppliger, P. Regnault, L. Baraldi, M. T. Gale: Electron beam writing of continuous resist profiles for optical applications, Proc. EIPB92, Orlando, USA (1992)

71. N. Feldstein, T. S. Lancsek: A technique for selective electroless plating, RCA Rev. **32**, 306-310, (1971)

72. B. Kluepfel, F. Ross, eds., *Holography Market Place* (Ross books, Berkeley, CA, USA, 1991)

73. N. Streibl: Beam shaping with optical array generators, J. Mod. Opt. **36**, 1559-1573 (1989)

74. J. R. Leger, M. L. Scott, W. B. Veldkamp: Coherent addition of AlGaAs lasers using microlenses and diffractive coupling, Appl. Phys. Lett. **52**, 1771-1773 (1988)

75. T. Shino, K. Setsune, O. Yamazaki, K. Wasa: Rectangular-apertured micro-Fresnel lens arrays fabricated by electron-beam lithography, Appl. Opt. **26**, 587-591 (1987)

76. S. J. Walker, J. Jahns: Array generation with multilevel phase gratings, J. Opt. Soc. Am. A **7**, 1509-1513 (1990)

77. M. R. Taghizadeh, J. I. B. Wilson, J. Turunen, A. Vasara, J. Westerholm: Optimization and fabrication of grating beamsplitters in silicon nitride, Appl. Phys. Lett. **54**, 1492-1494 (1989)

78. P. Ehbets, H. P. Herzig, D. Prongué, M. T. Gale: High efficiency continuous surface-relief gratings for two-dimensional array generation, Opt. Lett. **17**, 908-910 (1992)

79. M. Chen: "Is phase-shift mask technology production worthy?", Proc. SPIE **1463**, 2-5 (1991)

80. S. V. Babin, A. I. Erko: Fabrication of diffraction X-ray elements, Nucl. Instrum. and Meth. in Phys. Res. **A282**, 529-531 (1989)

6 Characterization of Interconnection Components

J. Schwider
Physikalisches Institut Universität Erlangen-Nürnberg, Angewandte Optik

6.1 Introduction

Free space optical interconnects are used in optoelectronic and optical networks as it is discussed in the second section of this book. A very promising scheme uses optoelectronic switching planes and optical interconnects between those planes. In the most general approach this is a more or less regular 3D-fabric of elements, this is why such a structure can be used as a symbolical example for free space interconnects. Here, we will deal with passive optical elements only which are necessary to interconnect emitters on one plane with opto-electronic detectors on the plane of the following stage.

So, mainly two functions have to be realized by microoptical elements: focussing/collimating and deflection [1,2]. If also a fanout of light from one point in the first plane to several or all points in the following plane shall be implemented either amplitude or aperture dividing elements are necessary. In a similar manner fanin geometries can be realized.

The flexibility of free space optical interconnects rests mainly on the space-variance of the interconnect approach. This suggests the use of micro-optical elements. These elements are produced on substrates (in most cases being transparent). These substrates carry lenses, mirrors, prisms, or they are the mechanical support of diffractive elements either in the form of surface profiles or in the form of thick phase holograms [3] (e.g. made in dichromated gelatin). Similar assumptions have to be made for refractive elements: either of the diffusion-type [4] or molten resist structures [5].

The optical performance of these elements or of an array of such elements depends on one or several of the following characteristica: 1) surface quality (flatness) of the supporting structure, 2) the quality of the curved surface of a micro-lens, or of the flat surfaces of prisms or mirrors in case of refractive/reflective elements, 3) on the type and the quality of the surface corrugation for surface relief gratings, and 4) on the quality of the corrugation of a voluminous holographic material (e.g. on the modulation strength).

Therefore, the need for the characterization of the following features is obvious: 1) characterization of the surface quality of the substrates (homogeneity is assumed), i.e., flatness in the macroscopical and the microscopical range, 2) characterization of the surface form of lenses, 3) characterization of the surface relief profile of diffracting elements, 4) characterization of volume gratings, 5) characterization of the deflection angles of microoptic deflectors.

The optical performance of the micro-optical elements be it the focussing power or the deflection accuracy are characterized by such merit functions as are: 1) wave aberrations, 2) point spread function (PSF), 3) modulation transfer function (MTF), 4) deflection angle, and 5) efficiency (e.g. diffraction efficiency).

Accordingly, the measurement of surface parameters is described in Sect 6.2. We start with the macro-geometry of flat and of spherical surfaces considering methods of varied sensitivity. Then the measurement of surface profiles of binary as well as arbitrary form is described. In Sect. 6.3 we deal with aberration measurements, where other merit functions can be calculated from the wave aberration data as are PSF and MTF. Other tests of the optical performance in the field have to be carried out if segmented space-invariant imaging is considered.

6.2 Characterization of the surface structure

Each surface can be disassembled into a series of "surface waves". Therefore, it is straightforward to characterize the surface by the spectral content of these surface frequencies which are present in the geometry of the surface under test. It is common usage to discern gross surface defects as e.g. curvature, from surface waves with periods below 1 mm. The highest frequency of interest is determined by the experiment to be carried out, since every optical experiment acts as a low-pass filter limiting in this way the test needs for passive optical elements as e.g.: for lenses, HOE, or similar elements. In all cases the resolution of the test equipment should be superior to the requirements of the experiment at least by one order of magnitude.

6.2.1 Macrostructure measurements

At first, we will discuss methods for testing the gross quality of the possible boundary surfaces. Depending on the surface deviations different interferometric methods are applicable. Here, we will only discuss two extreme cases: high accuracy planeness/sphericity tests using Fizeau- or Twyman-Green Interferometers and grazing incidence interferometers to obtain deviation pictures with strongely deviating plane surfaces.

Testing of flats. Figure 6.1 shows a typical planeness test setup [6]. This is a Fizeau interferometer having laser illumination. Interference occurs between the sample and the reference surfaces, i. e., between the two surfaces facing each other. The collimator supplies plane wave illumination necessary for quantitative interferometry. The reference surface has to be calibrated to obtain an absolute instrument. Modern versions are equipped with the technical means for the phase shifting technique giving the planeness deviations in suitable form. The technical means are: a photoelectrical area detector (e.g. a CCD array), a phase shifter (commonly a piezo-driver), and an on-line computer with interfaces to the detector and PZT. Calibration data can be subtracted since the interferometer data are stored in the on-line computer. Typical accuracies range from $\lambda/100$ to $\lambda/1000$, with λ as the used wavelength in the interferometer. The lateral resolution is mainly determined by the sampling rate due to the pixel content of the detector array. Typical values are a few hundred points per diameter of the sample under test. Interferometers of this type are commercially available [7]. The optical limitations due to the restricted detection aperture are less stringent.

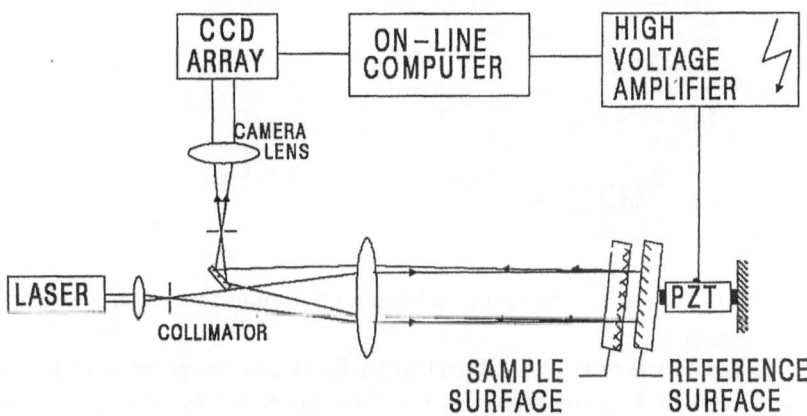

Fig. 6.1 Scheme of a planeness Fizeau interferometer with laser illumination and phase shifting facilities.

Grazing incidence instruments. If arrays of lenses having small numerical aperture are used then bigger surface deviations can be tolerated. The supporting structure must be sufficiently flat only across a single subaperture. In this case it is more suitable to measure with reduced sensitivity. Therefore, grazing incidence test setups would be more appropriate because of the big deviations across the whole substrate or holographic plate or similar. With grazing incidence upon the test sample the sensitivity can be varied within broad limits. While for normal incidence the fringe distance represents a height variation by $\lambda/2$ in case of grazing incidence one fringe represents a height variation of $\lambda/2\cos\alpha$ where a is the incidence angle of the light beam on the sample under test. A prism Fizeau interferometer has been proposed [8] with the test beam originating from the hypotenuse surface under near grazing exit angle. This type of interferometer (Fig. 6.2) has been used for wafer flatness tests [9]. Birch [10] proposed a grating interferometer with the feature that fringe and grating period are identical. The obvious disadvantage of all grazing incidence instruments is the inevitable anamorphic distortion of the object geometry.

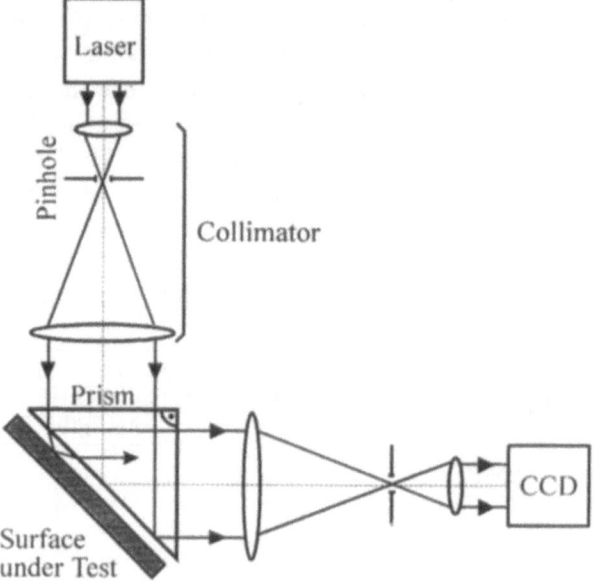

Fig. 6.2 Scheme of a grazing incidence Fizeau prism interferometer.

Testing of spherical surfaces. Spherical surfaces can be tested with the help of Twyman-Green or Fizeau arrangements. The probe wave is made to impinge perpendicularly on the surface under test (Fig. 6.3) which assumes special beam shaping optics. In this way one obtains a constant sensitivity across the whole surface. The other conditions are very similar to the situation for planeness tests. While beam shaping optics for concave spheres is easily obtained since microscopic objectives fulfill the correction requirements the situation is more complicated for convex ones. For these surfaces special and expensive beam shaping optics are necessary. The objectives have to be corrected for spherical aberration and should fulfill the sine condition. These corrections are necessary to avoid recalibrations for each radius of curvature on the one hand and on the other enable fringe adjustments which occur either accidentally or on purpose.

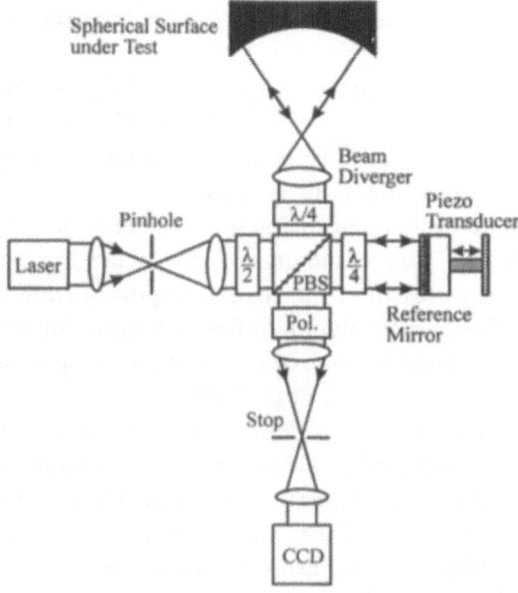

Fig. 6.3 Scheme of a Twyman-Green interferometer for testing spheres. The interferometer is a polarisation type instrument making intensity balancing within wide margins (rotatable $\lambda/2$ plate) and phase shifting techniques possible. The $\lambda/4$ plates and the polarizing beam splitter (PBS) enable an insulation of the interferometer from the illuminatoring laser.

6.2.2 Microstructure measurements

While the macroscopic deviations from planeness or sphericity mainly affect the global function of an array of elements the midrange deviations contribute to the aberrations of the single element. Furthermore, the microstructure of the surface is very essential for the performance of diffractive and also of refractive elements. The form of the groove profile of a period of the diffractive elements determines the amount of light being diffracted into the desired order. Deviations from the ideal form are the cause for losses in diffraction efficiency on the one hand and on the other for parasitic stray light which will contribute to crosstalk in optical interconnects. Also refractive lenses will be deteriorated in their optical function if corrugations of the surfaces in the microregion occur. Within this section methods for the measurement of surface profiles shall be dealt with.

To characterize the scope of possible surface profiles some examples shall be given. One group of typical examples for elements which have to be characterized are binary phase gratings such as Dammann gratings [11] or smooth relief gratings [12]. With binary gratings mainly the depth and the aspect ratio of the structure have to be measured. The smoothness of the upper and of the lower level areas is essential for the amount of stray light generated in the application of the element.

Another group of elements are approximations of sawtooth profiles by staircase designs [13,14]. The optical characterization of such profiles with sufficient accuracy and reliability is a difficult task. For interferometry in the microregion microscopic imaging is mandatory to obtain a sufficient spatial resolution. Computer generated diffractive elements have a broad spectrum of spatial frequencies. A very common element is the Fresnel Zone Plate (FZP) which shows a linear growth of the spatial frequency from the center to the rim of the FZP. Therefore, the resolution of the measuring interference microscope is determined by the resolution requirements in the rim region and the field of view by the extension of an inner zone of the FZP.

Interference microscopes were described by several authors; here only one monography shall be referred to [15]. There is a multitude of different working principles. For small numerical apertures (smaller than 0.3) Fizeau interferometers are in use as well as the Mirau-interferometer (Fig. 6.4). A special derivate of this principle is the interference microscope after Schulz [16,17]. In this instrument (Fig. 6.5) the reference surface is unsharply imaged together with the object surface. Therefore the microstructure of the reference surface is averaged if partially coherent illumination is applied. The aperture of the Mirau-type instruments is limited for two reasons: high aperture objectives have short working distances and the adjustment of a fringe pattern is only possible by tilting the object. This results for high numerical apertures in intolerable defocusing of the object across the field of view.

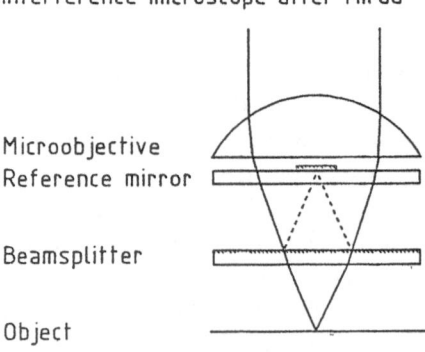

Interference microscope after Mirau

Microobjective
Reference mirror

Beamsplitter

Object

Fig. 6.4 Scheme of Mirau micro interferometer.

Beside the high spatial frequency content of the surface profile there is an additional problem if the structure is "deep", i.e. if the depth of the structure exceeds $\lambda/4$ and even more so if it exceeds the depth of focus of the imaging optics. As an example for such structures a staircase profile for the CO_2-Laser wavelength Fig. 6.6 is given.

For high numerical apertures the Linnik-interferometer (Fig.6.7) is established in the testing community [18,19]. Since the imaging optics is incorporated in the object arm there are no limitations concerning aperture and working distance. For reasons of symmetry the reference arm has an identical structure. Because of the symmetry and the usage for surface inspection this instrument allows for white-light illumination. This is a very essential advantage since it allows for the identification of fringe orders across unresolved step profiles.

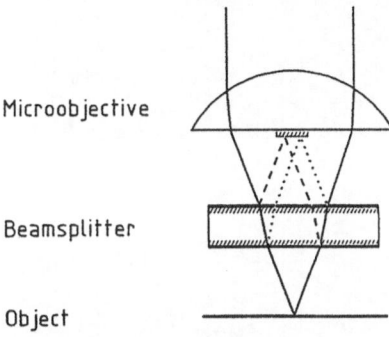

Fig. 6.5 Scheme of three-beam micro interferometer after Schulz.

The interferometers enumerated so far are reflected light instruments, i.e., the object surface is tested via the reflected wavefront. In most of the applications concerning interconnects treated in the literature the diffractive element is used in transmitted light as e.g. a lens. Therefore, the measurement of the phase delays in transmitted light delivers the relevant information more directly than reflected light methods. For structures etched into glass substrates the measuring results can be easily interpreted for both interferometer types. In case of layers on glass, as e.g. photoresist on glass, the measured phase delays are influenced also by the layer structure [20] and the situation may be especially troublesome in reflected light if the element is designed for transmitted light. The best way out would be to measure the element in the same configuration as in the intended application and with the same wavelength.

For this reason also transmitted light interferometers are necessary to get unambiguous results for transmitted light optical elements. In Erlangen we built a Mach-Zehnder interferometer (MZ) for transmitted light interferometry. The scheme of the instrument is shown in Fig. 6.8. Since high numerical apertures are necessary for the microscopical imaging, two identical microscopes are incorporated into the Mach-Zehnder, one in the test arm and the other in the reference arm of the interferometer. The interferometer allows for realtime evaluations with the phase shifting technique. A measuring result is shown in Fig. 6.9. Beside the above advantage, the MZ is not so severly hampered by spurious fringes as reflected light instruments with laser illumination.

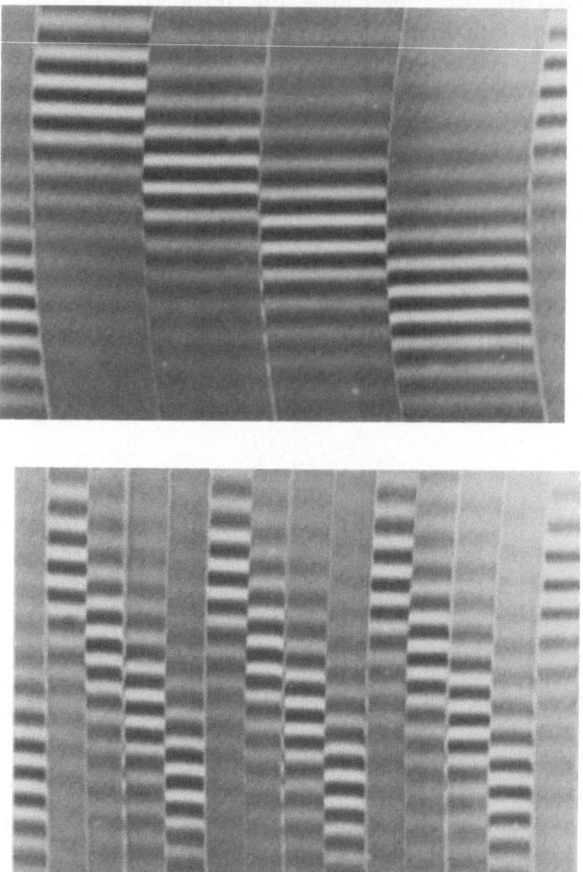

Fig. 6.6 Interferogram of a staircase profile for a CO_2-Laser, above: inner zone, below: rim region, field of view: 227 μm x 293 μm.

Corrections of the thickness values of layers on glass have to be calculated with the help of a structure model. The main problem for such calculations is the uncertainty of the refractive index of the layers, since there may be a big difference between the indices of the bulk material and the layer.

Nearly all modern interferometers have phase shifting capability [21] and allow the display of a whole host of merit functions be it as simple merit numbers as e.g. rms , peak to valley (p/v) figures, or Strehl definition; or surface structures or statistical functions as correlation length and power spectra which can be used as measure for the scatter caused by the surface under test.

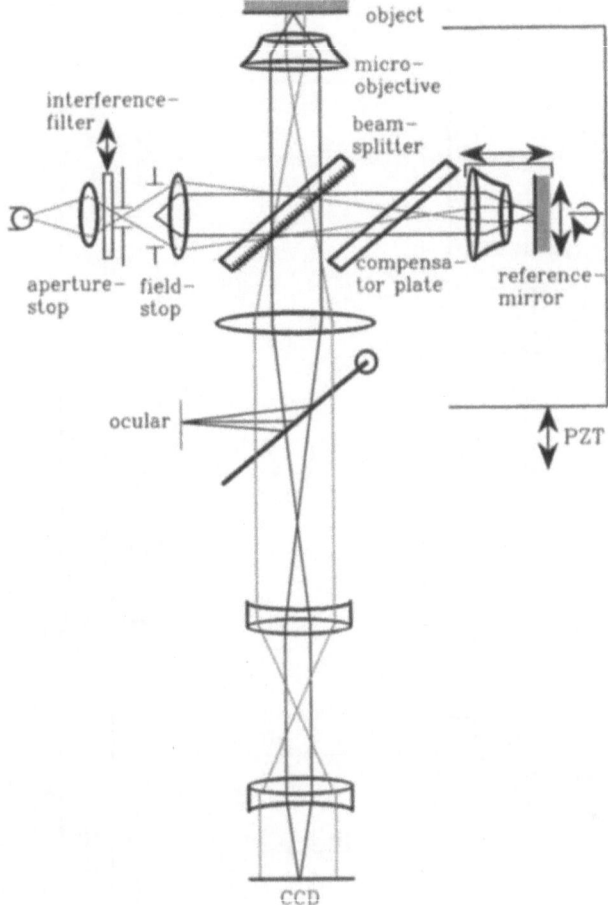

Fig. 6.7 Scheme of Linnik micro interferometer. The reference beam is positioned on a PZT-device enabling phase shifting techniques.

One of the general problems in surface inspection with the help of interferometers is the fact that only wavefronts can be compared. This means that the surface deviations are transferred to wavefront deviations via reflection or transmission. The surface to be tested and the wavefront are exact copies of each other only under the following assumptions:

- the surface to be tested is sharply imaged onto the detector and
- the wavefronts are free from other defects.

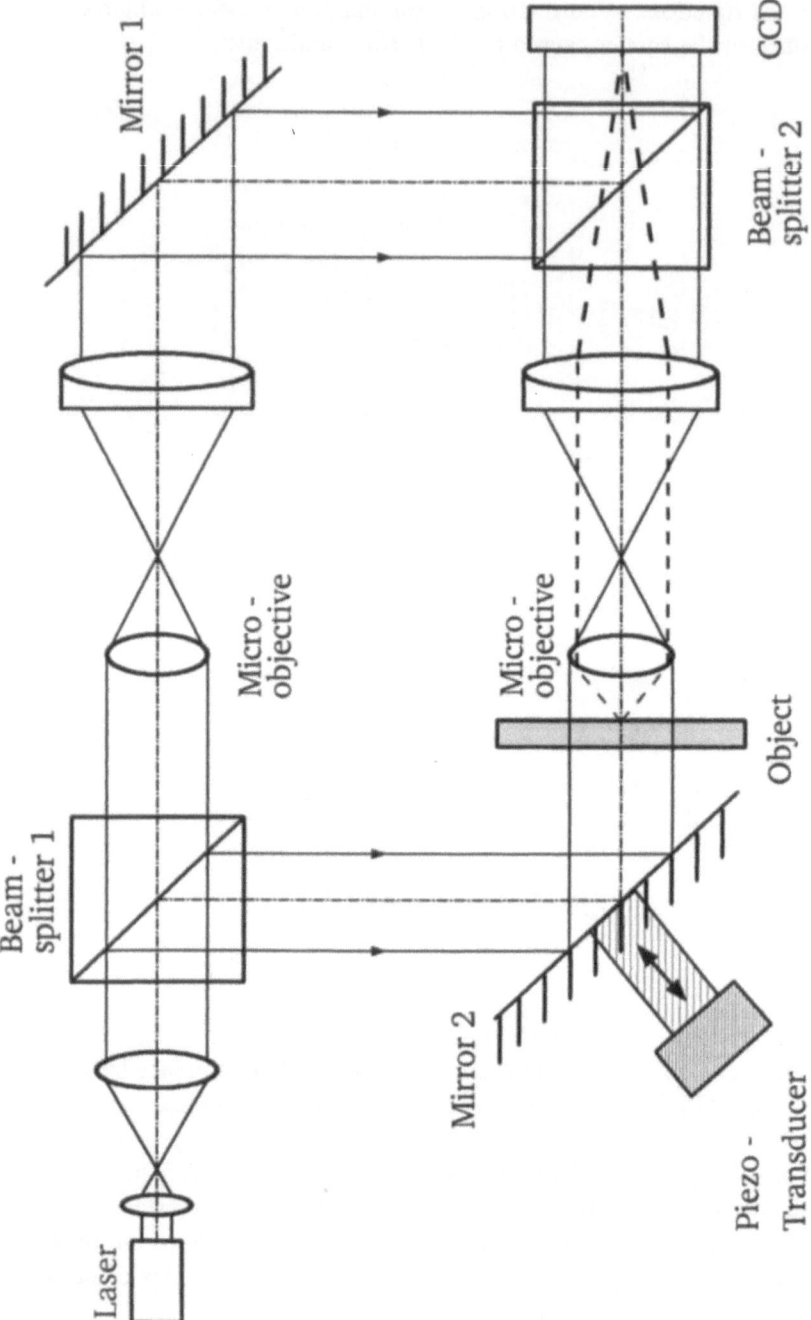

Fig. 6.8 Scheme of Mach-Zehnder micro interferometer for measurements in transmitted light.

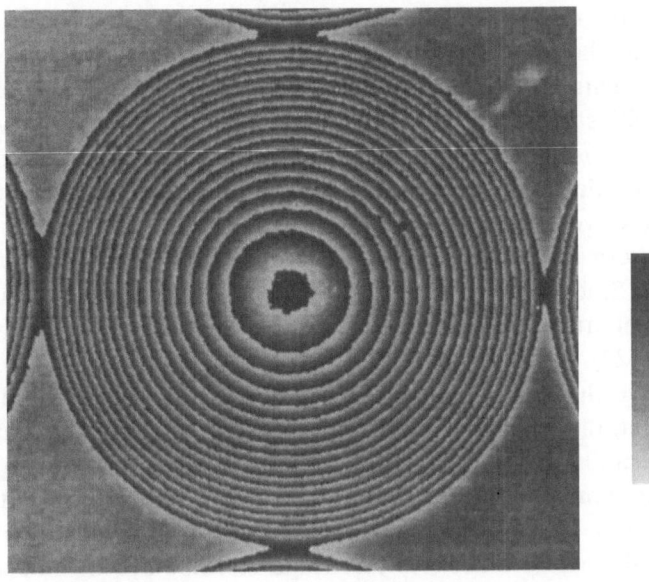

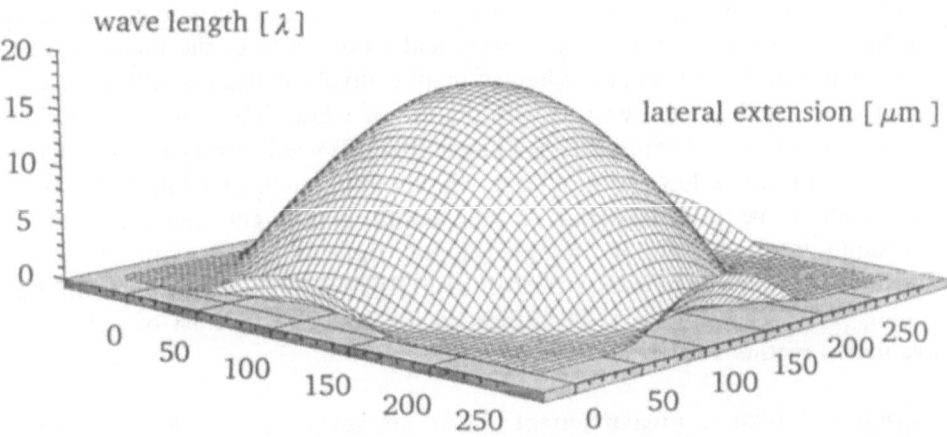

Fig. 6.9 Interference evaluation of a graded index lens (Nippon Sheet Glass) with the interferometer after Fig. 6.8, above: interferogram (λ = 633 nm), below: pseudo-3D-plot. Diameter D = 250 μm, focal length f = 700 μm.

While the first assumption can be fulfilled in most cases rather easily the second assumption is more stringent than it seems at first glance.

This can be understood in the following way: the waves entering the interferometer have, in the case of coherent illumination, in no way smooth wavefronts. This is the result of the diffraction at dust particles and other inhomogeneities in the ray path. By these disturbances secondary wavelets are generated which interfere with the strong background wave. These statistically distributed particles generate a stochastic modulation of the modulus and the phase of the waves in the interferometer. Consequently, the interferometer shows a fringe pattern having stochastic fringe displacements from equidistancy and straightness. Since the difference of two stochastic ensembles results again in a stochastic pattern the true surface profile is screened by a noisy background information [22].

To improve the quality of the interferogram it is necessary to use partially coherent light. In the case of interference microscopes this means an illuminating aperture of similar order of magnitude as the imaging aperture and the use of a spatially and temporally incoherent light source; commonly a white light source in combination with interference filters. But one has to be careful with the increase of the aperture because the fringe distance becomes dependent on the aperture angle as has been shown by several authors [23-26] and especially by Schulz and Elssner [27]. For numerical apertures well below 0.4 one can correct the fringe distance by a global factor, i. e., $2/(1+\cos u)$, where u is the maximum aperture angle. For higher aperture values it may happen that the height difference is no longer unequivocally dependent on the wavelength.

The evaluation of interference patterns obtained with interference microscopes can be done manually either in situ by use of ocular micrometers or using interference colors or fringe adjustments and hard copies of the interferogram. Since with visual evaluations the human brain is involved also complicated fringe patterns are easily identified and with the use of white light fringes the correct fringe orders can be determined. The human intelligence is a very decisive factor if e.g. step profiles have to be measured. The high contrast white light fringe maximum can be followed across the step and the step height can be measured if a second interferogram in monochromatic light is made maintaining the adjustment of the interference microscope. The accuracy is of course limited (order of 1/20 of a fringe). Therefore, in the last years a host of automated evaluation methods has been developed.

Automated profile measurement. There are several review articles on this subject [21,28,29]. For the characterization of interconnect components in the microregion the interest is mainly concentrated on profile sections through the element or better through one period of the element structure and on the uniformity of the structure in the two lateral dimensions. Therefore, we may concentrate here on two forms of automated interferometry, i.e. heterodyne and phase shifting interferometry.

Heterodyne interferometry delivers the phase to be measured modulated on a carrier frequency in the time domain which can be generated within or outside of the interferometer by means of acousto-optic modulators. A solution relying on differential interferometry has been proposed by Makosch and Solf [30]. The probe beam is split by a Wollaston prism into two slightly displaced beams having orthogonal polarisations. Due to the different polarisations it is possible to generate the carrier frequency outside of the interference microscoope. Here the high accuracy of electronic phase difference measurement between the interference pattern and the local oscillator is the main advantage. The object is scanned across the selected section and the phase difference between the two sheared light spots on the object is measured.

The role of the local oscillator beam can alternatively be assumed by an unsharp spot around the sharp probing spot [31]. Due to the close neighborhood of the two beams also this interferometer is rather stable against shock and vibration. The unsharp imaging of the local oscillator beam provides an averaging over the microstructure of the sample under test. Due to the symmetry of the interferometric sensor head small tilt movements are averaged and in this way eliminated. Piston movements are eliminated anyways because reference wave and probe wave experience identical phase variations. It is not immediately clear whether binary objects with steps higher than $\lambda/4$ can correctly be measured. The main working field is supersmooth surface topography.

Phase shifting interferometry. Interference patterns from a two-beam interferometer have the form:

$$I(x,y) = I_0(x,y)(1 + V(x,y) \cos (\Phi(x,y) - \varphi)), \tag{6.1}$$

where $I(x,y)$ is the measured intensity, $I_0(x,y)$ the mean intensity, $V(x,y)$ the visibility, and $\Phi(x,y)$ is the phase to be measured. The three unknowns $\Phi(x,y)$, $I_0(x,y)$, and $V(x,y)$ have to be separated from the measured intensity $I(x,y)$. For this purpose the arbitrary reference phase φ can be varied (in widespread use is a piezoelectrically driven mirror) in order to obtain several intensity values in each point of the test sample enabling the separation of $\Phi(x,y)$ the phase to be measured. The phase shifting method is characterized by an equidistant stepping of the reference phase through at least one period of the interference pattern [21]. In each sample point more than 3 intensity values are generated which enables a least squares approximation of a cos-type interference function which provides the phase $\Phi(x,y)$ mod (2π) in the point (x,y):

$$\tan \Phi(x,y) = \sum_{r=1}^{R} I_r \sin \varphi_r \Big/ \sum_{r=1}^{R} I_r \cos \varphi_r \tag{6.2}$$

where $\phi_r = 2\pi(r - 1)/R$, $r = 1,2,...,R$ (≥ 3) holds. A rather simple algorithm follows if 4 reference phase values: 0, $\pi/2$, π, $3\pi/2$ are chosen:

$$\tan \Phi(x,y) = (I_2 - I_4)/(I_1 - I_3), \tag{6.3}$$

which follows immedeately if the 4 intensity distributions are written down and resolved to get $\tan \Phi$.

The phase resulting from equ.(6.2) is discontinuous and can be unwrapped using a priori knowledge [21]. The phase unwrapping algorithm works rather well as long as the sampling theorem (more than 2 samples per period of the local interference pattern are necessary to reconstruct the cos-type intensity distribution) is fulfilled in the whole field of view. However, microscopic objects show very often small spots where the phase varies very rapidly violating in this way the sampling theorem. Such pixels or even regions have to be discerned and discarded from the unwrapping procedure. It is rather difficult to define a sufficient criterion which allows a secure detection of such spots.

Phase shifting evaluations for binary objects. A further difficulty occurs with discontinuous phase distributions if the phase difference between neighboring pixels exceeds p since then the order number of the fringes gets lost at the discontinuity. In this case only a priori knowledge can help.

Typical structures are meander-shaped ones. The interference fringes may show a discontinuity of many interference fringes across the edge (see e.g. Fig. 6.10). The normal way out is the use of a white light interferometer. The white light fringe maximum indicates the locations where the optical path difference becomes zero on both sides of the edge. To find the corresponding locations it is mandatory to use a strong fringe adjustment. Furthermore, a rather regular support surface is assumed to make evaluations easy [15]. Phase shift evaluations are only unambiguous within one period. To keep the phase change across the discontinuity below π the interferometer could either be illuminated with a long wavelength [32] or one makes two or more measurements with wavelengths λ_a, λ_b not so far apart from each other. In this way it is possible to measure at an effective wavelength λ_{eff} which is e.g. one order of magnitude larger than normal wavelengths of the visible region:

$$\lambda_{eff} = \lambda_a * \lambda_b / (\lambda_a - \lambda_b). \tag{6.4}$$

For λ_{eff} the phase difference of neighboring pixels across the step can be held below π which is necessary to follow the order number across the step.

Since two interference patterns are superimposed incoherently one gets an additive Moire. Typical additive Moires give agreeable contrast for the difference frequency only if the high frequency term is also sharply imaged at the same time.

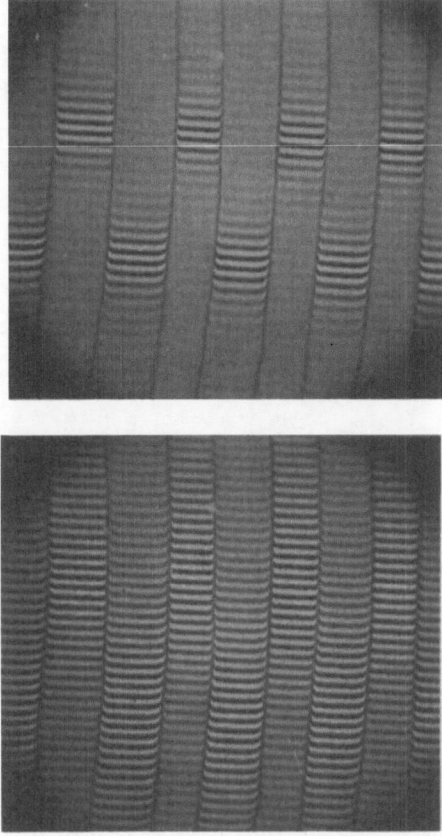

Fig. 6.10 Binary phase object in reflected light (Linnik interferometer), above:
white light, below: interference filter $\lambda = 533.5$ nm.

This is caused by the addition theoremes of the two cosine functions. Let the two
cosine functions be cos a and cos b then the additive Moire delivers:

$$2 \cos((a+b)/2) \cos((a-b)/2). \tag{6.5}$$

From this follows that the low frequency term of the interference pattern is only
visible if the high frequency term is sharply imaged. For the application of Phase
Shifting Interferometry (PSI) this assumes that for all regions of the field of view
the phase unwrapping procedure must be carried out at both wavelength λ_a, λ_b
starting from the same pixels and with the same offset.

The phase jumps at the effective wavelength across the discontinuity are held
below π. These methods are borrowed from the wellknown length measuring
algorithms using different wavelengths and the Koesters interferometer [33].
Cheng and Wyant [34], and Creath [35] applied these methods to a two-beam-
Mirau interference microscope. The object is measured at two wavelengths
whose quotient is not a rational number.

The principle of phase unwrapping can be best understood for an one-dimensional section of the data. We consider a detector line with pixels 1,...,r,r+1, ... The phase at the r-th pixel for the two wavelength λ_a, λ_b shall be Φ_r^a, Φ_r^b:

$$\Phi_r^a = OPD*2\pi/\lambda_a$$

$$\Phi_r^b = OPD*2\pi/\lambda_b,$$

where OPD is the optical path difference encountered by the light rays (for an extensive discussion of phase unwrapping problems please consult [29]). For single wavelength interferometry the condition

$$|\Phi_r^a - \Phi_{r+1}^a| \gg \pi \tag{6.6}$$

is tested and in the case of a phase jump p corrected. Here we have to test the phase difference $\Phi_r^a - \Phi_r^b = \Phi_r^{eff}$ for phase discontinuities. If the effective wavelength has properly been selected, the difference phase Φ_r^{eff} will per definition change by less than p, although at the discontinuities of the object the phase will discontinuously change for both wavelengths. This means that the OPD across the step keeps below $\lambda_{eff}/2$ or the object surfaces varies by less than $\lambda_{eff}/4$ for reflected light interferometers. With the help of the phase information for λ_{eff} it is possible to find the change in the order number of the fringes at λ_a, or λ_b, respectively. The maximum step height which can be measured depends strongly on the accuracy of the phase measurement of the single wavelength. If the accuracy at the single wavelength λ_a is λ_a/q then the ratio λ_{eff}/λ_a has to be smaller than $q/2$. That means above 10 µm step height the requirements for the accuracy become rather stringent. Due to the additive Moire character of this method the phase unwrapping problems are not made easier.

Mechanical profile measurements. The microgeometry may be measured with the help of mechanical profilometers, where a diamond needle scans the sample surface. The vertical movement of the tip is measured electro-dynamically with great accuracy. The ambiguities resulting from the mechanical stage have to be averaged by the use of an integrating air bearing stage [36]. A test example for a binary diffractive element is shown in Fig. 6.11. Limitations of this method are: (i) mechanical contact with the surface of the object with the danger of scratches or other contaminations, and (ii) the radius of curvature of the diamond tip limits the lateral resolution more than the light spot of an optical method.

The big advantage of this method is, that it measures the geometrical height of the surface structure independently of the layer structure of the diffracting element or other object features and the scan can follow also rather deep structures without suffering from defocus problems.

Alternatively, the mechanical tip of the Talystep may be replaced by an optical scanning head from a CD-player [37]. Beside the indication by optical means all other problems are similar to the mechanical scanning device. The accuracy is of the order of 1/50 of a wavelength.

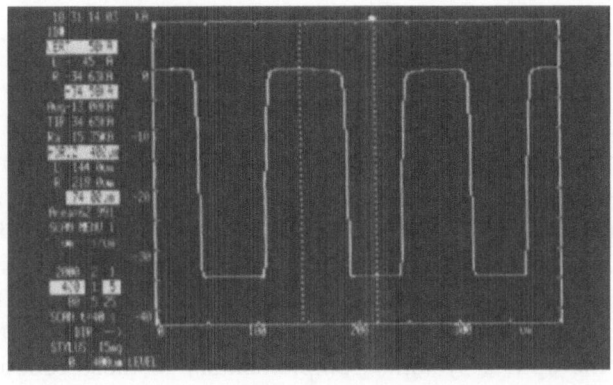

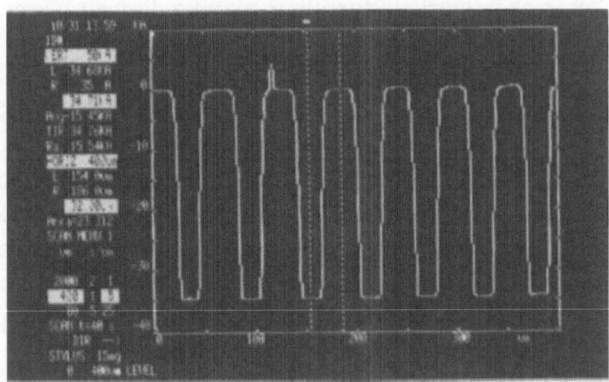

Fig. 6.11 Talystep trace of a binary etched fused silica surface (see Fig. 6.10).

Atomic force microscope. Atomic force microscopes can in a similar manner as the mechanical profilometer be used to measure surface profiles. The sensitivity is of the order of a few Angstroms in the height measurement and nm's in lateral dimensions [38]. The atomic forces acting on the scanning cantilever tip are considerably smaller than in mechanical profilometers. Therefore, the resolution is by some orders of magnitude higher and the surface load smaller. The resolution depends strongly on the tip geometry. The tip cone angle has been reduced to 10° and the tip radius to 100 Å. The atomic forces are indicated by the bending of the cantilever. The bending is optically measured with laser diode illumination and a position sensitive diode which senses the lateral movement of the laser beam reflected at the backface of the cantilever. The cantilever is made either from Si_3N_4 or a single Si-crystal. An overview over the field is given by Sarid [39].

6.3 Characterization of optical performance

The optical performance of micro optical imaging elements can be characterized with the help of different methods. The optical channel is characterized by the pulse response or in optical terms the point spread function (PSF). Fourier transform of the pulse response is the transfer function, i.e., it gives the modulation degree as a function of the spatial frequency: the so-called modulation transfer function (MTF). This has to be measured for each field point of the lens or the optical system. Complementary to those functions the wave aberrations can be measured delivering a more direct impression of the deviations of the element. This can be of great value for the production process since the location of the disturbances can be detected in a more direct way. From the wave aberrations the other characteristic functions may be derived by means of Fourier transformation [20].

6.3.1 Wave aberration measurement and related merit functions

The overall optical function of a microoptic element depends more on the macroscopic distribution of the refractive index or the geometrical location of diffractive phase features than on the microscopic form of the substructures used to generate the deflection of the light rays. The test of the global parameters such as: wave aberrations, optical transfer function and optical point spread function can be carried out with different methods known from the testing of macrooptic elements. Cline and Jander [40] used the phase shifting technique [21] for the measurement of the wave aberrations of GRIN lenses made by Nippon Sheet Glass Co. In this case the aim of the evaluations was the determination of the Zernike coefficient for spherical aberration from the wave aberration pattern measured with the help of a Twyman-Green interferometer (TWG). The coupling efficiency of such GRIN lenses for coupling light from a laser into a fiber depends rather strongly on just this aberration.

The problem with the Twyman-Green interferometer is on the one hand the double passage of the wave through the lens under test. On the other hand the TWG offers the advantage of simple adjustment and straightforward calibration procedures, i.e. absolute measurements [41].

With the same interferometer also holographical lenses (HOE) may be tested. The setup is shown in Fig. 6.12. The test result for an aberration-optimized HOE with a numerical aperture of N.A.=0.25 (see Fig. 6.13). Aberration-optimized means that the astigmatism has been diminished by choosing an appropriate recording geometry [42].

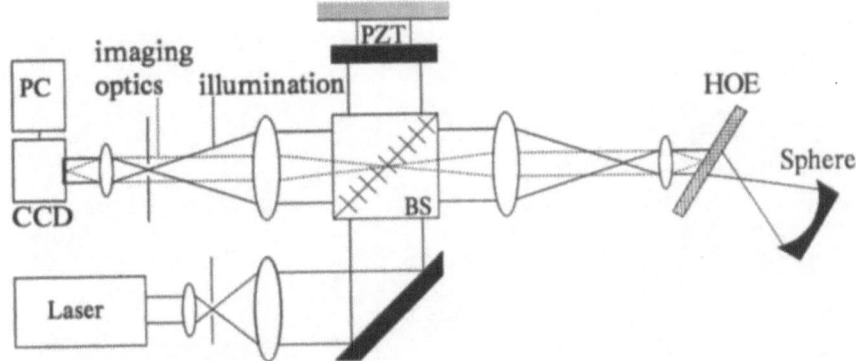

Fig. 6.12 Scheme of a Twyman-Green interferometer for the test of holographic lenses in DCG (volume hologram). In this setup the lens is passed twice and the light path is bent due to the carrier frequency of the HOE.

Another production technology for micro-optical lenses uses melting processes in photoresist to obtain lenslet arrays [43]. These lens arrays have been tested with the help of different interferometers, i.e. Twyman-Green, spherical Fizeau, and Mach-Zehnder interferometer. The tests showed good performance of the lenslets across the main part of the aperture. The constancy of the focal length for a tandem of two arrays could be demonstrated in a Mach-Zehnder interferometer (Fig. 6.14).

If the wave aberrations are measured, it is possible to fit a suitable set of orthogonal polynomials to the data from the interferometric test. In this way the main contributions to the aberration can be identified. Probably the best choice are the Zernike circle polynomials. The lower order contributions may be identified with spherical aberration, coma, astigmatism etc.

The coefficients of the series expansion are attributable to the optical performance of the element under test.

Once the wave aberrations have been measured also other characteristic functions can be calculated. The point spread function can be calculated by Fourier transformation of the pupil function kW:

$$PSF = |Four\{exp(jkW(x,y))\}|^2 \tag{6.7}$$

The inverse Fourier transform of the point spread function delivers the optical transfer function which is a complex function with the MTF (modulation transfer function) and the PTF (phase transfer function). In most cases the MTF is the essential merit function. Only with strong nonsymmetric aberrations also the PTF becomes essential [44].

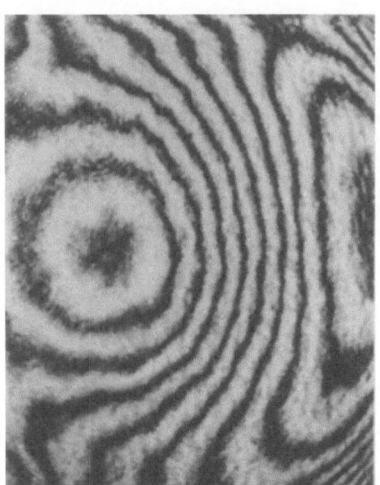 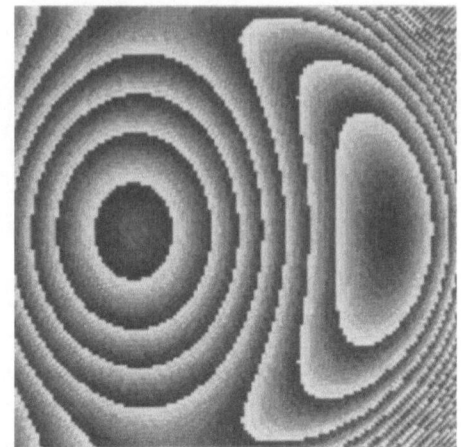

Fig. 6.13 Interferometric test of a holographic lens, optimized aberration with the help of a third order theory, numerical aperture sin(u) = 0.25, recording wavelength λ = 488 nm, reconstruction wavelength λ = 633 nm, left: interferogram at λ = 633 nm, right: synthetic interferogram calculated with the help of ray trace program.

An alternative to the TWG is the Mach-Zehnder interferometer. On the one hand the MZ offers single pass geometry which is advantageous in case of large wave aberrations. A further drawback of the TWG is its sensitivity to spurious reflections. In interferometry complex amplitudes are superimposed.

Therefore, the intensity of spurious reflections should be at least 3 orders of magnitude smaller than the signal intensities. This condition can be met rather sufficiently in transmitted light but only very poorly in reflected light. The Mach-Zehnder test geometry is therefore a very powerful alternative. Fig. 6.15 shows a Mach-Zehnder arrangement for measuring the wave aberrations of microlenses. The use of a He-Ne-laser allows for this unsymmetric setup. With spatially partially coherent light a more symmetric setup (Fig. 6.19) must be used. A whole set of aberration data for a lens made by melting photoresist is given in Fig. 6.16.

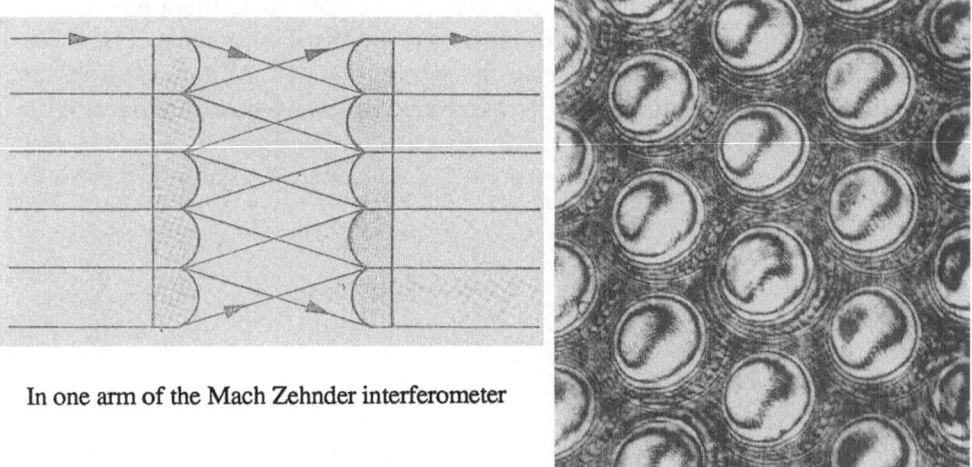

In one arm of the Mach Zehnder interferometer

Fig. 6.14 Interferometric test of the uniformity of the focal length of a tandem of lens arrays made by melting photoresist (courtesy M. Hutley, NPL, Teddington UK).

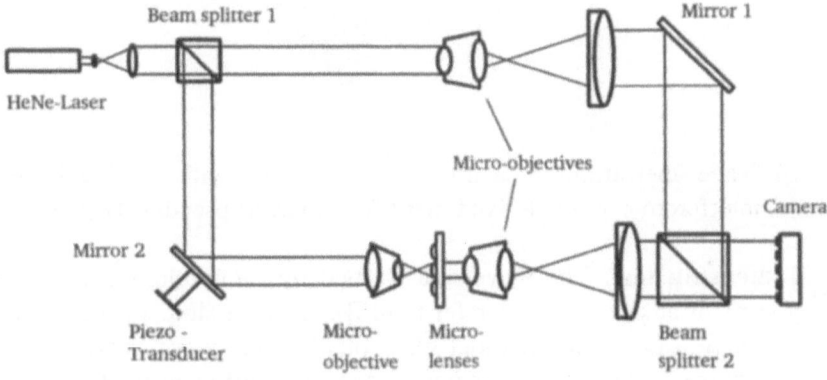

Fig. 6.15 Mach-Zehnder interferometer adapted to the measurement of wave aberrations. He-Ne-Laser illumination allows for an unsymmetric setup.

On the other hand the calibration of an interferometer containing a beam diverger is not at all simple and presupposes other interferometric tests (see Sect. 6.3.2).

The disadvantage of the interferometric methods is that laser light has to be used in order to cope with different optical path differences. Noninterferometric methods can also be used to determine the point spread function and the MTF

[45]. In this case even white light sources are admissible which might be useful if simultaneous measurements at different wavelengths are at stake. For more details the reader is adviced to consult the special literature Murata [45]. So far only micro-optical elements (MOE) in transmitted light have been discussed. Also reflection type MOE's can be used in optical interconnects. A possible test geometry is shown in Fig. 6.17. Here a TWG arrangement is assumed in order to test an optical lens element in reflection.

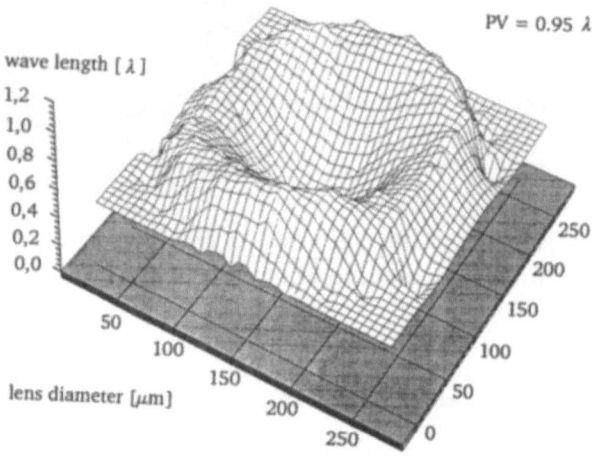

Fig. 6.16 Wave aberrations of a microlens measured with the Mach-Zehnder interferometer and derived merit functions. a) pseudo-3D-plot,.

Lateral shearing test. The wavefront formed by a microlens can undergo a shearing test (for an overview we refer to [46]). A rather simple shearing test for laser illumination is a plane parallel glass plate. If the wave aberrations show rotational symmetry then one shear interferogram is sufficient for the evaluation of the lens (comp. Fig. 6.18). Polychromatic illumination requires a test geometry which enables optical path difference zero. One possibility is a Michelson interferometer. The width of the illuminating slit determines the possible lateral shear.

6.3.2 Calibration problems

Interferometers show all phase differences which the light suffers from any optical component of the interferometric setup. Therefore, it is mandatory to calibrate the aberrations of the empty interferometer and the reference surfaces to generate the reference wave or which are used to redirect the light as in the case of the TWG.

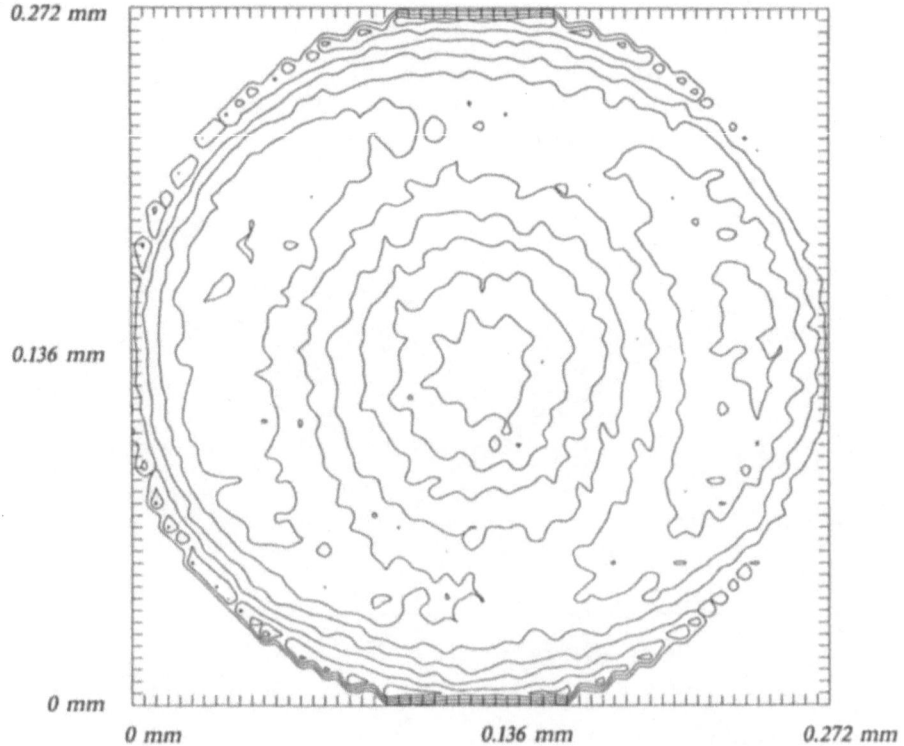

Wave Aberration

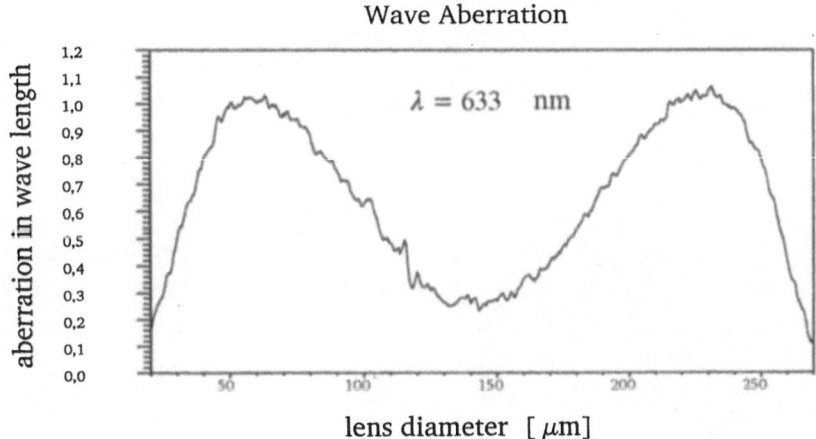

Fig. **6.16** Wave aberrations of a microlens measured with the Mach-Zehnder
interferometer and derived merit functions. **b)** contour line plot: height
difference λ/8 between contour lines, **c)** section through the wave
aberrations,

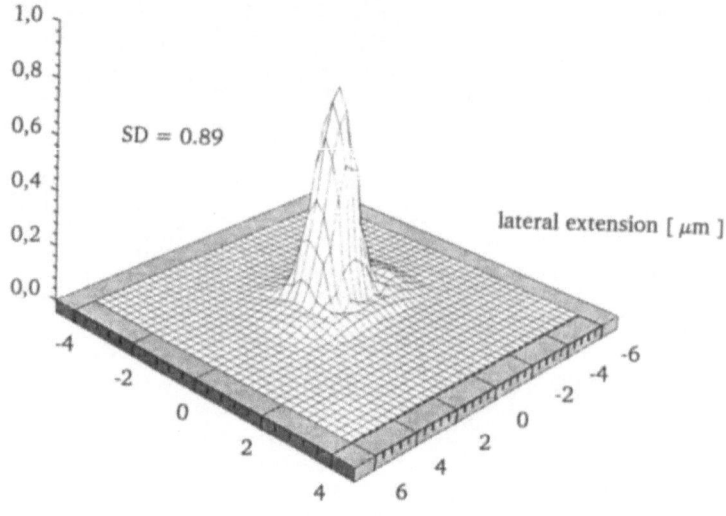

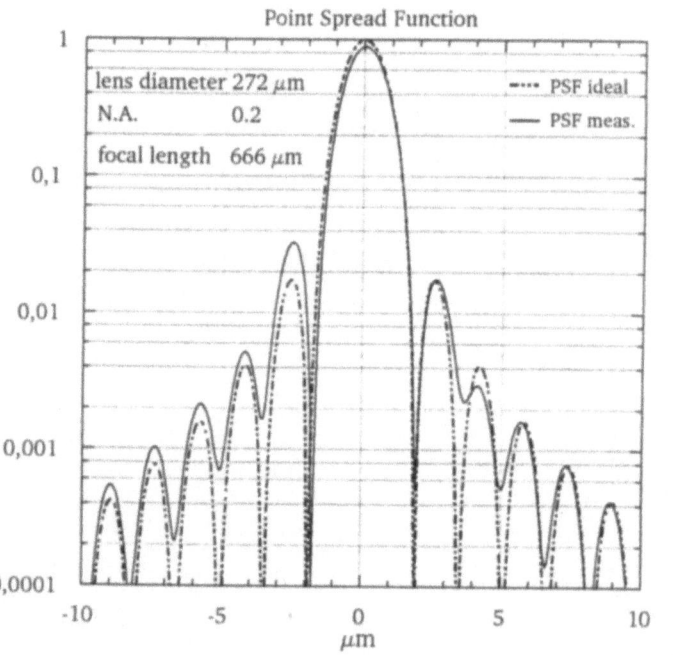

Fig. 6.16 Wave aberrations of a microlens measured with the Mach-Zehnder
interferometer and derived merit functions. **d)** point spread function
(SD: strehl definition), **e)** cross section through the PSF in logarithmic
scale,

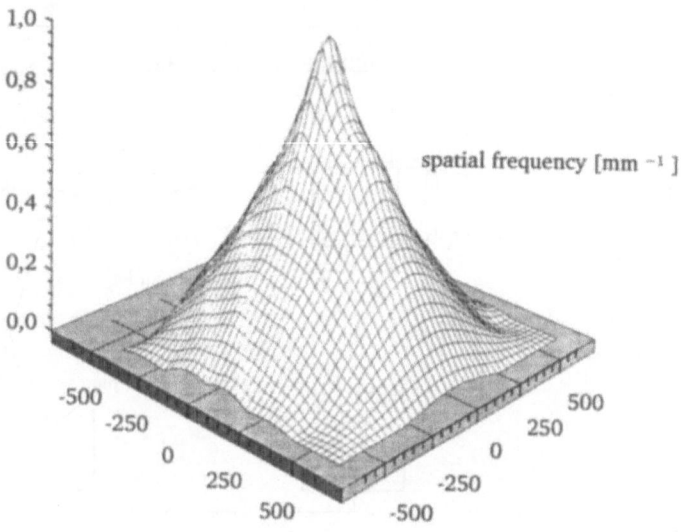

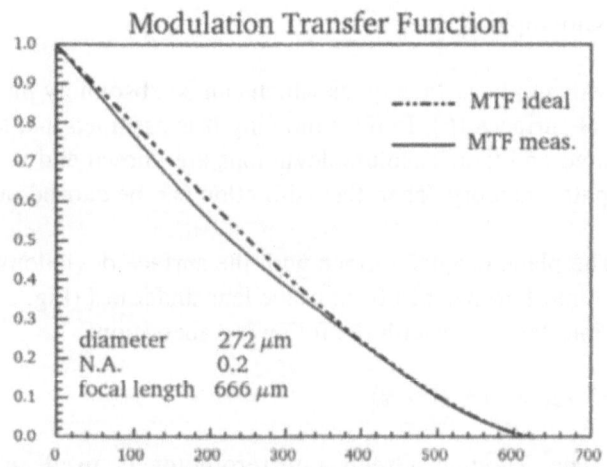

Fig. 6.16 Wave aberrations of a microlens measured with the Mach-Zehnder interferometer and derived merit functions. f.) modulation transfer function, g.) central section through MTF.

Twyman-Green interferometer. Calibration of interferometers means absolute measurements of surfaces or other optical elements. Otherwise one has to be sure that the calibrating element be it a surface or an optical system has known or negligible aberrations.

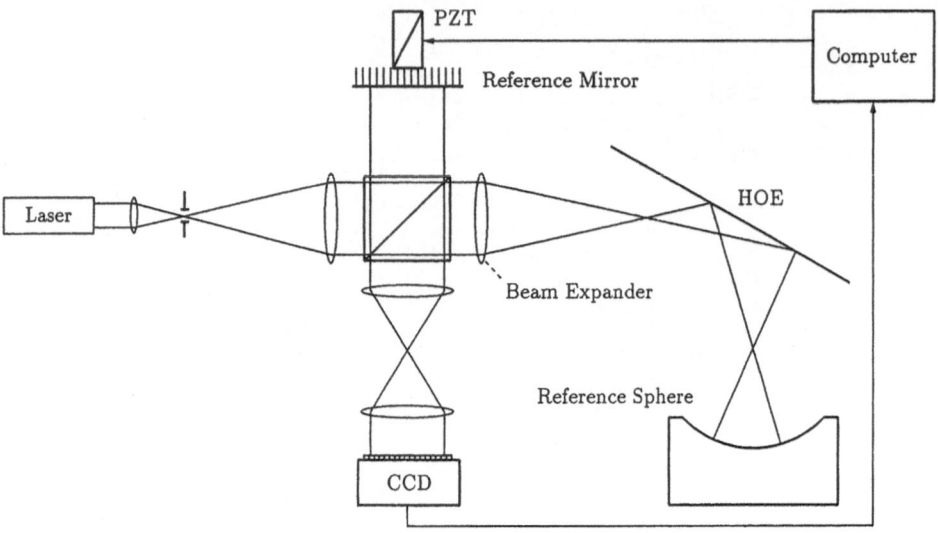

Fig. 6.17 Scheme of a Twyman-Green interferometer to test reflected light holographic lenses.

There are two simple surface types which can be absolutely measured, i. e., plane and spherical surfaces [6]. In the following it is assumed that such surfaces have been measured and their absolute deviations are known and as data map resident in the computer memory. Then the calibration can be carried out in the following way.

Step 1: The plane normal surface with the surface deviations $P(x,y)$ is inserted in the TWG interferometer in front of the lens under test (Fig. 6.3). One has then a Michelson interferometer with the following aberrations:

$$W_1(x,y) = -W_{ref.}(x,y) + 2P(x,y) \qquad (6.8)$$

Step 2: The Twyman-Green interferometer is used in the normal test configuration with a known spherical normal having the deviations $S(x,y)$. The wave aberrations measured by the TWG are:

$$W_2(x,y) = -W_{ref.}(x,y) + 2W_{lens}(x,y) + 2S(x,y) \qquad (6.9)$$

Since $P(x,y)$ and $S(x,y)$ are assumed to be known the lens aberrations $W_{lens}(x,y)$ can be calculated as:

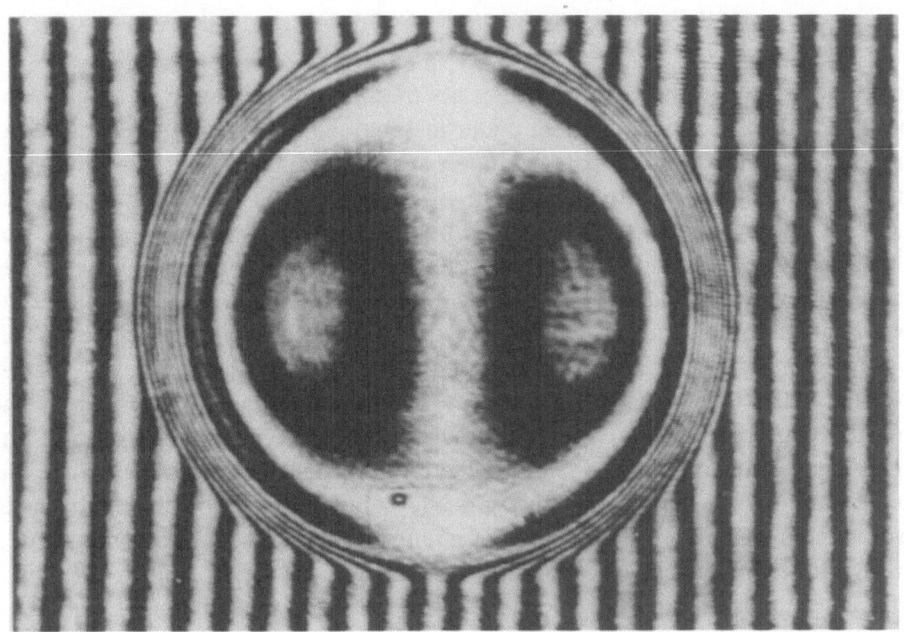

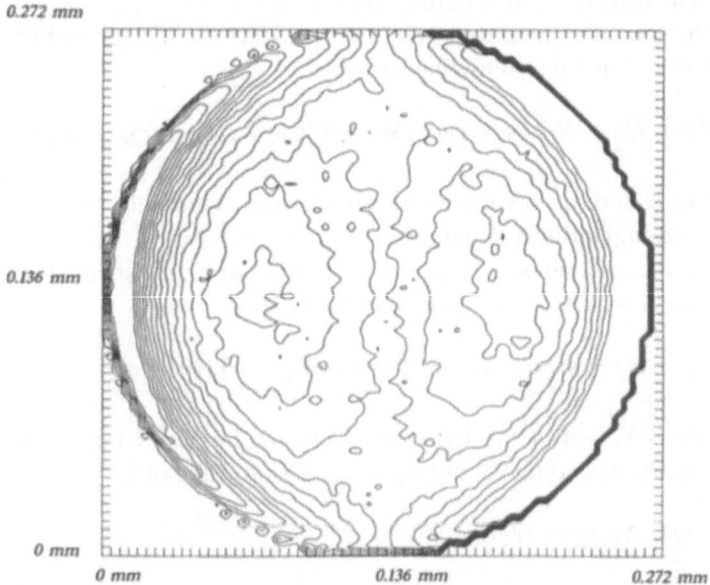

0.272 mm

0.136 mm

0 mm

0 mm   0.136 mm   0.272 mm

Fig. 6.18 Shearing interferogram of a microlens, above: interferogram, below:
contour line plot of the wave aberration difference measured with
phase shifting technique, focal length: $f = 900$ μm, diameter: 272 μm.
The difference between two lines in the contour line plot is $\lambda/10$,
respectively.

$$W_{lens}(x,y) = 1/2(\ W_2(x,y) - W_1(x,y) - 2[S(x,y) - P(x,y)])$$ (6.10)

Mach-Zehnder. In the case of strong aberrations double passages through the optical lens or the optical element under test may be inconvenient because a ray trace through the double pass geometry becomes necessary, and the high sensitivity to adjustment errors might be troublesome. It is a common feature that the beam splitters should only be passed by plane waves if strong aberrations due to the glass thickness shall be avoided. For the test of MOE the test object has to be magnified which makes the use of some auxiliary objectives of appropriate numerical aperture for a compensation of the spherical wavefront necessary. This objective could be calibrated by means of a TWG as discussed above.

Usually, the deviations introduced by the mirrors can be corrected by the measurement of the empty interferometer and by storing those data in the memory of the computer. Here, however, one has to be careful since the combination of the lens under test with the compensating objective introduces an inversion of the coordinate system about the optical axis. To arrange for a throughout symmetrical test situation one needs 4 auxiliary objectives with a sufficient numerical aperture and same focal length as the lens under test. Total symmetry is also required to avoid radial shear of the test beam in relation to the reference beam. The calibration can again be performed by a two-step procedure. The test situation is outlined in Fig. 6.19.

Step 1: The setup for measuring the reference data set is given in Fig. 6.19. The wave aberrations have the following form:

$$W_{MZ1} = W_{1'} + W_3 + W_{O3} + W_{O4} - (W_1 + W_2 + W_{O1} + W_{O2} + W_{4'})$$ (6.11)

This equation is chosen in such a manner that the coordinate systems remain in coincidence through all test situations.

Step 2: In the second step e.g. the objective with the aberrations W_{O4} is replaced by the lens under test. The aberrations are:

$$W_{MZ2} = W_{1'} + W_3 + W_{O3} + W_{Test} - (W_1 + W_2 + W_{O1} + W_{O2} + W_{4'})$$ (6.12)

The aberrations W_{Test} of the lens under test can be derived by subtracting the two equations (6.11) and (6.12) of step 1 and 2 from each other:

$$W_{Test} = W_{MZ1} - W_{MZ2} - W_{O4},$$ (6.13)

where W_{O4} has to be measured by a procedure discussed in connection with the TWG interferometer. For the second step it is assumed that the objectives 1 through 4 have a sufficient numerical aperture and same focal length in order to guarantee the absence of shear effects which could cause severe measuring errors.

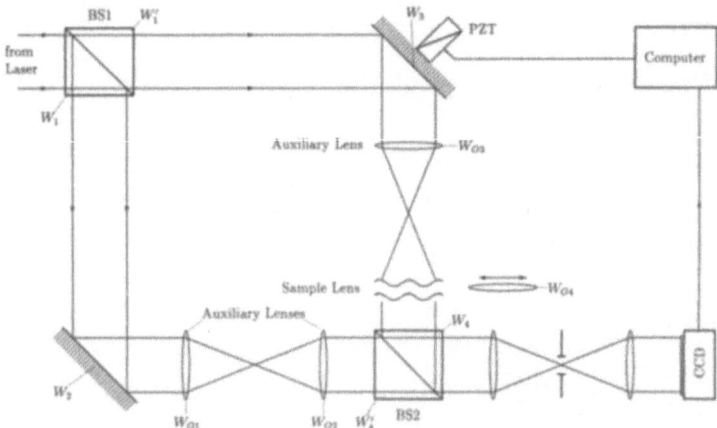

Fig. 6.19 Scheme of a Mach-Zehnder: Calibration configuration , W_{Oi}: wave aberrations of the auxiliary objectives Oi.

Since it might be difficult to find an auxiliary objective having sufficient numerical aperture and the same focal length as the lens under test absolute methods known from planeness testing could be used in a modified form. For this purpose three lenses of the same type on different substrates could be combined in pairs in the Mach-Zehnder after Fig. 6.19, where the pair of lenses replaces the combination of the lenses O3 and O4. Then one obtains the sum of the aberrations of the pairs which can be resolved for the aberrations of a single lens by solving the linear system of 3 equations. It has been shown that the solution can only be obtained on a section and additional procedural measures are necessary to obtain the two-dimensional wave aberration of the lens (for details consult [6]).

6.3.3 Direct PSF measurements

With the help of modern imaging devices it is possible to scan the intensity pattern in the focal plane of micro-optic lenses and also of other elements (see e.g. [47-49]). If the sampling density does not suffice the focal plane can be imaged with the help of a microscopic objective to the appropriate scale. Since such auxiliary objectives are available as diffraction-limited systems, the additional spread can be held within tolerable values by using a high aperture system. Fig. 6.20 shows an example of such a scan through the light spots in the focal plane for a graded index lens array made by Nippon Sheet Glass which shows the form of the PSF in one section providing information about the geometrical spot size.

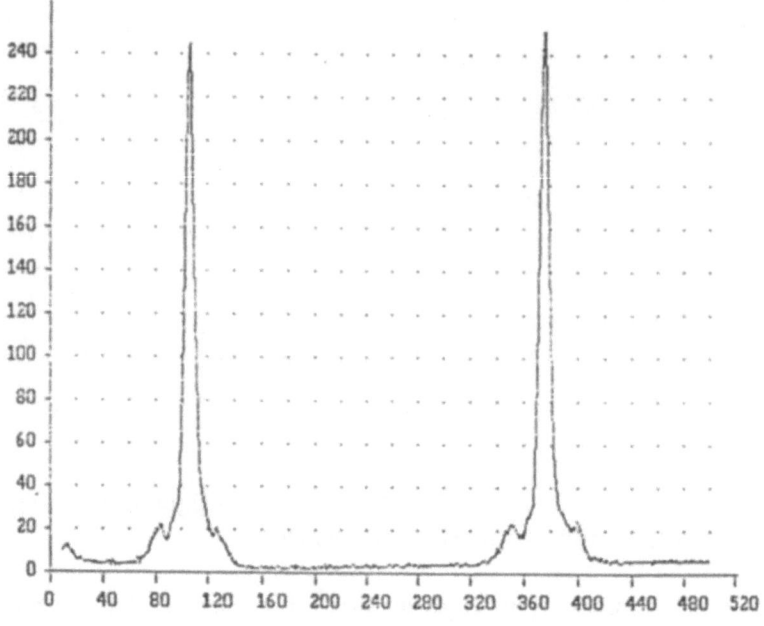

F1=HELP   Startpos.=68   Verst.=6   Offset=0   Triggerung=continous
27.9.1991   helles Austasten

Fig. 6.20 Scan through the focus of a graded index lens array (two focus spots).

For space invariant imaging by means of microlenses also off-axis aberrations may be of interest. In order to access such aberrations the element has to be turned about its nodal point to enable field measurements.

Although not directly correlated with PSF measurements, measurements of the locations of an array of spots can be carried out with the same technique. So, e.g., distortion measurements have been carried out in case of graded index lenses [4]. The distortion was measured through imaging a rectangular grid pattern.

6.3.4 Angle measurement of reflective/refractive and diffractive deflectors

Light deflectors are a very essential part of an optical interconnect. The main quantity of interest is of course the angle between the incident and refracted (or diffracted) rays. The angle between the incident and deflected ray can be determined with the help of a spectrometer table with a divided circle or a coded angle-measuring device. As indication one uses the maxima of the intensity distribution in the back focal plane of the observing telescope for the case of a measurement with and without the microoptic component in the ray path. For planar elements also the autocollimating properties of the plane entrance window may be used. But what remains is the measurement of the angles between two or more angular positions defined by the center of gravity of the light beams.

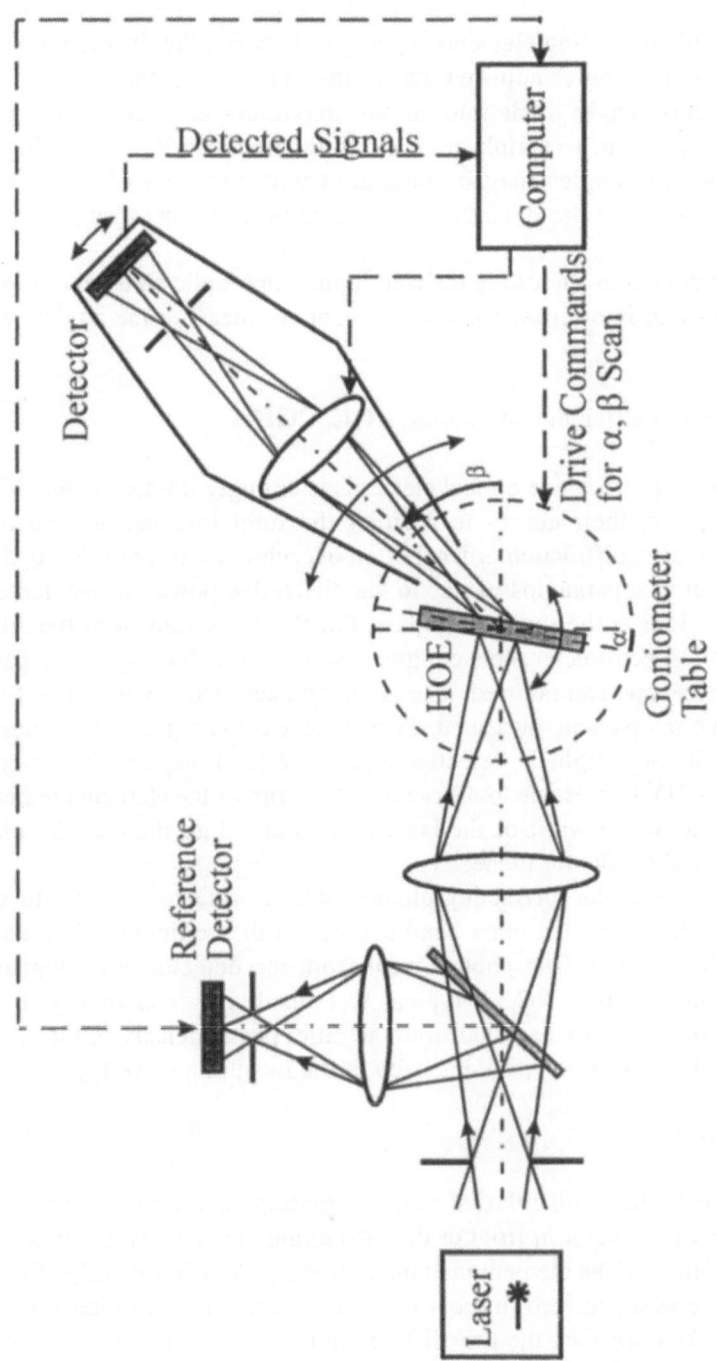

Fig. 6.21 Scheme of a goniometer stage for the measurement of the diffraction efficiency of holographic elements.

For holographic or grating elements the angles between the diffraction orders are determined by the angles adjusted during the exposure of the element. Volume holograms show angle deviations of the maximum efficiency rays from the adjusted geometry due to shrinkage or swelling effects during processing.

Small deviation angles may be measured with the help of autocollimation telescopes as long as the deviations remain within the measuring range of the telescope.

If discrete deviation angles are realized in the microoptic element one could use angle etalons and interferometric arrangements to measure the angles with high accuracy.

6.3.5 Diffraction efficiency of volume HOE, CGH

The efficiency of diffractive optical elements is strongly limited by the diffraction efficiency η, i.e. their ability to redirect the light into the wanted direction commonly the first diffraction order (For an overview we refer to [3, 50, 51]). It is rather difficult to separate losses due to the diffractive power of the element from surface reflections at the supporting glass. For the measurement of the diffraction efficiency η a spectrometer table equipped with a photodetector in the back focal plane of a telescope can be used. The angle characteristics without and with the element in the ray path are measured. To eliminate drifts of the light source a small part of the incident light is detected seperately [52] and is used as reference (comp. Fig. 6.21). Losses due to diffraction at the rim of the element are minimized by imaging the beam waist of the laser light source into the element plane and afterward into the detector plane.

On the one hand the diffracting element can be rotated in order to vary the incidence angle and on the other hand the arm of the goniometer is used to scan the diffraction pattern. The photocurrent from the detector is analog to digital converted and fed to a PC. A typical test result is given in Fig. 6.22. The diffraction efficiency is calculated from the ratio: peak intensity of the spot of the diffracted order I_{diff} to the peak intensity of the incoming wave I_{in}:

$$\eta = I_{diff} / I_{in}, \tag{6.14}$$

where the intensities are relative values, relating to the power of the source measured by the detector in front of the diffracting element. This efficiency is the overall efficiency of the element including all losses. Often, the ratio of the first to the sum of the first and zero-th order is given as diffraction efficiency for volume holograms. But even then the true diffraction efficiency can only be measured if the outcoupling of the light out of the HOE is done for all diffraction orders under equal conditions, e. g. by index-matching a suitable glass cylinder to the HOE-surface. The cylinder axis should coincide with the diffracting element, so that all diffracted orders are travelling along radii of the cylinder and hit the mantle of the cylinder perpendicularly.

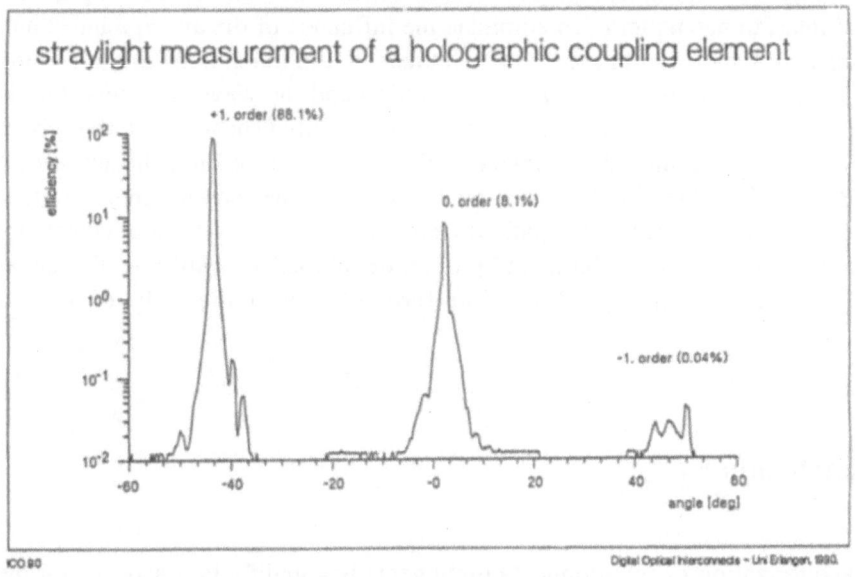

Fig. 6.22 Measuring result for the diffraction efficiency of a deflecting HOE.

6.4 Characterization of index profiles

Microoptic graded index imaging elements play a major role in optical communication lines [4]. Therefore, measuring methods for preforms of selfoc fibers, selfoc lenses and also for refractive index distributions in general have been developed. With interferometers either sections of an index profile or the integral effect of the index distribution can be measured. For the measurement of a section the lens element has to be cut into slices (and polished) which then can be placed into an interferometer to measure the optical path difference perpendicularly to the gradient of the index. Nondestructive measurements, i.e. measurements of the integral path difference allow only in case of rotational symmetry a reduction of the data to get the 3D-index distribution. For this purpose the sample is tested in a position perpendicularly to the working position. The influence of the cladding is commonly compensated for by immersing the sample in an index-matched fluid. Although there is the possibility to derive the 3D-index distribution from the measurement in different directions [53] it is very difficult to apply such techniques to samples with strong symmetries as grin rods or fibers.

To measure the path difference in transmitted light shearing interferometry [46] has been used. Small objects as fibers or preforms can be doubled in the shearing interferometer so that one compares the object with a smooth surrounding giving a normal interference pattern. To eliminate the influence of the average index and of the shape of the element the sample is immersed into index matching oil. If the sample is too large to get a true lateral separation and the wavefront curvature is too big, one may use small shears resulting in a differential shearing interferogram. In this case one measures the gradient of the wavefront and not the wavefront itself [54]. Provided the sample is rotational symmetric one shearing interferogram suffices to obtain the integral path difference and therewith the wavefront aberrations. Special procedures [55] were developed to calibrate the shear amount and the sensitivity of the interferometer with the help of known aberrations.

6.5 Conclusion

The characterization of interconnect components has mainly two aspects: on the one hand *the measurement of the form of the element,* (or of the supporting structure), or of a periodic substructure, and on the other hand *the measurement of the optical function.*

Two characteristic functions of the interconnnect components can be identified: *focusing and deflection.* The microoptic elements are either of the refractive/reflective or of the diffractive type.

Optical interconnects require a great deal of design freedom. This freedom is obtained by using preferably space-variant components which provide one or a few interconnect channels. High packing density of space-variant channels is only possible with microoptics. The test equipment has to take this fact into account. So, e.g. the exact imaging of the microoptic component onto the detector is mandatory which makes microscope optics necessary.

One part of the Section 6 dealt with the characterization of the surface structure, since a great part of the elements is based on diffraction. The optical test equipment for these structures has its problems. On the one hand the resolution requirements are very high since the structures are of the order of a few wavelength. So, the interpretation of interference patterns with respect to the surface structure becomes questionable. This is especially the case if in addition the structures are more or less discontinuous and deep. Measurements in transmitted light are therefore necessary since the test configuration is in many cases compatible to the application. Automated interferometers need further development of the measuring procedure and software to cope with the problems posed by the complexity of such elements. Furthermore, the interference pattern is rather involved in case of deep discontinuous structures and under partially incoherent illumination. Here comparisons with results from atomic force microscopy might help to clear up the limits of such measuring interferometers. AFM has the problem that it is no real noncontact method and that the field of view is rather small.

The characterization of the optical function of microoptic components is among other things difficult since the diameter of the element is small (a few hundred wavelength). Diffraction effects of the rim region determine the aberrations and the function to a higher degree than usual in optics. High numerical apertures cause small adjustment tolerances which make the measurement a difficult task and the use of such elements expensive.

The measurement of deflection angles and diffraction efficiencies is also limited by diffraction effects and the limited surface quality of the supporting structure. The verification of the results of the coupled wave theory e. g. is hampered by Fresnel reflections and scattering effects in the glass support. So, it is obvious that the development of testing methods and equipment is not at all a closed case.

References

1. N. Streibl, K.-H. Brenner, A. Huang, J. Jahns, J. Jewell, A. Lohmann, D. A. B. Miller, M. Murdocca, M. E. Prise, Th. Sizer, "Digital Optics", Proc. IEEE 77, 1989, p. 1954-1969
2. N. Streibl, "Beam shaping with optical array generators" J. of Mod. Opt. 36, 1989, p. 1559-1573
3. L. Solymar, J. Cooke, "Volume holography and volume gratings", Academic Press 1981
4. K. Iga, Y. Kokubun, M. Oikawa, "Fundamentals of microoptics", Academic Press, Tokyo 1984
5. D. Daly, R.F. Stevens, M.C. Hutley, "The manufacture of microlenses by melting photoresist", Meas. Sci. Technl. 1, 1990, p. 759-766
6. G. Schulz, J. Schwider, "Interferometric testing of smooth surfaces", Progress in Optics 13, 1976, p. 92-167, editor E. Wolf, North Holand Publ. Amsterdam
7. Zygo description of phase shifting interferometer, Zygo Corp., Middlefield CT, USA
8. N. Abramson, "The Interferoscope a new type of interferometer with variable fringe separation", Optik 30, 1969, p. 56
9. J. Schwider et al., "Semiconductor wafer and technical flat planeness testing interferometer", Measurement 5, 1987, p. 98-101

144

10. K.G. Birch, "Oblique incidence interferometry applied to non-optical surfaces", J. Phys. E. Sci. Instr. 1973, Volume 6, p. 1045-1048
11. N. Streibl, U. Nîlscher, J. Jahns, S. Walker, "Array generation with lenslet arrays", Appl. Opt. 30, 1991, p. 2739-2742
12. H.-P. Herzig, D. Prongue, R. Dändliker, "Design and fabrication of highly efficient fan-out elements", Proc. of the Int. Top. Meeting on "Optical Computing" in Kobe, Japan, April 1990
13. J. A. Cox, "Overview of diffractive optics at Honeywell", Proc. SPIE 884, 1988 p. 127-131
14. J. R. Leger, M.L. Scott, P. Bundman, M.P. Griswold, "Astigmatic wavefront correction of gain-guided laser diode array using anamorphic diffractive microlenses", Proc. SPIE 884, 1988, p. 82-89
15. W. Krug, J. Rienitz, G. Schulz, "Contributions to Interference Microscopy", Hilger & Watts, 1964, London
16. G. Schulz, "Ein einfaches Interferenzmikroskop für Auflicht", Naturwissenschaften 48, 1961, p. 565-566
17. U.-C. Minor, G. Schulz, "Über ein Auflicht-Interferenzmikroskop ohne Vergleichsfläche", Wiss. Zeitschr. Hochsch. Elektrotechn. Ilmenau 8, 1962, p. 475-479
8. C. Koliopoulos, "Interferometric optical phase measurement techniques", Thesis, 1981, Tucson OSC
19. Topo 3D-Interferometer-prospectus, Wyko Corp.
20. M. Born, E. Wolf, "Principles of Optics", 6-th edition, Pergamon Press, 1980
21. J. Schwider, "Advanced evaluation techniques in interferometry", Progress in Optics 28, 1990, p. 272-359, editor E. Wolf, North Holand Publ. Amsterdam
22. J. Schwider, R. Burow, K.-E. Elßner, J. Grzanna, R. Spolaczyk, K. Merkel, "Digital wavefront measuring interferometry: some systematic error sources", Appl. Opt. 22, 1983, p. 3421-3432
23. G. Schulz, "Über Interferenzen gleicher Dicke und Längenmessung mit Lichtwellen", Ann. Phys. (6) 14, 1954, p. 177-187
24. G. Schulz, "Über Interferenzen gleicher Dicke bei größeren Keilwinkeln und größerer Apertur", Optik 16, 1959, p. 280-287
25. K. Creath, "Calibration of numerical aperture effects in interferometric microscope objectives", Appl. Opt. 28, 1989, p. 3333-38
26. J. Biegen, "Calibration requirements for Mirau and Linnik microscope interferometers", Appl. Opt. 28, 1989, p. 1972-1974
27. G. Schulz, K.-E. Elßner, "Errors in phase measurement interferometry with high numerical apertures", Appl. Opt. 30, 1991, p. 4500-4506
28. K. Creath "Phase-Measurement Interferometry Techniques", Prog. in Optics 26, 1988, p. 349-393
29. Osten, "Digitale Verarbeitung und Auswertung von Interferenzbildern", Akademie Verlag, Berlin 1991

30. G. Makosch, B. Solf, "Surface profiling by electro-optical phase measurements", Proc. SPIE Vol. 316, 1981

31. C. C. Huang, "Optical heterodyne profilometer", Opt. Eng. 23, 1984, p. 365

32. J. C. Wyant, K. Creath, "Recent advances in interferometric optical testing", Laser Focus/Electrooptics 21, Nov. 1985, p. 118-132

33. W. Kösters, "Anwendung der Interferenzen zu Messzwecken", Handbuch der physikalischen Optik Bd.1, p. 471-498, ed. E. Gehrcke, Leipzig 1927, J.A. Barth

34. Y.-Y. Cheng, J.C. Wyant, "Two-wavelength phase shifting interferometry", Appl. Opt. 23, 1984, p. 4539-4543

35. K. Creath, "Step height measurement using two-wavelength phase-shifting interferometry", Appl.Opt. 26, 1987, p. 2810-2816

36. Talystep, Taylor & Hobson, Instrument description, 1987

37. Rodenstock, Laser Stylus RM 600

38. D. Rugar, P Hansma, "Atomic force microscopy", Phys. Today, Oct 1990, p. 23-30

39. D. Sarid, "Scanning force microscopy with appliocations to electric, magnertic, atomic forces", Oxford University Press, New York, 1991

40. T. W. Cline, R. B. Jander, "Wavefront aberration measurements on Grin-rod lenses", Appl. Opt. 21, 1982, p. 1035-1041

41. J. Schwider, K.-E. Elssner, J. Grzanna, R. Spolaczyk, "Results and error sources in absolute sphericity measurements", IMEKO Laser Measurement Working Group Symposium Budapest, November 1986

42. O. Falkenstörfer, H. Koboll a, U. Krackhardt, N. Lindlein, J. Schwider, N. Streibl, R. Völkel, H. Weißmann, "Optimization of holographic lenslets and their measurement", IOP Short meeting on Microlens Arrays, Teddington 1.May 1991, p. 53-60

43. M.C. Hutley, D. Daly, R.F. Stevens, "The testing of microlens arrays", Proc. IOP-Short-meeting on "Microlens arrays", 1.May 1991 Teddington, Institute of Physics Series No. 30, p. 67-81

44. O'Neill, "Introduction to statistical optics", 1963, Read. Mass.

45. K. Murata, "Instruments for the measuring of optical transfer functions", Prog. in Optics vol. V, ed. E. Wolf, North Holland publ. House, 1966

46. D. Malacara, "Optical shop testing", J. Wiley & Sons, 1978, New York

47. D. Prongue, H. P. Herzig, "Design and fabrication of HOE for clock distribution in integrated circuits", Conference Proceedings "Holographic Systems, Components and Applications", Bath, Sept. 1989

48. H.-P. Herzig; "Holographic optical elements (HOE) for semiconductor lasers", Opt. Comm. 58, 1986, p. 144-148

49. K. Hamanaka, H. Nemoto, M. Oikawa, E. Okuda, T. Kishimoto, "Multiple imaging and multiple Fourier transformation using planar microlens arrays", Appl. Opt. 29, 1990, p. 4064-4070

50. R. R. A. Syms, "Practical volume holography", Clarendon Press, Oxford 1990

51. H. Kogelnik, "Coupled wave theory for thick hologram gratings", Bell Syst. Techn. Journ. 48, 1969, p. 2909

52. G.R. Chamberlin, D.E. Sheat, A.M. Hill, "Holographic transmission gratings for use in the 1250-1600 fibre window", Conference Proceedings "Holographic Systems, Components and Applications", Bath, Sept. 1989

53. G. Birnbaum, C.M. Vest, "Holographic nondestructive evaluation: status and future. Int. Adv. in Nondestr. Test. 9, 1983, p. 257-282

54. Y. Kokubun, K. Iga, "Index profiling of distributed-index lenses by a shearing interference method", Appl. Opt. 21, 1982, p. 1030-1034

55. Y. Kokubun, T. Usui, M. Oikawa, K. Iga, "Wave aberration testing system for micro-lenses by shearing interference method", Jap. J. Appl. Phys. 23, 1984, p. 101-104.

First section: Components

Part 1.2 Active interconnect components

7 Optoelectronic semiconductor devices

F. Lozes-Dupuy, H. Martinot, S. Bonnefont
LAAS-CNRS, Toulouse, France

7.1 Introduction

Over the last years considerable progress has been made in semiconductor optoelectronics because of the needs of optical telecommunications and the emergence of new fields. Today, semiconductor optical devices are used in fiber-optic systems and for satellite communication, for optical data communication, storage, reading and writing, for optical sensors and measurements, and as solid-state laser pumps. These devices are expected to be more extensively applied in emerging areas such as optical interconnects, optical signal processing and computing or optical memory. This development has been made possible by the greater maturity of material growth and device fabrication techniques, and by an increased knowledge of semiconductor materials and device structures. Ongoing research in this field is reported in the literature and enhanced fabrication tends to make high-yield and low-cost semiconductor devices available.

The purpose of this chapter is to give an introduction to the main trends in the field of optoelectronic semiconductor devices that could play a key role in future optical interconnections. The first Section describes advances in epitaxial growth technologies that form the basis of the development of new semiconductor materials and device structures. The second Section indicates those trends which in the field of technological processes have led to new types of devices with enhanced performance. Rather than report on the worldwide research covering the various types of devices, this paper will focus on a few examples illustrating the great deal of success achieved during the last decade in the design and fabrication of devices suitable for one-dimensional (1D) or two-dimensional (2D) integration. The third Section deals with 2D laser arrays and 2D optical modulator arrays which are particularly interesting for a wide variety of possibilities offered by the parallel processing architecture in optical information processing. A discussion of detectors is not included in this text as they represent a mature

150

topic and are not a limiting technology in current optical interconnect developments [1].

7.2 Semiconductor materials and growth techniques

7.2.1 III-V compound semiconductors

Unlike Ge and Si exhibiting an indirect bandgap, various III-V compounds possess a direct energy bandgap and can be used as efficient active materials in light emitting devices. Therefore, common optoelectronic structures consist of a substrate upon which several crystalline layers of III-V materials are grown. Fig. 7.1 shows the relationship between the lattice constant and the bandgap energy in possible binary or ternary solid solutions of III-V compounds. The solid lines correspond to the direct bandgap and dotted ones to indirect bandgaps. Quaternary compounds cover the area defined by the four relevant binary compounds.

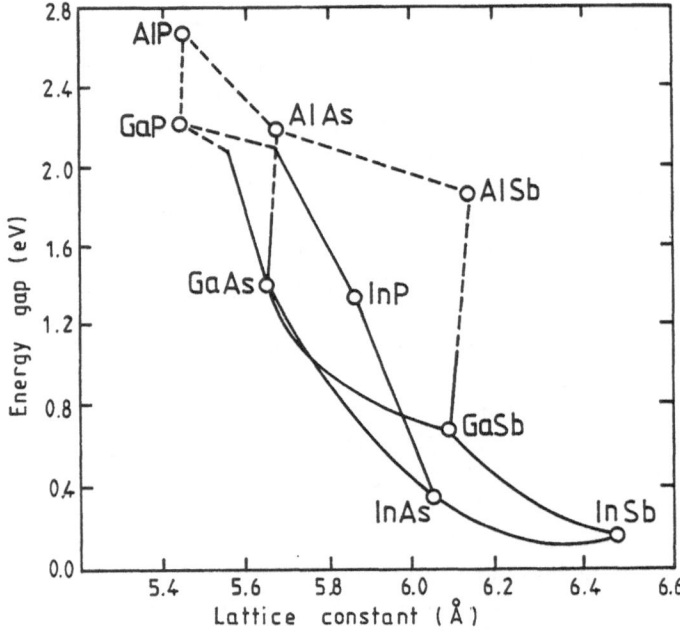

Fig. 7.1. Bandgap energy vs lattice constant for several III-V compounds. The boundaries joining the binary compounds give the ternary or quaternary bandgap energy and lattice constant.

By carefully controlling the alloy composition, it is possible to grow high-quality heterostructures which alternate III-V semiconductor layers containing different compositions but identical lattice constants. The almost ideal natural lattice match between GaAs and AlAs has led to a successful ternary combination of $Al_xGa_{1-x}As$ on GaAs substrates, with bandgap energies corresponding to wavelengths between 0.6 and 0.9 μm. Among lattice-matched material systems, the one most extensively developed has been the $In_xGa_{1-x}As_yP_{1-y}$ system grown on InP substrate, to take advantage of the zero dispersion (1.3 μm) and minimum loss (1.5 μm) of the silica single mode optical fiber.

More recently, however, there has been considerable interest in extending the wavelength range in the visible and near-infrared regions. In particular, the quaternary alloy $(Al_xGa_{1-x})_yIn_{1-y}P$ lattice-matched to GaAs with emission wavelengths in the 0.6-0.7 μm wavelength range is the most attractive material for optical-disk memory systems, laser printers, bar code readers, plastic fiber communication or HeNe laser replacement. Much attention is also devoted to quaternary alloys $Al_xGa_{1-x}As_ySb_{1-y}$ and $In_xGa_{1-x}As_ySb_{1-y}$ lattice-matched to GaSb, since they are suitable for lasers and detectors operating in the 2-4 μm region where fluoride glass fibers are predicted to have less losses than those of silica fibers.

7.2.2 Crystal growth techniques

Although the first semiconductor lasers were reported in 1962, it was not until 1970 that continuous room temperature laser operation was demonstrated. This required implementing lattice-matched heterostructures, which, in these early development stages, were fabricated by liquid-phase epitaxy (LPE). In this near equilibrium growth technique, an initially saturated solution of the layer constituents is in contact with the surface of the substrate onto which it precipitates while maintaining crystalline quality [2,3]. This technique has long been particularly useful for the fabrication of high-quality optoelectronic devices in various III-V materials. Also it is rather inexpensive for exploring novel devices in which the thickness of the layers, the morphology of the surface and the size of the wafer are not severe constraints. Moreover, the unique regrowth properties of LPE have proven extremely useful for buried heterostructure fabrication resulting in a planar structure [4]. However, redissolution of the solid surface by the melt and growth rate in the 1 μm min^{-1} range are real limitations to accurate layer thickness control and good surface morphology [5]. The fabrication of high performance optoelectronic devices has therefore led to advanced epitaxial growth techniques such as metal-organic chemical vapor deposition (MOCVD) and molecular beam epitaxy (MBE).

In MOCVD, which is a vapour phase epitaxial layer technique, the layers are grown from chemically reactive gases at atmospheric or low pressure [6]. A low-pressure MOCVD system is shown in Fig. 7.2. [7]. The group III and V alkyls and/or hydrides of the constituent elements are transported in the reactor by flowing hydrogen. The substrate is placed on a graphite susceptor heated by radio frequency to growth temperatures in the 500-800°C range and growth rates are usually between 1-4 μm h^{-1}. MOCVD is very advantageous for large-area high-quality uniform multilayered structures that can then be grown with reproducibility. However, a major disadvantage lies in the high toxicity of the hydride gases (arsine and phosphine).

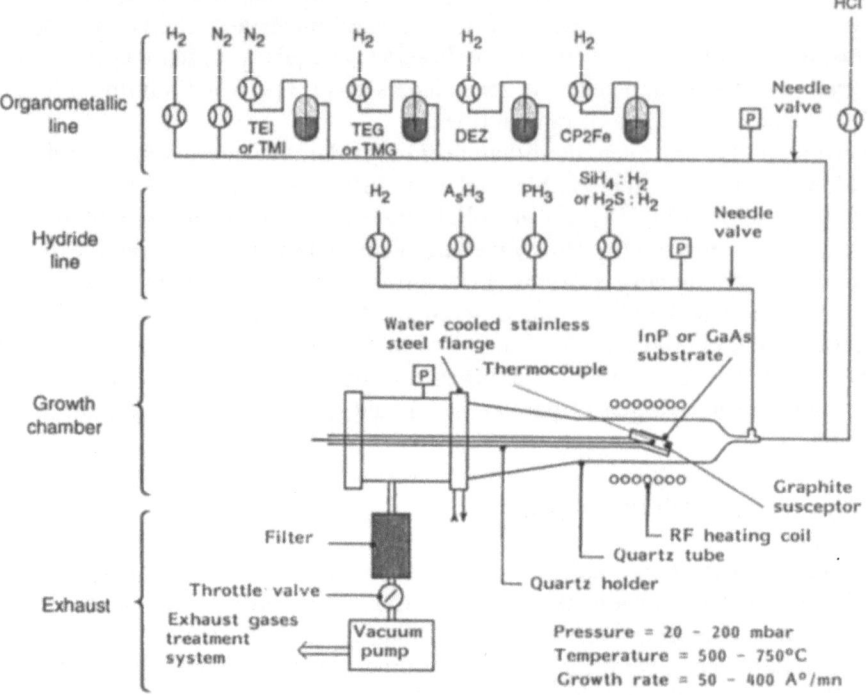

Fig. 7.2. Schematic illustration of low-pressure MOCVD reactor (Courtesy of D. Rondi - THOMSON-CSF LCR).

MBE is a vacuum evaporation process, where molecules or atoms are evaporated from effusion cells onto a heated substrate kept in an ultra-high vacuum (UHV) chamber [8, 9]. A typical system is shown in Fig. 7.3. The substrate is mounted on a heated rotating molybdenum block to a temperature low enough for the group III elements to condense on the substrate. The growth process is governed by kinetics and thermodynamics. The main growth parameters are the substrate temperature, the flux densities of the group III beams and the V/III flux ratio. The molecular beams are monitored by appropriate shutters and precise control of the temperature of the effusion cells. Temperature must remain the same across the whole substrate to ensure uniform layer thickness. High-performance GaAs/AlGaAs devices are typically grown in the 580-720°C range with a growth rate of ~ 1 μm h^{-1}. The MBE technique offers the flexibility required to prepare various mixed alloys and structures involving alternating layers with compositional profiles that are sufficiently accurate to achieve monolayer dimensions. Since MBE is a UHV process, a useful surface analysis can be performed during growth by *in situ* diagnostic tools such as reflection high-energy electron diffraction (RHEED) and mass spectrometry. MBE is therefore a

153

very useful process for basic physics studies and device applications, even if the ultra-high vacuum system results in a costly technique.

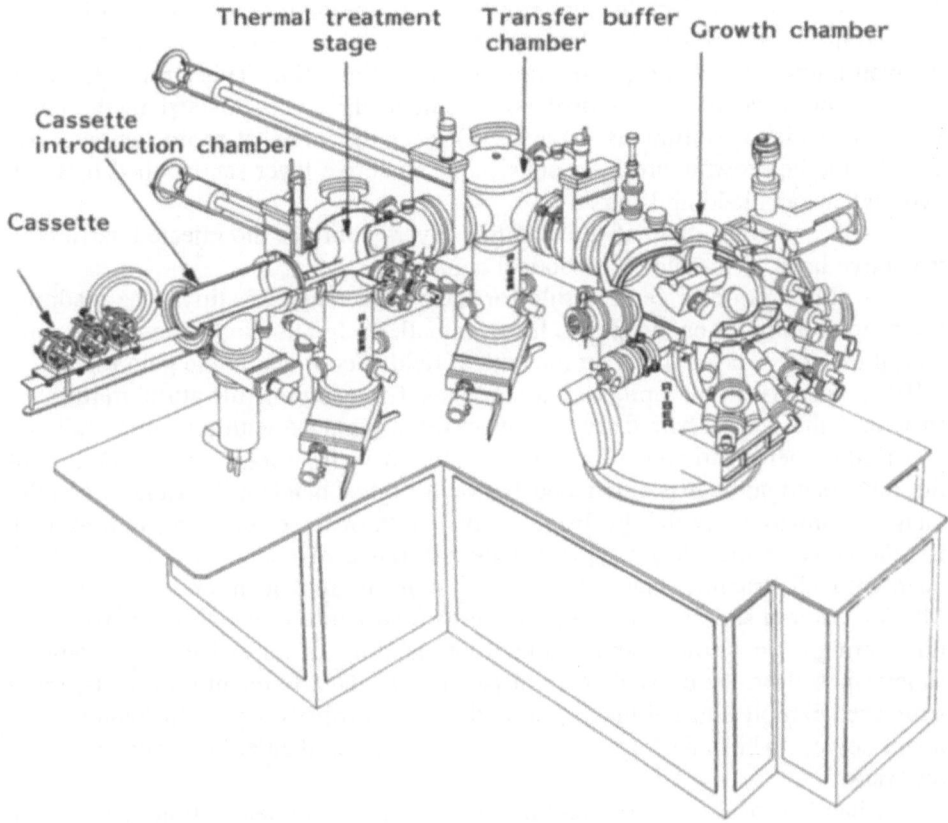

Fig. 7.3. Schematic of a typical MBE system (Courtesy RIBER).

Other growth techniques, referred to as gas-source MBE (GSMBE), chemical beam epitaxy (CBE), metal-organic molecular beam epitaxy (MOMBE), have recently emerged [10, 11]. Basically, MOCVD conditions are achieved in a MBE system by choosing metal-organic and hydride sources for the III or V elements. This is very useful for growing alloys containing phosphorus because it is difficult to handle a solid phosphorus source in MBE. They provide InP-based structures with very sharp interfaces and high control of thickness and composition over large substrate area. Another very attractive feature is that selective area epitaxy (SAE) of III-V materials can be achieved with a high degree of perfection [12]. In the future, these techniques appear highly promising for the fabrication of high-quality optoelectronic integrated circuits.

7.3 Basic device trends

7.3.1 Basic heterostructures

The development effort in semiconductor optoelectronic devices primarily focused on the physics and technology of various discrete components based on the well-known double heterostructure (DH) configuration. This structure, which provides both carrier and optical confinement [13], was devised in the early 1970s to achieve continuous wave (CW) laser operation at room temperature. The double heterostructure laser consists of an active layer sandwiched between two higher-gap cladding layers :
 - the wide bandgap of the cladding layers confines the injected carriers in the active layer to produce high optical gain ;
 - although not a general rule for the whole range of alloys, the cladding layers with a higher bandgap also have a smaller refractive index compared with that of the active layer such that the optical field is confined in the gain region.
 With the advent of epitaxial techniques for growing ultrathin multilayer structures the double heterostructure were often replaced with quantum-well and superlattice heterostructures [14]. In a quantum well structure, the thickness of the low -bandgap well sandwiched between higher bandgap barriers is smaller than or comparable to the de Broglie wavelength or the electron mean free path, i.e. the active region has a typical thickness range of about 20 Å to 200 Å. A quantum well structure may consist of single or multiple quantum wells (SQW or MQW), while a superlattice is a periodic well structure in which the barriers are thin enough to allow carrier tunneling, leading to an artificial periodicity superimposed on the crystal's natural periodicity. Due to quantum effects, these structures exhibit unusual electronic and optical properties [15,16], which yield novel devices whose performance far exceeds that of standard devices with bulk materials.
 Confinement of the free carriers in the well causes a quantization of the wave functions and the energy levels in the growth direction of the layers [17]. The first advantage of quantum-well structures is the ability to tune the emission by varying the well thickness or to achieve wavelength switching from the lowest energy state to higher ones [18-20]. Secondly, the specific "staircase" density of states leads to finite absorption and laser optical gain even at low electronic energies and low temperatures [21, 22]. The third feature is strong exciton effects, even at room temperature, which provide unique large nonlinear optical effects compared to bulk materials. Near the optical absorption edge, the QW exciton absorption saturates at lower optical intensity than that required for bulk material, and application of even a moderate electric field normal to QW layers also causes enhanced absorption change [23, 24]. The magnitude of the electro-absorption effect is about fifty times larger than in bulk materials.
 Quantum-well lasers have several advantages over conventional double heterostructure. In addition to the tunability of the wavelength, the threshold current density is significantly reduced : it approximately falls by a factor of five

in GaAs/AlGaAs lasers, and milliampere threshold currents are currently reported [25, 26]. Other advantages are a reduced temperature dependence, a high output power, a high speed response and a narrow linewidth [27-32]. The excitonic nonlinear absorption and the high nonlinear refractive index of QW structures prompt intense research in optical switches, modulators and directional couplers [33-39]. Of course, attempts are made to perform more complex functions by integrating these devices with quantum well lasers owing to their modulation bandwith and their low power consumption [40, 41].

7.3.2 Towards lower-dimensional structures

Significant performance improvement should be obtained by further reducing the dimension of the structures to one or zero, by developing quantum wires or boxes. As the density of states becomes sharper and narrower, optical gain calculations show the possibility of reductions in threshold current of one or two orders of magnitude as compared to the best quantum well lasers [42-45]. These calculations predict lasers with higher differential slope efficiencies allowing high output power in a large operating-temperature range and also indicate higher modulation rates and improved noise characteristics. Another attractive possibility is to achieve shorter laser wavelengths by taking advantage of the shift to higher emission energies associated with the reduced dimension [46]. Larger optical non-linearities are obviously highly attractive for high performance optical switch since this large electric field-induced refractive index variation would allow the required optical path to be reduced to length of a few mm and would lead to ultra-high speed devices with modulation capabilities beyond a few tens of GHz [47,48]. However the development of quantum wire or quantum box structures depends to a great extent on the development of advanced material processing techniques. Even if low dimensional quantum structures can be grown with atomic layer precision, their lateral size and shape are highly sensitive to the technique used to achieve lateral confinement. The best microlithographic techniques currently yield resolutions in the 10 nm range, and several approaches, such as growth on nonplanar substrates, focused ion beam implantation technique, electron-beam lithography are being investigated to evaluate their potential [49-54].

7.3.3 Lattice-mismatched heterostructures

Bandgap engineering is an overall concept to the design of the transport and optical properties of a heterostructure for a specific device application [55]. Initially, it was mainly limited to lattice-matched layered structures, alternating doped and undoped layers, combining a smaller and a wider bandgap material or using continuously graded gap materials. Recently, there has also been considerable interest in the growth of layer materials with lattice-constant mismatches in the percent range.

The main interest is the possible integration of optoelectronic and high speed III-V devices into silicon very large scale integrated (VLSI) circuit technology since for example optical interconnection may provide practical solutions by replacing wire connection in board-to-board or chip-to-chip interconnection. Silicon has a different thermal expansion coefficient than those of III-V materials and this causes a poor structural quality of the epilayers when cooling down from growth temperature to room temperature. So far, the high defect density introduced in the lasers still compromises their performance and reliability [56-58]. Room-temperature CW operation for over 1 000 h has been demonstrated [59]. Even if GaAs-on-silicon heteroepitaxy has not been successful for such integrations, some promising results have been reported for GaAs on InP heteroepitaxy, allowing integration of GaAs MESFET technology on long wavelength InP based OEICs [60-62].

The second promising area concerns the use of lattice mismatched strained layers where the layer thickness is thin enough to elastically accommodate lattice mismatch, without the formation of dislocations. The motivation for using these new semiconductor structures results from the ability to engineer the band structure through the increased selection of semiconductor materials and from the possibility of combining the influence of quantum size effects and strain effects. One of the main advantages is the wide range of bandgaps and lattice constants made possible by the growth of strained layers. Strained quantum well lasers exhibit enhanced performance compared to unstrained devices, i.e lower threshold currents, higher modulation bandwith, high reliability [63-69]. They open up new commercial applications for pumping erbium-doped optical amplifiers or for extending the range of available wavelength from frequency-doubled diode lasers [70-72]. They are also successfully applied to the fabrication of surface-emitting lasers or transmission optical modulators since the GaAs substrate is transparent at wavelength greater than 0.9 μm. Another attractive feature of strained structures is that certain indirect-gap bulk material can be converted to direct gap through zone-folding effects. Enhancement of the optical absorption coefficient and photoluminescence measurements stimulate research in some systems such as GeSi strained layer superlattices but these structures are still a long way from practical use for optoelectronic applications [73, 74].

7.4. Optoelectronic device arrays

7.4.1 Introduction

The development of diode lasers has resulted in a variety of device configurations and semiconductor materials being produced for numerous applications. It is likely that this trend will continue, even in the field of telecommunications where second generation devices are used to achieve complex functionalities with optoelectronic integrated circuits (OEICs).

On the other hand, massively parallel interconnections and information processing, based on light beam propagation are one of the most promising areas of future optoelectronic systems. To fulfil these expectations, surface emitting lasers and optical functional devices that can memorize or perform logical operations have to be developed. These devices must be compact and low power consuming and must also feature reproducible characteristics so that they can be integrated into a 2D array.

7.4.2 Progress in edge-emitting diode lasers

The basic structure of a laser diode is a double heterostructure with a Perot-Fabry cavity achieved by simple cleavage planes at both ends of the device. Following the first demonstration of CW room temperature operation the aim of semiconductor laser research was twofold.

First and foremost 1.3 and 1.5-μm InGaAsP- based lasers needed to be developed for optical fiber applications. Single longitudinal mode operation was achieved by incorporating a Bragg grating element into the cavity. Two cavity structures can commonly be used : in the distributed Bragg reflector (DBR) laser, the grating region is separated from the gain region, and in a distributed feedback (DFB) laser, the grating is etched into a waveguide layer adjacent to the active layer [75]. Over the last decade, considerable progress has been made in high-bandwith with low frequency chirping and large wavelength tuning range with narrow linewidth [32]. Advanced semiconductor laser structures, including multielectrode tunable DFB and DBR lasers and multiple quantum well layers, have only begun to be explored in photonic integrated circuits (PICs) where they are expected to offer dramatic simplification in terms of optical interconnections technology, in particular in high-density multichannel systems [76-78].

Secondly, high-power diode lasers were required. The reliability and performance of semiconductor lasers used in fiber-optic or optical memory systems suggested that they could also be used for applications such as solid-state laser pumps, green or blue radiation emission by second harmonic generation, space communications or optical processing systems. Several approaches to higher powers have been investigated : optimizing the design of single stripe devices, broadening the active width, or multistripe monolithic laser diode arrays.

Major advances in the growth of very uniform quantum well lasers and in the prevention of mirror degradation have supported the fabrication of high-power lasers emitting in stable, single-lobed far-field patterns. High output CW powers have been achieved, in the 200-500 mW range from single stripe lasers, and a few Watts from broad-area lasers. With their low threshold current densities of about 100 A/cm^2, and emission wavelengths from 0.8 to 1.1 μm, AlInGaAs/AlGaAs and InGaAs/AlGaAs strained single-quantum-well diode lasers have recently been the focus of much attention. InGaAs/AlGaAs laser diodes emitting at 0.98 μm are particularly suitable for pumping erbium-doped fiber amplifiers, and InGaAs/AlGaAs diodes operating at 1.06 μm could replace the neodymium:yttrium aluminium garnet (Nd:YAG) lasers when high optical quality is not required [70-72, 79].

Due to their high power capability, several monolithic arrays of diode lasers have been explored. Most of the research effort has focused on one-dimensional arrays of mutually coupled diode lasers that can operate as a coherent array [80-82]. However most of the structures reported to date suffer from gain-spatial hole-burning and poor discrimination between the lateral modes of the array, giving rise to undesirable beam broadening [83]. Among the various types of arrays, leaky-mode coupling [84,85] has enabled corresponding arrays to exhibit 350 mW CW and 1.5 W pulsed coherent emission. Interest in higher output powers has led to partially coherent devices designed to replace flashlamps for pumping solid state lasers. Conventional diode laser pumps [86] are 1cm- wide laser arrays, which typically consist of ten to thirty 10-stripe lasers spaced on a monolithic GaAs bar, with a total active aperture width of 1 to 3 mm. "Bars" of QW lasers, operating at the nominal wavelength of 808 nm, are today commercially available for pumping Nd:YAG lasers. They produce up to 10 W or more of reliable CW output power, and a few tens of Watts with "quasi-CW" optical pulses (typically 200 µs pulses at 100 Hz), the pulse width being comparable to the lifetime of the excited states of solid-state laser media [87, 88]. Even higher output power levels have been achieved by mounting several bars on plates and closely arranging them to create a stacked laser array : average output power as high as $120W/cm^2$, with quasi-CW peak power in excess of 1200W, have been reported [89]. Visible laser bars are now becoming available for shorter-wavelength applications such as printing, displays or medical applications : AlGaInP laser bars, operating to 8.5 W CW at 680 nm have recently been reported [90].

7.4.3 Surface-emitting laser arrays

Just as photonics and diode pumped solid state lasers are becoming a mature technology, semiconductor laser-research has shifted over the last years in the direction of surface emitter technology. Because of their potential for forming 2D laser arrays, surface-emitting lasers will have important applications, especially at a very high power, and in optical parallel signal processing, optical data storage and optical interconnects. In addition, surface-emitting lasers offer the advantage of wafer-level processing and testing and the ability for optoelectronic integration without the need for a cleavage process.

Since applications will require the best device configuration, several types of surface-emitting lasers are currently being developed. These structures can be classified into three categories :
- conventional edge-emitting lasers with 45° deflecting mirrors ;
- DBR and DFB lasers using second-order diffraction gratings ;
- short vertical cavity laser.

Various surface-emitting lasers with horizontal cavities coupled with 45° beam deflectors have been demonstrated. Mirrors etched at 45° are used either as an intracavity mirror for total reflection or an external mirror to deflect the beam from the laser facets by 90° (Fig. 7.4.). Deflecting mirrors for InGaAsP/InP lasers were fabricated by selective chemical etching followed by a mass transport process

[91]. Such a technique is not known for the AlGaAs/GaAs system, and dry etching techniques, such as reactive ion etching (RIE) or ion beam etching (IBE) at oblique incidence are mostly used to form the mirrors. The etched angle and the smoothness of the surface of the mirrors must be precisely controlled to achieve performance standards comparable to that of conventional cleaved devices [92-96]. High-peak power up to 1.5 kW/cm^2 has been demonstrated from monolithic 2D AlGaAs/GaAs incoherent laser arrays [97]. High-power CW operation of these arrays is restricted by thermal dissipation, and heat sinks containing microchannels for cooling fluid flow are used for heat extraction [98].

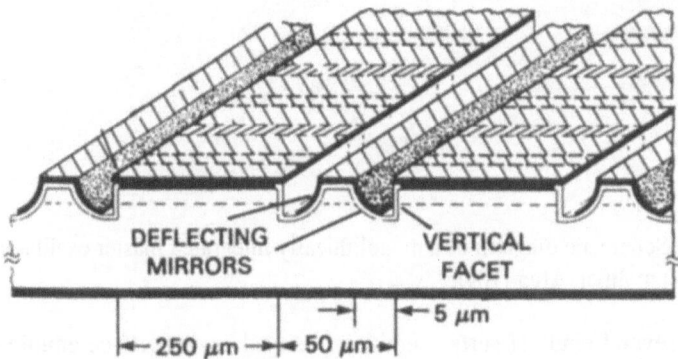

Fig. 7.4. Schematic diagram of a monolithic 2D surface-emitting array with parabolic deflectors. After [97].

An alternative approach to achieve a surface-emitting laser consists of replacing the Perot-Fabry cavity with a grating that provides feedback for laser oscillation in second order and couples the light out perpendicular to the surface in first order. The grating has the additional capability of transmitting light in zeroth order to an adjacent colinear device to achieve mutual injection locking in the longitudinal direction [99]. Such grating surface emitting (GSE) lasers, consisting of alternating quantum well gain and grating sections, have been used to form coherent linear GSE array, providing both dynamic single mode operation and narrow beam divergence. A linear GSE array has demonstrated CW single longitudinal mode operation with spectral linewidth less than 300 kHz [100], and the angular divergence of the output beam can be as narrow as 0.01° in one direction since the source size is as large as a few millimeters on one side. Linear arrays of GSE lasers have also demonstrqted pulsed output powers as high as 16 W [101]. Two-dimensional coherent arrays have been obtained by combining evanescent field overlap or Y branches to achieve lateral coherence and injection coupling through second-order distributed Bragg reflectors in the longitudinal direction. [102]. The use of GaInAs strained quantum wells emitting at ~1 μm with transparent GaAs substrates has allowed high power CW operation to more than 3 W [103]. Another approach to obtaining higher coherent output power is given by monolithically integrated master-oscillator power amplifier (M-MOPA) GSE arrays. A DBR laser is used as a master oscillator, and its output is amplified by a linear chain of amplifiers separated by detuned second-order grating output couplers as shown in Fig. 7.5. [104]. Single frequency operation was reported to an output pulsed power of greater than 4.5 W [105].

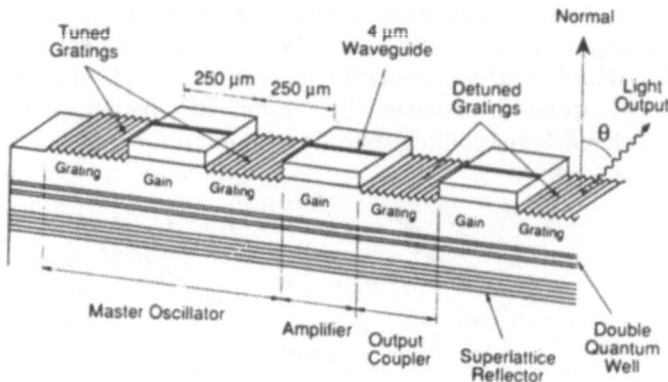

Fig. 7.5. Schematic diagram of a monolithically integrated master oscillator/power amplifier. After [104].

Among several kinds of surface emitters, vertical cavity surface emitting lasers (VCSELs) are a subject of considerable interest since their compact size and their low threshold currents make these devices well suited for large scale 2D arrays of arbitrary configuration. VCSELs offer other advantages such as single longitudinal mode operation due to their short cavity length, and low-divergence circular output beam allowing efficient coupling without additional optics.

Although VCSELs were first introduced by Iga et al more than ten years ago [106], these lasers did not receive widespread interest until the first room-temperature continuous-wave was reported, and the performance has improved significantly with the demonstration of threshold currents in the 1mA range [107-110]. Due to the very short gain length, the key elements of these lasers are the high mirror reflectivities (~ 99%) required at both ends of the cavity, and the tight lateral current confinement to reduce the threshold current [111-114]. Among the various VCSELs recently reported, one of the most promising structures is depicted in Fig. 7.6. It consists of an InGaAs quantum well active region sandwiched between epitaxially grown, doped distributed Bragg reflectors, composed of alternate AlGaAs/AlAs quarter wavelength layers. The laser output is taken either through the transparent substrate or from the top surface using a window in the electrode. The entire structure is grown by MBE. High packing density, of over two million devices per square centimeter have been demonstrated. Recently, there have been several advances on VCSELs, such as the fabrication of wavelength-tunable lasers with wavelength tunability over 40 nm [115], phase-locked 2D arrays [116-119], multiple wavelength 2D laser arrays [120]. Several array architectures were demonstrated, based on an independently addressable scheme or arranged in a matrix addressing architecture [121-123].

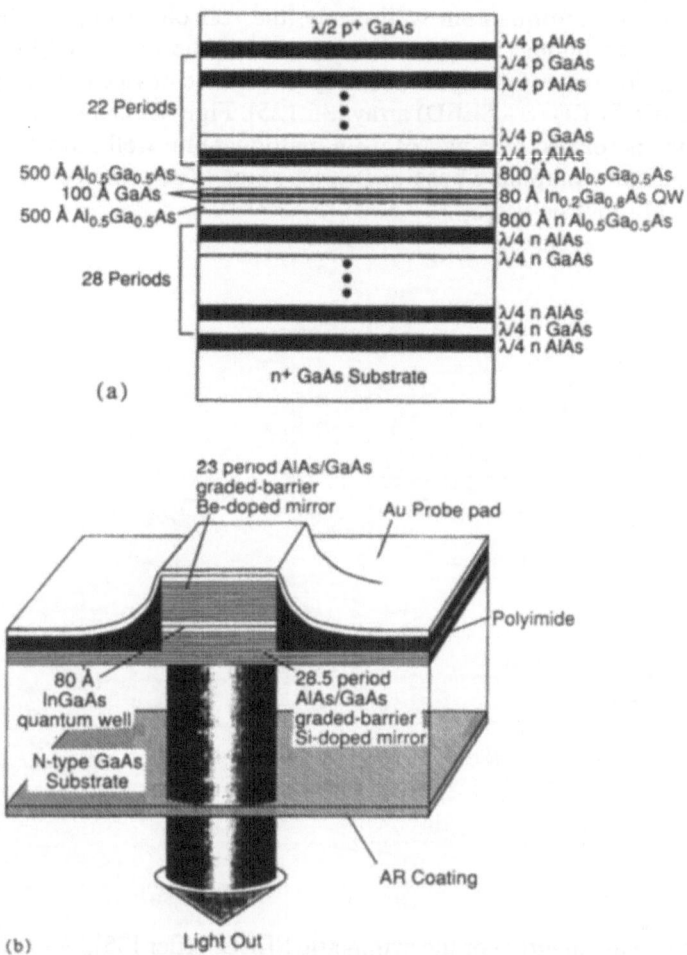

Fig. 7.6. Schematic diagram of a vertical-cavity laser structure. After [109].

7.4.4 Self-electrooptic effect device arrays and vertical-to-surface transmission electrophotonic devices

The growth technologies of quantum wells can be used to achieve new types of optoelectronic devices that differ from laser diodes. The first such mechanism investigated to obtain optical modulation is the quantum-confined Stark effect (QCSE) [124]. Applying an electric field perpendicular to the quantum wells layer gives an actual shift of the absorption exciton peak at ambient temperature. The most important aspect of QCSE is that the absorption variation can give useful light modulation with low voltage even in a single pass through the thickness of the multiple quantum well material. By combining the QCSE with optical

162

detection in p-i-n multiquantum well diode, the self electro-optic effect device (SEED) acts as an optical bistable element and can be processed in a 2D array configuration. For large logic systems the most suitable device currently available is the symmetric SEED (S - SEED) array [35,125]. Fig. 7.7. shows the single array element which is formed by a pair of p-i-n multiquantum well diodes. This device has complementary outputs whose switching point is determined by the ratio of the two optical input powers.

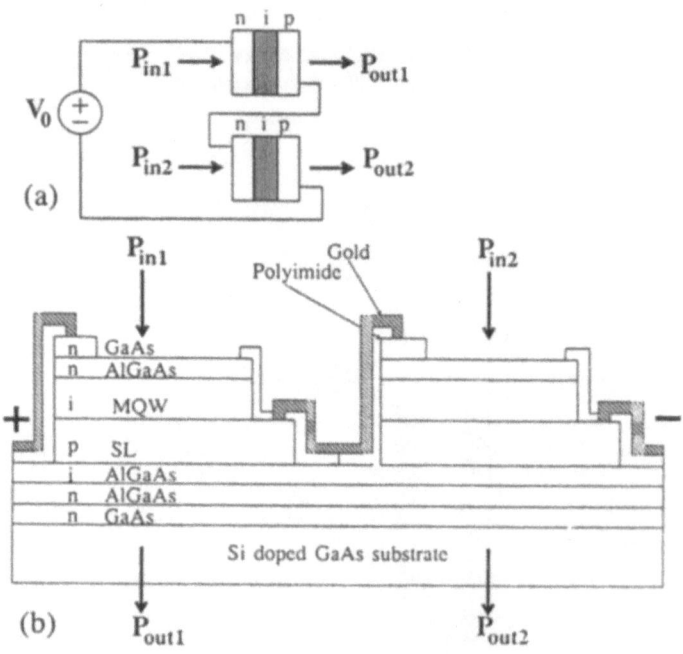

Fig. 7.7. Schematic diagram of the symmetric SEED. After [35].

The most improved GaAs/GaAlAs S - SEED are designed to work in reflection by including in the epitaxy process the growth of a dielectric mirror stack. Arrays as large as 256 x 128 GaAs/GaAlAs S - SEED elements have been reported and cascaded parallel operation of two arrays has been demonstrated [126-128]. The operating characteristics of these devices are limited by the incident optical power and future improvements can be achieved by using higher power source or integrating field effect transistor into each modulation device (F - SEED). It is also to be noticed that the SEED concept can lead to various potential functions : for example, an oscillator circuit operation at frequencies exceeding 100 MHz has been reported. The SEED system experiments are promising on the whole but improvements in integration processes are still necessary to assess the ultimate importance of these devices.

The vertical-to-surface transmission electrophotonic device (VSTEP) is another device concept proposed to meet the requirements of 2D optical signal processing and interconnections [129]. In a VSTEP device low-power-consuming electronic processes are inserted within a photonic device which provides light detection and emission. Surface normal optical switches based on the monolithic integration of a heterojunction phototransistor [130] or a pn-pn photothyristor [131] with a light emitting diode or an edge emitting laser diode have been extensively studied. High sensitivity and low holding power have been achieved by decreasing surface generation-recombination currents at device perimeter [131]. However, light emitting diode based switches are relatively inefficient in conversion efficiency from electrical to optical energy and in optical gain terms, while 2D arrays of edge emitting laser based photothyristors are difficult to achieve. These problems have been circumvented by vertical cavity surface emitting laser diodes (VCSEL). VC - VSTEP (vertical cavity - Vertical-to-surface transmission electrophotonic) devices have been fabricated by the monolithic integration of an InGaAs quantum well VCSEL with a GaAs/GaAlAs heterojunction phototransistor (Fig. 7.8.) [132] or by embedding a pn-pn photothyristor structure between the two distributed Bragg reflector mirrors of an InGaAs multiquantum well VCSEL [133]. The only disadvantage of these devices is that cascading is not possible because of the poor spectral overlap between the source and the detector. Recently, cascadable switches have been demonstrated by integrating a GaAs VCSEL into a self contained GaAs photothyristor (Fig. 7.9.) [134, 135]. All these devices exhibit low switching power (tens of nW), high optical gain and high constrast. However, in spite of recent developments, VCSEL characteristics, including lower series resistance, threshold voltage and current density, higher quantum and power efficiencies, have to be significantly improved so that the VC - VSTEP concept can solve the problem of compact 2D arrays manufacturing with sufficiently low thermal dissipation.

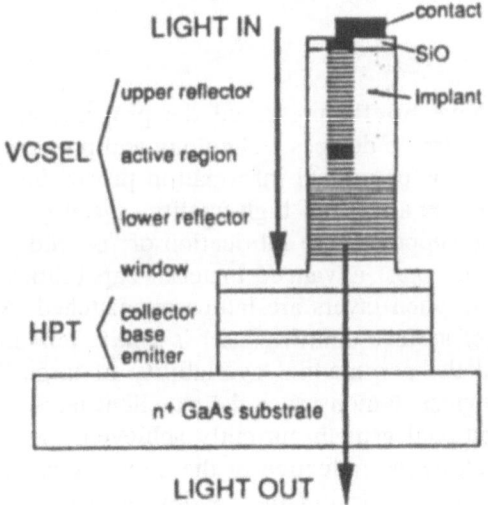

Fig. 7.8. Device structure of vertically integrated VCSEL and HPT. After [132].

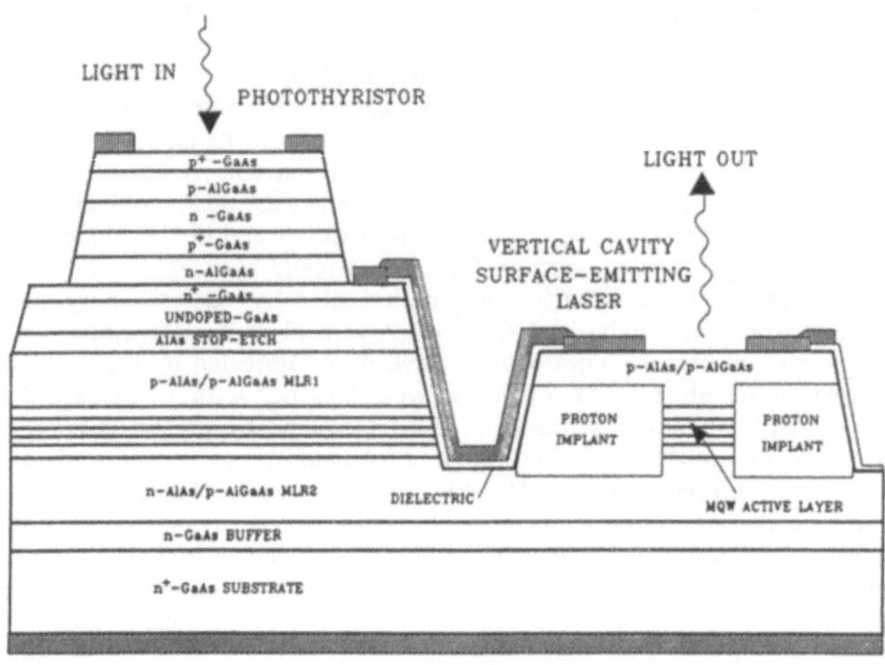

Fig. 7.9. Cross sectional structure of monolithic integrated VCSEL device. After [135].

7.5 Summary

In this review, we have briefly presented the progress made in the field of semiconductor optoelectronic devices technology and provided several examples focusing on its potential for optical information processing systems. One key development has been the advent of high quality crystal growth and processing technology which has supported the fabrication of "second generation" devices taking advantage of the most advanced material capabilities in quantum wells and superlattices even when layers are lattice-mismatched. Among these novel devices, the success of surface-emitting laser is a milestone in the emergence of 2D laser arrays, and the commercial availability of large 2D arrays of self-electrooptic effect devices demonstrates the excellent large-area uniformity and reproducibility of epitaxial growth currently achieved. The emergence of new physical concepts such as the reduction of the dimensionality of semiconductor devices, innovations in the design of the device structures to achieve new functionalities, and the advances in material growth combined with the processing technology will further improve the characteristics of the devices and their usefulness in systems. Again, there is a need for tunable lasers, narrowed linewidth lasers, high power lasers and short pulse lasers and the sources must be integrable in large arrays with high yields and low-cost. Significant progress continues in switches and modulators, directional couplers, optical amplifiers, tunable filters, based on the requirements for OEIC technologies. In addition, major strides have already been made in heteroepitaxy, opening up many exciting opportunities to integrate silicon electronics and GaAs and/or InP optoelectronics and to provide hybrid systems. Research in the field of optical interconnects would take advantage of the best of both technologies.

References

1. Useful reviews on detectors can be found, e.g., in Tsang W.T.: Semiconductors and Semimetals, Willardson, R.K., and Beer, A.C., eds., Academic Press, 22D and 22E, 1985 ; Scribner D.A., Kruer M.R., Killiany J.M. : Infrared focal plane array technology, Proc. IEEE, 79, 1991, pp 66-85.
2. Casey Jr, H.C., Panish, M.B. : Liquid phase epitaxy. In Heterostructure Lasers, Part B, Academic Press, 1978, pp 109-132.
3. Nakajima, K. : The liquid phase epitaxial growth of InGaAsP. In Semiconductors and semimetals, , 22 A, 1985, pp 1-93.
4. Botez, D. : Liquid phase epitaxy over channelled substrates. J. Cryst. growth, 70, 1984, pp 150-154.
5. Thulke, W. : Can liquid-phase epitaxy still be useful for optoelectronic devices. In Materials for optoelectronic devices, OEICs and photonics, Proc. of E-MRS Conference, Strasbourg, Nov. 27-30, 1990, pp 61-67.
6. Stringfellow, G.B. : Organometallic vapor-phase epitaxial growth of III-V semiconductors. In Semiconductors and semimetals, Willardson, R.K., and Beer, A.C., eds., Academic Press, 22A, 1985, pp 209-259.
7. Razeghi, M. : Low-pressure metallo-organic chemical vapor deposition of GaInAsP alloys. In Semiconductors and semimetals, Willardson, R.K., and Beer, A.C., eds., Academic Press, 22A,1985, pp 298-375.
8. Cho, A.Y., Arthur, J.R. : Molecular beam epitaxy. J. Progress in solid state chem., 10, 1975, pp 157-191.
9. Tsang, W.T. : Molecular beam epitaxy for III-V compound semiconductors. In Semiconductors and semimetals, Willardson, R.K., and Beer, A.C., eds., Academic Press, 22A,1985, pp 95-207.
10. Perales, A., Goldstein, L., Accard, A., Fernier, B., Leblond, F., Gourdain C., Brosson, P. : High performance DFB-MQW lasers at 1.5 µm grown by GSMBE. Electron. Lett., 26, 1990, pp 236-237.
11. Tsang, W.T. : Progress in chemical beam epitaxy. J. Cryst. growth, 105, 1990, pp 1-29.
12. Davies, G.J., Duncan, W.J., Skevington, P.J., French, C.L., Foord, J.S. : Selective area growth for opto-electronic integrated circuits. In Materials for

optoelectronic devices, OEICs and Photonics, Proc. of E-MRS Conference, Strasbourg, Nov. 27-30, 1990, pp 93-100.

13. Casey Jr, H.C., Panish, M.B. : Heterostructure lasers, Part A, Academic Press, 1978.

14. Baets, R. : Heterostructures in III-V optoelectronic devices. Sol. St. Electron., 30, 1987, pp 1175-1182.

15. Weisbuch, C. : Fundamental properties of III-V semiconductor two-dimensional quantized structures : the basis for optical and electronic device applications. In Semiconductors and semimetals, Willardson, R.K., and Beer A.C., eds., Academic Press, 24, 1987, pp 1-133.

16. Okamoto, H. : Semiconductor quantum-well structures for optoelectronics - Recent advances and future prospects. Jap. J. of Appl. Phys., 26, 1987, pp 315-330.

17. Dingle, R., Wiegmann, W., Henry, C.H. : Quantum states of confined carriers in very thin AlGaAs-GaAs-AlGaAs heterostructures. Phys. Rev. Lett., 33, 1974, pp 827-830.

18. Holonyak, N., Kolbas, R.M., Dupuis, R.D., Dapkus, P.D. : Quantum-well heterostructure lasers. IEEE J. Quantum Electron., 16, 1980, pp 170-185.

19. Mittelstein, M., Arakawa, Y., Larsson, A., Yariv, A. : Second quantized state lasing of a current pumped single quantum well laser. Appl. Phys. Lett., 45, 1986, pp 1689-1691.

20. Tokuda, Y., Tsukada, N., Fujiwara, K., Hamanaka, K., Nakayama, T. : Widely separate wavelength switching of single quantum well laser diode by injection-current control. Appl. Phys. Lett., 49, 1986, pp 1629-1631.

21. Dutta, N.K., Hartman, R.L., Tsang, W.T. : Gain and carrier lifetime measurements in AlGaAs single quantum well lasers. EEE J. Quantum Electron, 19, 1983, pp 1243-1246.

22. Asada, M., Kameyama, A., Suematsu, Y. : Gain and intervalence band absorption in quantum-well lasers. IEEE J. Quantum Electron., 20, 1984, pp 745-753.

23. Chemla, D.S., Miller, D.A.B., Smith, P.W. : Nonlinear optical properties of multiple quantum well structures for optical signal processing. In Semiconductors and semimetals, Willardson, R.K., and Beer, A.C., eds., Academic Press, 24, 1987, pp 279-318.

24. Wood, T.H. : Direct measurement of the electric-field-dependent absorption coefficient in GaAs/AlGaAs multiple quantum wells. Appl. Phys. Lett., 48, 1986, pp 1413-1415.

25. Lau, K.Y., Derry, P.L., Yariv, A. : Ultimate limit in low threshold quantum well GaAlAs semiconductor lasers. Appl. Phys. Lett., 52, 1988, pp 88-90.

26. Kapon, E., Simhony, S., Harbison, J.P., Florez, L.T., Worland P. : Threshold current reduction in patterned quantum-well semiconductor lasers grown by molecular beam epitaxy. Appl. Phys. Lett., 56, 1990, pp 1825-1827.

27. Tsang, W.T. : Quantum confinement heterostructure semiconductor lasers. In Semiconductors and semimetals, Willardson, R.K., and Beer, A.C., eds., Academic Press, 24, 1987, pp 397-458.

28. Chin, R., Holonyak Jr, N., Vojak, B.A., Hess, K., Dupuis, R.D., Dapkus, P.D. : Temperature dependence of threshold current for quantum-well AlGaAs-GaAs heterostructure laser diodes. Appl. Phys. Lett., 36, 1980, pp 19-21.

29. Hayakawa, T., Suyama, T., Kondo, M., Hosoda, M., Yamamoto, S., Hijikata, T. : High-power (2.2 W) CW operation of (111)-oriented GaAs/AlGaAs single-quantum-well lasers prepared by molecular-beam epitaxy. J. Appl. Phys., 64 (5), 1988, pp 2764-2766.

30. Arakawa, Y., Yariv, A. : Theory of gain, modulation response, and spectral linewidth in AlGaAs quantum well lasers. IEEE J. Quantum Electron, 21, 1985, pp 1666-1674.

31. Lang, H., Wolf, H.D., Korte, L., Hedrich, H., Hoyler, C., Thanner, C. : GaAs/AlGaAs quantum well laser for high-speed applications. IEE Proc., part J, 138, 1991, pp 117-121.

32. Lee, T.P. : Recent advances in long-wavelength semiconductor lasers for optical fiber communication. Proc. IEEE, 79, 1991, pp 253-276.

33. Scherer, A., Jewell, J.L., Lee, Y.H., Harbison, J.P., Florez, L.T. : Fabrication of microlasers and microresonator optical switches. Appl. Phys. Lett., 55, 1989, pp 2724-2726.

34. Walker, R.G. : High speed III-V semiconductor intensity modulators. IEEE J. Quantum Electron., 27, 1991, pp 654-667.

35. Lentine, A.L., Hinton, H.S., Miller, D.A., Henry, J.E., Cunningham, J.E., Chirovsky, M.F. : Symmetric Self-Electrooptic Effect Device : Optical Set-Reset Latch, Differentiel Logic Gate and Differential Modulator/Detector. IEEE J. Quantum Electron., 25, 1989, pp 1928-1936.

36. Law, K.K., Whitehead, M., Merz, J.L., Coldren, L.A. : Simultaneous achievement of low insertion loss high contrast and low operating voltage in asymmetric Fabry-Perot reflection modulator. Electron. Lett., 27, 1991, pp 1863-1865.

37. Huang, T.C., Chung, Y., Dagli, N., Coldren, L.A. : GaAs/AlGaAs multiple quantum well field-induced optical waveguide. Appl. Phys. Lett., 57, 1990, pp 114-116.

38. Ozeki, Y., Johnson, J.E., Tang, C.L. : Polarisation bistability in semiconductor lasers with intracavity multiple quantum well saturable absorbers. Appl. Phys. Lett., 58, 1991, pp 1958-1960.

39. Jin, R., Chuang, C.L., Gibbs, H.M., Koch, S.W., Polky, J.N., Pubanz, G.A. : Picosecond all-optical switching in single-mode GaAs/AlGaAs strip-loaded nonlinear directional couplers. Appl. Phys. Lett., 53, 1988, pp 1791-1793.

40. Tarucha, S., Okamoto, H. : Monoliithic integration of a laser diode and an optical waveguide modulator having a GaAs/AlGaAs quantum well double heterostructure. Appl. Phys. Lett., 48, 1986, pp 1-3.

41. Hernandez-Gil, F., Koch, T.L., Koren, U., Gnall, R.P., Burrus, C.A. : Tunable MQW - DBR laser with monolithically integrated GaInAsP/InP directional coupler switch. Electron. Lett., 25, 1989, pp 1271-1272.

42. Asada, M., Miyamoto, Y., Suematsu, Y. : Gain and the threshold of three-dimensional quantum-box lasers. IEEE J. Quantum Electron., 22, 1986, pp 1915-1921.

43. Yariv, A. : Scaling laws and minimum threshold currents for quantum-confined semiconductor lasers. Appl. Phys. Lett., 53, 1988, pp 1033-1035.

44. Miyamoto, Y., Miyake, Y., Asada, M., Suematsu, Y. : Threshold current density of GaInAsP/InP box lasers. IEEE J. Quantum Electron., 25, 1989, pp 2001-2006.

45. T. Takahashi, Y. Arakawa : Nonlinear gain effects in quantum well,quantum well wire, and quantum well box lasers. IEEE J. Quantum Electron., QE-27, 1991, pp 1825-1829.

46. Maile, B.E., Forchel, A., Germann, R., Straka, J., Korte, L., Thanner, C. : Lateral quantization induced emission energy shift of buried GaAs/AlGaAs quantum wires. Appl. Phys. Lett., 57, 1990, pp 807-809.

47. Shimomura, K., Suematsu, Y., Arai, S. : Analysis of semiconductor intersectional waveguide optical switch modulator. IEEE J. Quantum Electron., 26, 1990, pp 883-892.

48. Aizawa, T., Shimomura, K., Arai, S., Suematsu, Y. : Observation of field-induced refractive index variation in quantum box structure. IEEE Photon. Technol. Lett., 3, 1991, pp 907-909.

49. Petroff, P.M., Gossard, A.C., Logan, R.A., Wiegmann, W. : Toward quantum well wires : fabrication and optical properties. Appl. Phys. Lett., 41, 1982, pp 635-638.

50. Kapon, E., Simhony, S., Bhat, R., Hwang, D.M. : Single quantum wire semiconductor lasers. Appl. Phys. Lett., 55, 1989, pp 2715-2717.

51. Simhony, S., Kapon, E., Colas, E., Hwang, D.M., Stoffel, N.G., Worland, P. : Vertically stacked multiple-quantum-wire semiconductor diode lasers. Appl. Phys. Lett., 59, 1991, pp 2225-2227.

52. Fukui, T., Ando, S., K. Fukai, Y. : Lateral quantum well wires fabricated by selective metalorganic chemical vapor deposition. Appl. Phys. Lett., 57, 1990, pp 1209-1211.

53. Izrael, A., Sermage, B., Marzin, J.Y., Ougazzaden, A., Azoulay, R., Etrillard, J., Thierry-Mieg, V., Henry, L. : Microfabrication and optical study of reactive ion etched InGaAsP/InP and GaAs/GaAlAs quantum wires. Appl. Phys. Lett., 56, 1990, pp 830-832.

54. Vieu, G., Schneider, M., Mailly, D., Planel, R., Launois, H., Marzin, J.Y., Descouts, B. : Optical characterization of selectively intermixed GaAs/ GaAlAs quantum wires by Ga$^+$ masked implantation. J. Appl. Phys., 70, 1991, pp 1444-1450.

55. Capasso, F. : Graded-gap and superlattice devices by bandgap engineering" In Semiconductors and semimetals, Willardson, R.K., and Beer, A.C., eds., Academic Press, 24, 1987, pp 319-393.

56. Hall, D.C., Deppe, D.G., Holonak Jr, N., Matyi, R.J., Shichijo, H., Epler, J.E. : Thermal behavior and stability of room-temperature continuous AlGaAs-GaAs quantum well heterostructure lasers grown on Si. J. Appl. Phys., 64, 1988, pp 2854-2860.

57. Choi, H.K., Wang, C.A., Fan, J.C.C. : Room-temperature continuous operation of GaAs/AlGaAs lasers grown on Si by organometallic vapor-phase epitaxy. J. Appl. Phys., 68, 1990, pp 1916-1918.

169

58. Egawa, T., Soga, T., Jimbo, T., Umeno, M. : Room-temperature continuous-wave operation of AlGaAs-GaAs single-quantum-well lasers on Si by metalorganic chemical-vapor deposition using AlGaAs-AlGaP intermediate layers. IEEE J. Quantum Electron., 27, 1991, pp 1798-1803.

59. Sugo, M., Mori, H., Itoh, Y., Sakai, Y., Tachikawa, M. : 1.5 µm long-wavelength multiple quantum well laser on a Si substrate. Jap. J. Appl. Phys., 30, 1991, pp 3876-3878.

60 Van Ackere, M., Ackaert, A., Moerman, I., Lootens, D., Demeester, P., Van Daele, P., Baets, R., Lagasse, P. : GaAs single-quantum well GRIN-SCH ridge lasers grown on InP by MOVPE. Electron. Lett., 25, 1989, pp 47-48.

61. Lo, Y.H., Caneau, C., Bhat, R., Florez, L.T., Chang, G.K., Harbison, J.P., Lee, T.P. : High-speed GaAs-on-InP long wavelength transmitter OEICs. Electron. Lett., 25, 1989, pp 666-667.

62. Pollentier, I., Buydens, L., Demeester, P., Van Daele, P., Enard, A., Lallier, E., Glastre, G., Rondi, D. : Monolithic integration of GaAs MESFET and InP/InGaAsP 2 x 2 optical switch. Electron. Lett., 27, 1991, pp 2339-2340.

63. Eng, L.E., Chen, T.R., Sanders, S., Zhuang, Y.H., Zhao, B., Yariv, A., Morkoç, H. : Submilliampere threshold current pseudomorphic InGaAs/ AlGaAs buried-heterostructure quantum well lasers grown by molecular beam epitaxy. Electron. Lett., 55, 1989, pp 1378-1379.

64. Wang, C.A., Walpole, J.N., Missagia, L.J., Donnelly, J.P., Choi, H.K. : AlInGaAs/AlGaAs separate-confinement heterostructure strained single quantum well diode lasers grown by organometallic vapor phase epitaxy. Electron. Lett., 58, 1991, pp 2208-2210.

65. Tanbun-Ek, T., Logan, R.A., Chu, S.N.G., Sergent, A.M., Wecht, K.W. : Effects of strain in multiple quantum well distributed feedback lasers. Electron. Lett., 57, 1990, pp 2184-2186.

66. Yasaka, H., Takahata, K., Yamamoto, N., Naganuma, M. : Gain Saturation coefficients of strained-layer multiple quantum-well distributed feedback lasers. IEEE Photon. Technol. Lett., 3, 1991, pp 879-882.

67. Blez, M., Kazmierski, C., Quillec, M., Robein, D., Allovon, M., Gloukhian, A., Sermage, B. : First DFB GRIN-SCH GaInAs/AlGaInAs 1.55 mm MBE MQW active layer buried ridge structure lasers. Electron. Lett., 27, 1991, pp 93-94.

68. Thijs, P.J.A., Tiemeijer, L.F., Kuindersma, P.I., Binsma, J.J.M., Van Dongen, T. : High-performance 1.5 mm wavelength InGaAs-InGaAsP strained quantum well lasers and amplifiers. IEEE J. Quantum Electron., 27, 1991, pp 1426-1439.

69. Zah, C.E., Bhat, R., Menocal, S.G., Favire, F., Lin, P.S.D., Gozdz, A.S., Andreadakis, N.C., Pathak, B., Koza, M.A., Lee, T.P. : Reliable InGaAs quantum well lasers at 1.1 mm. Electron. Lett., 27, 1991, pp 552-553.

70. Murison, R.F., Moore, A.H., Lee, S.R., Holehouse, N., Dzurko, K.M., Cockerill, T.M., Coleman, J.J. : High power continuous operation of laser diode at 1064 nm. Electron. Lett., 27, 1991, pp 1979-1981.

71. Major, J.S., Plano, W.E., Welch, D.F., Scifres, D. : Single-made InGaAs-GaAs laser diodes operating at 980 nm. Electron. Lett., 27, 1991, pp 539-540.

72. Welch, D.F., Cardinal, M., Streifer, B., Scifres, D. : High-power single mode InGaAs/AlGaAs laser diode at 910 nm. Electron. Lett., 26, 1990, pp 233-235.
73. Pearsall, T.P. : Si-Ge alloys and superlattices for optoelectronics. In Materials for optoelectronic devices, OEICs and Photonics, Proc. of E-MRS Conference, Strasbourg, Nov. 27-30, 1990, pp 225-231.
74. Turton, R.J., Jaros, M. : Linear and nonlinear optical properties of direct gap Si-Ge superlattices. IEE Proc. J, 138, 1991, pp 323-329.
75. Agawal, G.P., Dutta, N.K. : Long wavelength semiconductor lasers. Van Nostrand Reinhold Company, 1986.
76. Suematsu, Y., Arai, S. : Integrated optics approach for advanced semi-conductor lasers. Proc. IEEE, 75, 1987, pp 1472-1487.
77. Forrest, S. : Optoelectronic integrated circuits. Proc. IEEE, 75, 1987, pp 1488-1497.
78. Koch, T.L., Koren, U. : InP-based photonic integrated circuits. IEE Proc. J, 138, 1991, pp 139-147.
79. Choi, H.K., Wang, C.A., Kolesar, D.F., Aggarwal, R.L., Walpole, J.N. : High-power, high-temperature operation of AlInGaAs-AlGaAs strained-single-quantum-well diode lasers. IEEE Photonics Technology Letters, 3, 1991, pp 857-859.
80. Welch, D.F., Cross, P., Scifres, D., Streifer, W., Burnham, R.D. : In-phase emission from index-guided laser array up to 400 mW. Electron. Lett., 22, 1986, pp. 293-294.
81. Goldstein, B., Carlson, N.W., Evans, G.A., Dinkel, N.A., Masin, V.J. : Performance of channelled-substrate-planar high-power phase-locked array operating in the diffraction limit. Electron. Lett., 23, 1987, pp 1136-1138.
82. Jansen, M., Yang, J.J., Ou, S.S., Botez, D., Wilcox, J., Mawst, L. : Diffraction-limited operation from monolithically integrated diode laser array and self-imaging (Talbot) cavity. Appl. Phys. Lett., 55, 1989, pp 1949-1951.
83. Thompson, G.H.B., Witheaway, J.E.A. : Analysis of the stability of the highest-order supermode in semiconductor laser arrays. Electron. Lett., 23, 1987, pp 444-446.
84. Botez, D., Jansen, M., Mawst, L.J., Peterson, G., Roth, T.J. : Watt-range, coherent, uniphase powers from phase-locked arrays of antiguided diode lasers. Appl. Phys. Lett., 58, 1991, pp 2070-2072.
85. Major, J.S., Mehuys, D., Welch, D.F., Scifres, D.R. : High power, high efficiency antiguide laser arrays. Appl. Phys. Lett., 59, 1991, pp 2210-2212.
86. Streifer, W., Scifres, D.R., Harnagel, G.L., Welch, D.F., Berger, J., Sakamoto, M. : Advances in diode laser pumps. IEEE J. Quantum Electron., 24, 1988, pp. 883-893.
87. Sakamoto, M., Endriz, J.G., Scifres, D.R. : 20 W CW monolithic AlGaAs (810 nm) laser diode arrays. Electron. Lett., 28, 1992, pp 178-180.
88. Sakamoto, M., Endriz, J.G., Scifres, D.R. : 120 W CW output power from monolithic AlGaAs (800 nm) laser diode array mounted on diamond heatsink. Electron. Lett., 28, 1992, pp 197-198.

89. Harnagel, G.L., Ahrabi, M., Browder, G.S., Worland, D.P., Endriz, J.G., Scifres, D.R. : High power, high-efficiency quasi-CW two-dimensional laser diode arrays. Electron. Lett., 27, 1991, pp 55-56.

90. Welch, D.F., Scifres, D.R. : High power, 8.5 W CW, visible laser. Electron. Lett., 27, 1991, pp 1915-1916.

91. Liau, Z.L., Walpole, J.N. : Large monolithic two-dimensional arrays of GaInAsP/InP surface-emitting lasers. Appl. Phys. Lett., 50, 1987, pp 528-530.

92. Kim, J.H., Larsson, A., Lee, L.P. : Pseudomorphic InGaAs/GaAs/GaAlAs single quantum well surface emitting lasers with integrated 45° beam deflectors. Appl. Phys. Lett., 58, 1991, pp 7-9.

93. Chao, C.P., Law, K.K., Merz, J.L. : Low threshold InGaAs/GaAs strained layer ridge waveguide surface emitting lasers with two 45° angle etched internal total reflection mirrors. Appl. Phys. Lett., 59, 1991, pp 1532-1534.

94. Takamori, T., Coldren, L.A., Merz, J.L. : Folded cavity transverse junction stripe surface-emitting lasers. Appl. Phys. Lett., 55, 1989, pp 1053-1055.

95. Ou, S.S., Yang, J.J., Jansen, M., Sargent, M., Mawst, L.J., Wilcox, J.Z. : High performance surface-emitting lasers with 45° intracavity micromirrors. Appl. Phys. Lett., 58, 1991, pp 16-18.

96. Goodhue, W.D., Donnely, J.P., Wang, C.A., Lincoln, G.A., Rauschenbach, K., Bailey, R.J., Johnson, G.D. : Monolithic two-dimensional surface-emitting strained-layer InGaAs/AlGaAs and AlInGaAs/AlGaAs diode laser arrays with over 50 % differential quantum efficiencies. Appl. Phys. Lett., 59, 1991, pp 632-634.

97. Goodhue, W.D., Rauschenbach, K., Wang, C.A., Donnely, J.P., Bailey, R.J., Johnson, G.D. : Monolithic two-dimensional GaAs/AlGaAs laser arrays fabricated by chlorine ion-beam-assisted micromaching. J. Electron. Mat., 19, 1990, pp 463-469.

98. Missagia, L.J., Walpole, J.N., Liau, Z.L., Philips, R.J. : Microchannel heat sinks for two-dimensional high-power-density diode laser arrays. IEEE J. Quantum Electron., 25, 1989, pp 1988-1992.

99. Hardy, A., Welch, D.F., Streifer, W. : Analysis of a dual grating-type surface emitting laser. IEEE J. Quantum Electron., 26, 1990, pp 50-60.

100. Carlson, N.W., Bour, D.P., Evans, G.A., Liew, S.K. : Spectral linewidth narrowing in monolithic grating-surface-emitting multiple-quantum-well distributed feedback lasers. IEEE Photon. Technol. Lett., 2, 1990, pp 242-243.

101. Welch, D.F., Parke, R., Hardy, A., Waarts, R., Streifer, W., Scifres, D.R. : High power, 16W, grating surface emitting laser with a superlattice substrate reflector. Electron. Lett., 26, 1990, pp 757-758.

102. Evans, G.A., Carlson, N.W., Hammer, J.M., Lurie, M., Butler, J.K., Palfrey, S.L., Amantea, R., Carr, L.A., Hawrylo, F.Z., James, E.A., Kaiser, C.J., Kirk, J.B., Reichert, W.F. : Two-dimensional coherent laser arrays using grating surface emission. IEEE J. Quantum Electron., 25, 1989, pp 1525-1538.

103. Evans, G.A., Bour, D.P., Carlson, N.W., Amantea, R., Hammer, J.M., Lee, H., Lurie, M., Lai, R.C., Pelka, P.F., Farkas, R.E., Kirk, J.B., Liew, S.K., Reichert, W.F., Wang, C.A., Choi, H.K., Walpole, J.N., Butler, J.K., Ferguson, W.F.,

DeFreez, R.K., Felisky, M. : Characteristics of coherent two-dimensional grating surface emitting diode laser arrays during CW operation. IEEE J. Quantum Electron, 27, 1991, pp 1595-1607.

104. Mehuys, D., Parke, R., Waarts, R.G., Welch, D.F., Hardy, A., Streifer, W., Scifres, D.R. : Characteristics of multistage monolithically integrated master oscillator power amplifiers. IEEE J. Quantum Electron., 27, 1991, pp 1575-1581.

105. Parke, R., Welch, D.F., Mehuys, D. : Coherent operation of 2D monolithically integrated master oscillator power amplifier. Electron. Lett., 27, 1991, pp 2097-2098.

106. Iga, K., Koyama, F., Kinoshita, S. : Surface emitting semiconductor lasers. IEEE J. Quantum Electron., 24, 1988, pp 1845-1855.

107. Lee, Y.H., Jewell, J.L., Scherer, A., McCall, S.L., Harbison, J.P., Florez, L.T. : Room-temperature continuous-wave vertical-cavity single-quantum-well microlaser diodes. Electron. Lett., 25, 1989, pp 1377-1378.

108. Clausen Jr, E.M., Von Lehmen, A., Chang-Hasnain, C., Harbison, J.P., Florez, L.T. : Improved threshold characteristics of air-post vertical-cavity surface-emitting lasers using unique etching process. Electron. Lett., 27, 1991, pp 2243-2245.

109. Geels, R.S., Coldren, L.A. : Submilliamp threshold vertical-cavity laser diodes. Appl. Phys. Lett., 57, 1990, pp 1605-1607.

110. Yang, Y.J., Dziura, T.G., Fernandez, R., Wang, S.C., Du, G., Wang, S. : Low-threshold operation of a GaAs single quantum well mushroom structure surface-emitting laser. Appl. Phys. Lett., 58, 1991, pp 1780-1782.

111. Geels, R.S., Corzine, S.W., Coldren, L.A. : InGaAs vertical-cavity surface-emitting lasers. IEEE J. Quantum Electron., 27, 1991, pp 1359-1367.

112. Jewell, J.J., Harbison, J.P., Scherer, A., Lee, Y.H., Florez, L.T. : Vertical-cavity surface-emitting lasers : design, growth, fabrication, charaterization. IEEE J. Quantum Electron., 27, 1991, pp 1332-1346.

113. Chang-Hasnain, C.J., Harbison, J.P., Hasnain, G., Von Lehmen, A.C., Florez, L.T., Stoffel, N.G. : Dynamic, polarization, and transverse mode characteristics of vertical cavity surface emitting lasers. IEEE J. Quantum Electron., 27, 1991, pp 1402-1409.

114. Ibaraki, A., Furusawa, K., Ishikawa, T., Yodoshi, K., Yamaguchi, T., Niina, T. : GaAs buried heterostructure vertical cavity top-surface emitting lasers. IEEE J. Quantum Electron., 27, 1991, pp 1386-1390.

115. Chang -Hasnain, C.J., Harbison, J.P., Zah, C.E., Maeda, M.W., Florez, L.T., Stoffel, N.G., Lee, T.P. : Multiple wavelength tunable surface-emitting laser arrrays. IEEE J. Quantum Electron., 27, 1991, pp 1368-1376.

116. Yoo, H.J., Hayes, J.R., Paek, E.G., Harbison, J.P., Florez, L.T., Kwon, Y.S. : Phase-locked two-dimensional arrays of implant isolated vertical cavity surface emitting lasers. Electron. Lett., 26, 1990, pp 1944-1946.

117. Van Der Ziel, J.P., Deppe, D.G., Chand, N., Zydzik, G.J., Chu, S.N.G. : Characteristics of single and two-dimensional phase coupled arrays of vertical cavity surface emitting GaAs-AlGaAs lasers. IEEE J. Quantum Electron., 26, 1990, pp 1873-1882.

118. Gourley, P.L., Warren, M.E., Hadley, G.R., Vawter, G.A., Brennan, T.M., Hammons, B.E. : Coherent beams from high efficiency two-dimensional surface-emitting semiconductor laser arrays. Appl. Phys. Lett., 58, 1991, pp 890-892.

119. Orenstein, M., Kapon, E., Stoffel, N.G., Harbison, J.P., Florez, L.T., Wullert, J. : Two-dimensional phase-locked arrays of vertical-cavity semi-conductor lasers by mirror reflectivity modulation. Appl. Phys. Lett., 58, 1991, pp 804-806.

120. Chang-Hasnain, C.J., Wullert, J.R., Harbison, J.P., Florez, L.T., Stoffel, N.G., Maeda, M.W. : Rastered, uniformly separated wavelengths emitted from a two-dimensional vertical-cavity surface-emitting laser array. Appl. Phys. Lett., 58, 1991, pp 31-33.

121. Morgan, R.A., Robinson, K.C., Chirovsky, L.M.F., Focht, M.W., Guth, G.D., Leibenguth, R.E., Glogovsky, K.G., Przybylek, G.J., Smith, L.E. : Uniform 64x1 arrays of individually-addressed vertical cavity top surface emitting lasers. Electron. Lett., 27, 1991, pp 1400-1401.

122. Orenstein, M., Von Lehmen, A.C., Chang-Hasnain, C., Stoffel, N.G., Harbison, J.P., Florez, L.T. : Matrix addressable vertical cavity surface emitting laser array. Electron. Lett., 27, 1991, pp 437-438.

123. Von Lehmen, A., Chang-Hasnain, C., Wullert, J., Carrion, L., Stoffel, N., Florez, L., Harbison, J. : Independently addressable InGaAs/GaAs vertical-cavity surface-emitting laser arrays. Electron. Lett., 27, 1991, pp 583-585.

124. Miller, D.A. : Photoelectronic applications of quantum wells. Opt. and Photon., 1, 1990, pp 7-14.

125. Giles, C.R., Wood, T.H., Burrus, C.A. : Quantum well SEED optical oscillators. IEEE J. Quantum Electron., 26, 1990, pp 512-518.

126. Mc Cormick, F.B., Lentine, A.L., Morrison, R.L., Walker, S.L., Chirovsky, L.M.F., d'Asaro, L.A. : Parallel operation of a 32 x 16 symmetric self electrooptic effect device array. IEEE Photon. Technol. Lett., 3, 1991, pp 232-234.

127. Mc Cormick, F.B., Tooley, F.A.P., Sasian, J.M., Brubaker, J.L., Lentine, A.L., Cloonan, T.J., Morrison, R.L., Walker, S.L., Crisci, R.J. : Parallel interconnection of two 64 x 32 symmetric self-electro-optic effect device arrays. Electron. Lett., 27, 1991, pp 1869-1871.

128. Chirovsky, L.M.F., Focht, M.W., Freund, J.M., Guth G.D., Leibenguth, R.E., Przybyblek, G.J., Smith, L.E., d'Asaro, L.A., Lentine, A.L., Novotny, R.A., Buchholz, D.B. : Large arrays of symmetric self electro-optic effect devices. Osa Top. Meet. Photon. Switch. Washington, D.C. : Opt. Soc. Amer. 1991, pp 150-153.

129 Lang, R., Kasahara, K., Sakaguchi, M. : VSTEP : a view on steps to optical inteconnects and processing. Proc. 16th European Conference on Optical Communication, Sept 16-20, 1990, pp 849-852.

130. Taylor, G.W., Mand, R.S., Simmons, J.G., Cho, A.Y. : Ledistor-a three terminal double heterostructure optoelectronic switch. Appl. Phys. Lett., 50, 1987, pp 338-340.

131. Kuijk, M , Heremans , P., Borghs, G. : Highly sensitive NpnP optoelectronic switch by AlAs regrowth. Appl. Phys. Let. 59, (5), 1991, pp 497-498.

132. Chan, W.K., Harbisson, J.P., von Lehmen, A.C., Florez, L.T., Nguyen, C.K., Schwarz, S.A. : Optically controled surface emitting lasers. Appl. Phys. Lett., 58, 1991, pp 2342-2344.

133. Numai, T., Sugimoto, M., Ogura, I., Kosaka, H., Kasahara, K. : Surface-emitting laser operation in vertical-to-surface transmission electrophotonic devices with a vertical cavity. Appl. Phys. Lett., 58, 1991, pp 1250-1252.

134. Zhou, P., Cheng, J., Schaus, C.F., Sun, S.Z., Hains, C., Zheng, K., Torres, A. : High performance latchable optical switch and logic gates based on the integration of surface emitting laser and photothyristors. Appl. Phys. Lett., 59, 1991, pp 2504-2506.

135. Zhou, P., Cheng, J., Schaus, C.F., Sun, S.Z., Hains, C., Zheng, K., Armour, E., Hsin, W., Myers, D.R., Vawter, G.A. : Cascadable, latching photonic switch with high optical gain by the monolithic integration of a vertical - cavity surface emitting laser and pn-pn photothyristor. IEEE Photon. Technol. Lett., 3, 1991, pp 1009-1012.

8 Spatial light modulators for interconnect switches

Andrew Walker
Dept. of Physics, Heriot-Watt University, Scotland.

8.1 Introduction

Spatial light modulators (SLMs) are optical elements which, under external control, can be induced to exhibit spatial variations of their optical properties within some overall working aperture. For applications in the field of optical interconnects these variations need usually to be local changes in transmission or reflection - although they can be based initially on induced changes in the polarisation or even the phase of the incident light.

One of the most widely discussed examples of an SLM-based optical interconnect is the optical cross-bar arrangement shown in figure 9.1. In this example the SLM is split into an array of independent sub-areas each of which can be switched between transmitting and opaque states. The light from multiple inputs is fanned out so that the information in each channel is broadcast across a corresponding column, while the output channels each collect light from one row in the matrix. In this way, by opening just one "window" per row each output channel can select any desired input.

This type of interconnect system, along with many others, is discussed in more detail elsewhere in this text. However it can be seen from this brief description that a wide variety of properties are required of the SLM: sufficient spatial resolution, high contrast switching, fast response, low insertion loss, etc. Similar demands are made by other SLM-based interconnect systems. For example, optical neural networks can be constructed using the configuration in figure 8.1 to provide the weighted interconnects between "neuron" devices. In this case, in addition to the above list of requirements, the SLM must have an analogue, rather than digital, response. The system acts as a vector-matrix multiplier with the matrix element values determined by the transmission coefficients set for each "pixel" in the SLM array. Finally it is also possible to make an SLM act as an externally controllable hologram capable of diffracting light into specified directions. For this application, phase modulation is preferred over amplitude (transmission) modulation, as it yields higher efficiency, while very high spatial resolution (and hence space-bandwidth product) is also demanded.

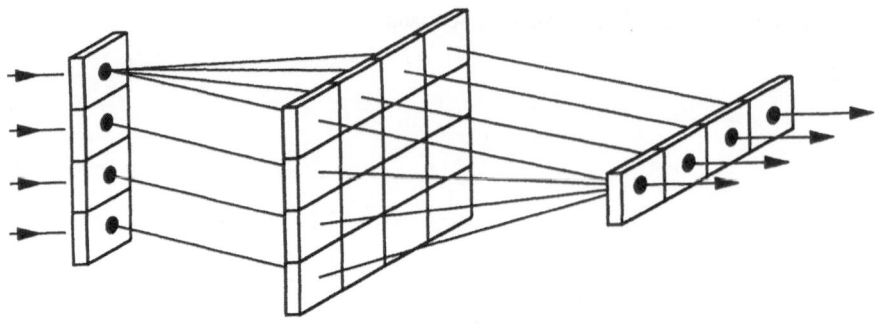

Fig. 8.1 Schematic of the optical cross-bar, or vector-matrix, interconnect architecture. The central component is an SLM, used to select from which input channel any particular output channel should receive data.

Another fundamental aspect of SLM technology is the mode of address - i.e. the method by which the required spatial variations in optical properties are input to the device. Two approaches exist: (i) optical address - where the control information is usually presented as an image (across the active area of the device); and (ii) electrical address - where serial or parallel electronic inputs specify the desired spatial pattern. Optically-addressed SLMs (OASLMs) are attractive in avoiding the problems electronics has in efficiently controlling, in parallel, discrete points within a large array (i.e. limited pin-out numbers). On the other hand an interface is often needed with electronics at some point in the system and the electrically-addressed SLM (EASLM) is then essential.

It can be concluded that no single SLM device will fit all the above requirements simultaneously. Indeed it has proven difficult to develop an ideal SLM for just one application! As a consequence a very large volume of research has been carried out with the aim of developing optimal SLM devices. The following sections review some of the approaches taken and indicates some of the different ways in which SLMs can be exploited to construct optical interconnects.

8.2 Mechanisms for Spatial Light Modulation

Any light modulation mechanism can, in principle, be exploited to develop an SLM device. In some cases the effect may naturally adapt to two-dimensional address of a uniform optical element, while in others an SLM can only be made by assembling independent microscopic modulator cells into a two-dimensional array. In both cases the mechanisms being exploited can be divided into three categories: (i) refractive (phase) modulation, (ii) absorptive (amplitude) modulation, and (iii) mechanical modulation. These correspond to either control of the material optical constants (as in i and ii) or mechanical movement of an optical surface. This section briefly discusses the phenomena that have been exploited to develop SLM devices. A summary of mechanisms and materials is provided in Table 8.1.

8.2.1 Refractive Modulation

This has proven to be the largest mechanism category for SLM research and development. Modulation of refractive index is an efficient method of controlling light - avoiding the energy dissipation of absorptive modulators.

The most widely utilised phenomenon is the linear electro-optical (Pockels) effect. Usually this is exploited to provide an electrically-controlled birefringent element, thus permitting manipulation of the polarisation of the incident "read" beam. The electric field which controls the birefringence of the modulator can derive either from a photoconductive input ("write") stage, in the case of an OASLM, or, for an EASLM, by a directly controlled array of electrodes or a scanned electron-beam. In most cases such SLMs are operated in reflection mode and require just a single polarising element, as shown in figure 8.2, to separate and analyse the output beam.

The most widely known example of an EASLM based on the electro-optic effect is the liquid-crystal display. These vary from coarse-pixel devices, used for example in watch displays, to the high pixel-density arrays for liquid-crystal televisions. Electro-optic materials, in addition to liquid crystals, that have been used to produce SLM devices include: $LiNbO_3$, KDP, BSO and PLZT.

Magneto-optic effects, in particular Faraday rotation, have also proven valuable in producing a polarisation rotating SLM. These are commonly operated in a digital mode, relying on magnetic domain switching effects to reverse the magnetic field and hence change the sign of the induced rotation of polarisation. Example materials are the iron garnets, with high bismuth content to maximise the Verdet constant.

Nonlinear refraction associated with optical excitation of a medium can be exploited to produce optically-addressed SLMs. Typical mechanisms are (i) interband electronic transitions in semiconductors, which lead to an irradiance-dependent refractive index (optoelectronic n_2), and (ii) optically induced temperature rises, which change the refractive index through the thermo-optic coefficient, dn/dT (optothermal n_2). The consequent variations in optical thickness of the medium can be translated into transmissivity or reflectivity changes by incorporating the nonlinear materials within a Fabry-Perot cavity. This can result in either a digital (optically bistable) or analogue response to an input "write" beam. If the underlying excitation of the medium can also be induced electrically, then equivalent EASLM devices can be constructed. This is particularly easy in the case of the optothermal devices where local heating can be induced by, for example, a scanned electron-beam. An example material that has been used to make optoelectronic nonlinear SLMs is GaAs, while optothermal devices have been based on materials such as ZnSe or liquid-crystals.

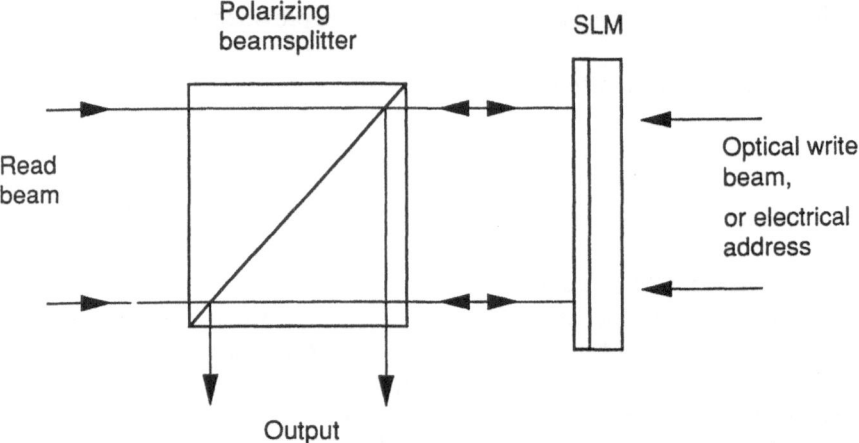

Fig. 8.2 Typical read arrangement for a controlled-birefringence SLM. Linearly polarised light is modified by the modulator stage (e.g. the polarisation direction is rotated) and consequently, on returning from the SLM surface, is reflected by the polarising beam splitter.The same technique can be used for non-birefringent reflective SLMs by adding a (fixed) quarter-wave phase retarder near the SLM plane.

The photo-refractive effect, which is another instance of an optical power dependence of refractive index, has also been used to develop SLM devices. The index changes, in this case, depend on the spatial variation of the illuminating pattern, rather than the local irradiance level. This is a consequence of the underlying mechanism - the redistribution of charge from lighter to darker areas and the electro-optical response of the medium to the consequent electric field variations. As a result, photorefractive materials are particularly good at reproducing interference fringe patterns. Example photorefractive materials, used for SLM applications, include BSO and $BaTiO_3$.

It should be noted, in addition, that (travelling-wave) 1-dimensional SLMs can be constructed using acousto-optic (Bragg) modulators or deflectors. By using multi-channel Bragg cells, these can also be extended to a 2-dimensional format. The prospects for developing programmable interconnects based on acousto-optic devices are covered in Chapter 10, and will therefore not be discussed any further in this chapter.

8.2.2 Absorptive Modulation

As with refractive modulation, the absorption coefficient of an optical material can also be controlled by applied fields or incident light. The first two effects described below are associated with the band edge of direct-gap semiconductors.

Electro-absorption is a term used generally to refer to electric field induced changes in absorption. Such effects tend to be largest at wavelengths in the vicinity of strong absorption features such as allowed atomic transitions and, in particular, semiconductor band edges. The earliest known band-edge electro-absorption phenomenon was the Franz-Keldysh effect. In this, as a result of the tilt of the band-gap (in real space) induced by an applied electric field and consequent tunneling from virtual states (below the conduction band) into real conduction-band states, the absorption edge is stretched out towards longer wavelengths. A second field-induced absorption phenomenon is the Stark effect. This was initially investigated in the context of induced shifts of atomic transitions but has proven to be particularly interesting with respect to exciton transitions in semiconductors. Isolated absorption resonances occur at wavelengths just longer than the absorption edge in direct-gap semiconductors as a result of the creation of excitons - i.e. mobile electron-hole pairs with insufficient energy to separate from each other and become fully-free charge carriers. Normally, such excitons are easily ionised (i.e. destroyed) by an applied electric field. However, by using quantum-well structures - where the electron and hole are trapped by the well potential within the same highly restricted volume - a strong Stark shift can be induced by an applied electric field without destroying the strength of the exciton absorption. This is known as the quantum-confined Stark effect (QCSE) and is discussed further in Chapter 7. SLM-type devices, based on both the QCSE and Franz-Keldysh effect, have been constructed using GaAs and InGaAs.

An alternative approach to exploiting the strong exciton or band-edge absorption in a semiconductor for the purposes of light modulation may be described as nonlinear absorption. In an OASLM type of device this would take the form of absorption saturation by a sufficiently intense "write" beam. It has been shown that exciton transitions can be saturated at power levels significantly less than those needed to induce useable changes in refractive index. The absorption saturation is a result of free carrier excitation and consequently the same result could be achieved by carrier injection using a forward biased p-i-n structure. In SLM format, this corresponds to the recently developed surface-emitting (vertical-cavity) laser arrays. Such a device, configured as an array of amplifiers (i.e. without resonant cavities) and with individual address of each element, would make an excellent high-contrast EASLM - going from a highly absorbing state to one giving gain as the current is increased. (Note that this could be less technically demanding than making a full laser array.)

As is the case with refractive optical modulation, temperature changes can also be used to induce variations in absorption. (The two optical constants are, of course, directly linked, via the Kramers-Kronig relations, and consequently changes in any one must imply a change in the other, in some spectral region.) The optical constants of a semiconductor at energies close to the band-gap are sensitive to the precise position of the absorption edge. Small increases in temperature can cause a significant shift in the absorption edge (commonly to longer wavelength) and consequently large increases in absorption can be seen in the tail of the absorption edge. This effect can be used to make an OASLM where the thermal input is supplied by the optical "write" beam. However it is clearly restricted to very specific band-tail wavelengths. Under the right conditions optically bistable switching can occur as a result of the increasing absorption causing a further increase in heating. This effect has been studied in semiconductors such as GaAs, InSb, ZnSe and CdS.

A rather different phenomenon, which could lead to the development of OASLMs, is photochromism. Photochromic materials can undergo a reversible transition between two structural states, each with a different absorption spectrum. This transition can be induced optically, in at least one direction, while the reverse process may occur as a result of thermal relaxation or illumination with light of a different wavelength. A well-known example of photochromism are the self-darkening sun-glasses. These are darkened by strong UV/blue light. If this illumination is structured spatially, then the pattern would be reproduced in the transmitted visible light. More useful, e.g. faster responding, photochromic media can be found amongst organic materials, e.g. fulgides, or even biological substances, e.g. bacterio-rhodopsin.

8.2.3 Mechanical Modulation

The mechanical, surface displacement, effects that are used to make SLM devices can be divided into two types: surface deformation and micro-mechanical deflection.

One of the earliest projection-display types of SLM, named "the Eidophor",. relied on the deformation of an oil-film (on the surface of a rotating glass disc) under the action of a scanned electron beam. The deformed oil surface deflected the light and hence modulated the proportion passing through the stop of the projection lens. In this way a TV image could be projected using a conventional high-power lamp. Both thermal and electrostatic techniques have been employed to induce surface deformation. EASLMs, with scanned e-beam as well as direct electrical address have been built using a variety of active layers: e.g. liquids, elastomer, thermoplastic, polymer-membrane.

An alternative approach relies on silicon processing technology to fabricate micro-mechanical elements controlled, by electrostatic forces, by the surrounding circuitry. Metallised or silica cantilevered mirrors can form the modulating elements, giving the option of not only modulating the incident light but also of deflecting it. The latter capability could clearly be valuable in the context of optical signal routing.

8.3 Examples of Spatial Light Modulators

8.3.1 Refractive SLMs

One of the better known SLMs is the liquid crystal light valve (LCLV) developed at Hughes Research Laboratories [1]. This is an electro-optic (controlled birefringence) device which has taken on many forms since it was first created. The generic structure for the optically-addressed LCLV, which has been retained throughout by Hughes and other laboratories [2, 3], is illustrated in figure 8.3. It consists of a photoconducting input stage, dielectric mirror, and a liquid crystal (LC) modulator. Local changes in conductivity of the photoconductor appear as variations in the field across the liquid crystal. Incoherent light can be used to write onto the input stage and a higher-power and/or coherent read beam, reflected from the LC-cell, has the spatial information then transferred onto it. Typical photoconductor layers are CdS [1], BSO [2, 3], and discrete silicon photo-diode/transistor arrays [4]. Nematic liquid crystals have been used in these devices, giving frame times of around 10-30 ms. Resolution and sensitivity are typically 10-40 lp mm^{-1} and 1-20 μJ cm^{-2} respectively.

Electrically addressed LCLVs have been developed by replacing the photoconductor in figure 8.3 by silicon circuitry. Two approaches can be distinguished. Firstly the Hughes group have used a CCD backplane to provide an interface between a serial electronic data input and the same type of MOS silicon-diode array used for their optically-addressed silicon-LCLV [5]. Secondly, a group at Edinburgh University, Scotland have developed a silicon-backplane LC-SLM[6]. This device uses an nMOS integrated circuit to control the voltage on an array of mirrors (in this case metal, rather than dielectric), above which the LC-cell is completed. The circuitry provides a latching switch for each pixel. The CCD device was configured as a 256 x 256 array on a 20 μm pitch, while silicon-backplane SLMs have been made in a number of formats: 16 x 16 (200 μm pitch), 50 x 50 (74 μm pitch) and 176 x 176 (30 μm pitch). Using twisted (45°) nematic liquid crystals operating in the "hybrid field effect" mode frame times of ~ 10 ms

182

were obtained with both approaches. Recent work with a ferro-electric liquid crystal by the Edinburgh group and STC Technology Ltd. (who had themselves worked independently on the development of Si-backplane SLMs) has shown 200 μs switching in test cells, compatible with a 1 ms frame-time for their 176 x 176 SLM array [7].

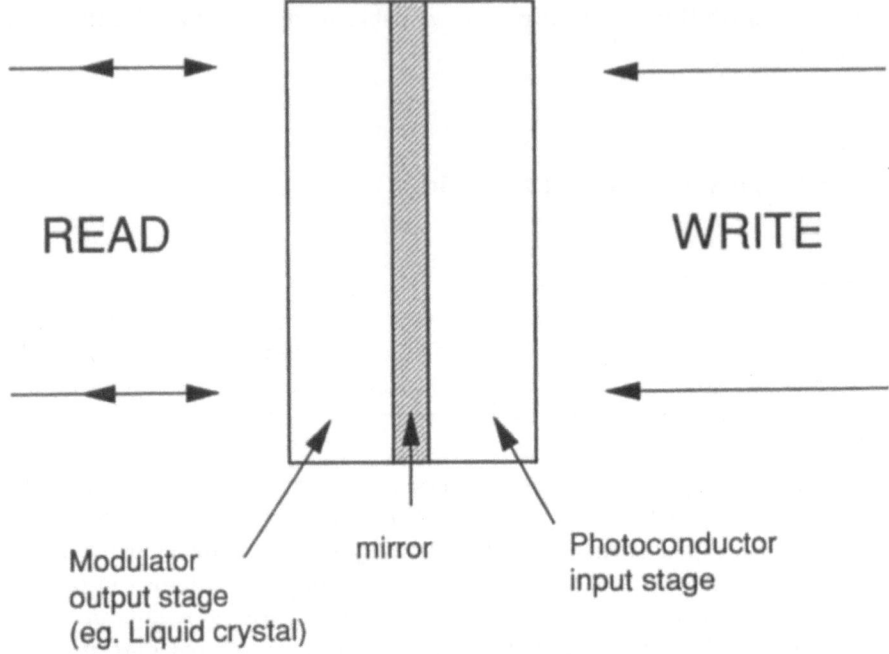

Fig. 8.3 Typical light-valve configuration. An optical write beam (e.g. an image) is detected by a photo-sensitive input stage which in turn controls the local voltage variations across the (reflective) field-sensitive modulator.

Ferroelectric liquid crystals, used in their chiral smectic C phase, are of considerable current interest - showing binary switching, depending on their structure in the cell, on timescales of 8-140 μs at a temperature of 25-30°C and using 20 volts [8]. For further increases in operating speed the Smectic A (electroclinic) phase holds out the prospect of switch times as short as a few microseconds. The construction of data-transparent (i.e. high transmission bandwidth) optical interconnect switches with microsecond reconfiguration times could be expected to have a significant impact.

An alternative approach to achieving higher frame-rates is to move away from the orientational electro-optic effects, exhibited by liquid crystals, to exploit the purely dielectric response of electro-optic crystals. The structure of the resulting SLM devices remains the same as in figure 8.3 (for the OASLM case) but with the

LC-cell replaced by a solid crystal layer. Examples of such OASLMs include: Photo-titus [9], which uses a KD*P modulator with an amorphous selenium input stage; and the PLZT-SLM, based on a PLZT modulator and a silicon photo-transistor array input. A closely related OASLM is the PROM [10], which uses a single crystal of BSO as both the input photoconductor and the electro-optic modulator. These devices have sensitivity and resolution comparable to, or slightly better than, the LCLV SLMs, 2-10 μJ cm^{-2} and 6-30 lp mm^{-1}, respectively, and exhibit significantly shorter frame-times: 20-100 μs.

Much higher sensitivity (< 1 μJ cm^{-2}) can be obtained by using a microchannel-plate charge amplifier stage, between the photo-conductor and the modulator. The MSLM [11] is an example of such a device and uses LiNbO$_3$ as the electro-optic crystal. Unfortunately the higher sensitivity is offset by a slower response time: ~ 30 ms.

The electrically addressed version of the PLZT-SLM [12] utilises direct electronic inputs to the silicon transistor array and achieves similar performance, in terms of single pixel address, as the OASLM. However, multiplexed full address circuitry significantly increases the total frame time. The alternative, e-beam, address scheme employed in the Titus SLM [13], also leads to longer frame-times (~ 30 ms) due to the difficulty of inducing rapid charge removal. Similar frame-times are also achieved in the EBSLM [14, 15], which incorporates e-beam address of a LiNbO$_3$ (rather than KD*P) crystal. An e-beam addressed version of the MSLM has also been developed [16] but again frame-times appear to be limited to > 25 ms.

An example of a magneto-optic EASLM is the LIGHT-MOD [17]. This has been developed as a commercial product and uses a matrix electrical address scheme, in which the magnetisation of the pixel at the cross-point is flipped between its two possible states. A 128 x 128 array of 76 x 76 μm pixels can be addressed in ~ 20 ms. A significant limitation is the low polarisation rotation that is induced (14°-32°) and optical absorption which, together, lead to a throughput efficiency of $< 6\%$ when operating in a high contrast mode.

Nonlinear refraction can be exploited to make very fast optically-addressed SLMs based on arrays of bistable nonlinear etalons. Pixellated GaAs arrays with sub-nanosecond frame times have been reported [18]. However such devices have low sensitivity (e.g. ~ 0.1-1 mJ cm^{-2}) and consequently, given their response speed, require the dissipation of high powers to maintain operation. As a result no arrays of significant size have yet been operated. It may, however, be possible to reduce the average input powers required by fabricating somewhat slower, microsecond response, devices (e.g. by utilising long carrier-lifetime material) and hence avoid the thermal dissipation problems in a large array.

Optically induced heating can be turned to good effect in the optothermal nonlinear interference filter devices. 16 x 16 pixellated-array SLMs have been fabricated using devices based on evaporated ZnSe layers [19]. Again, these are not sensitive devices - requiring a few milliwatts per pixel - but they can display switch times < 100 μs. An attractive alternative to providing the input power optically is to use e-beam address. This has been demonstrated in the ETIF device [20], in which a nonlinear interference filter has been included as the faceplate in a miniature CRT. Switching between bistable states in ~ 10 μs has been demonstrated [21].

A final example of a refractive SLM is the PICOC [22]. This is based on the photorefractive effect in BSO and is operated as an optically-addressed device. It relies on a volume phase grating initially being written into the photorefractive medium by two interfering beams and the subsequent selective erasure of this grating by the write image. The read beam diffracts only from the remaining grating regions and hence the desired image transfer is achieved. Because the write beam can cover a spread of wavelengths, this device (as with many OASLMs) can act as an incoherent-to-coherent image converter.

8.3.2 Absorptive SLMs

Electro-absorption modulator devices have been fabricated from a range of materials and in various configurations. SLMs are usually built up from arrays of discrete modulator elements. An example of an EASLM, based on the quantum-confined stark effect (QCSE) in GaAs MQW material, is the CCD-addressed one-dimensional SLM developed at MIT [23]. This device was designed to be used at 847 nm and consisted of a linear array of 16, 10 x 70 μm pixels with a frame-time of ~ 100 μs. Like all electro-absorption modulators the contrast was low - less than 1.5 : 1. For the longer telecommunications wavelengths (i.e. 1.3 μm, 1.55 μm) it is necessary to utilise the quarternary semiconductor InGaAsP. Two-dimensional QCSE modulator arrays of up to 10 x 10, 45 μm diameter pixels (125 μm pitch) have been made using InGaAs/InP MQW structures [24]. Again the modulation contrast was only ~1.5 : 1, but this was sufficient to demonstrate a 16-channel optical interconnect in which the EASLM provided a 100 Mbit/s parallel electronic-to-optical interface [25].

Work on optically-addressed electro-absorption devices has centred around the self electro-optic effect device (*SEED*). This device (described in more detail in Chapter 7) is a single-element detector-modulator and can provide a latching (bistable) response to incoming optical signals. Although the simplest SEED devices used the Franz-Keldysh effect in bulk GaAs [26], the largest arrays have been based on QCSE modulators fabricated from GaAs MQW material: e.g. 256 x 256 [27]. This latter device was actually configured as a 256 x 128 array of symmetric-SEEDs (S-SEEDs), achieved by electrically connecting adjacent pairs of SEEDs in series. This gives a latching optical switch which is sensitive to the ratio of light inputs to the two sensitive areas. The size of these optical windows was 5 x 10 μm with a pitch of 20 μm. The optical energy required to switch such devices is ~ 1 pJ per pixel, e.g. 1 μW of power for 1 μs. A similar, electrically-adddressed, S-SEED array has also been reported [27]. This was configured as an 8 x 16 array of switches with direct electronic address of each pair of elements. Better than 1 Gbit/s digital modulation was demonstrated.

A somewhat different class of electro-absorption modulator, mentioned in section 8.2.2 in the context of nonlinear absorption, is the surface emitting semiconductor laser. In this device, it is injected free-carriers which cause changes in the absorption - ultimately giving gain at sufficiently high carrier densities. Large arrays of micro-resonator lasers have been fabricated with emitting areas varying in size from 1 to 10 μm in diameter using both InGaAs (λ = 958 nm) and GaAs (λ = 845 nm) quantum-well structures [28]. Emitting devices of this type are of direct interest for applications in inter-chip optical connection systems [29]. However they would also appear to have significant SLM potential if they could be configured as electrically-addressed variable absorber/amplifier arrays.

Of the remaining absorptive modulation mechanisms - optically induced absorption saturation, temperature dependent absorption, and the photochromic effect - only the latter has been seriously considered for SLM development. Photochromic materials are of interest because of the essentially molecular-scale spatial resolution that they offer. This could permit their use in holographic SLM applications. A scanned-optical-input system, using a thermally-stable fulgide derivative, has been investigated [30]. Using 1 Watt of optical power a 256 x 256 pixel field could be addressed in about 30 ms. Better sensitivity is being sought from alternative photochromic substances such as the biological-origin material: bacterio-rhodopsin [31].

8.3.3 Mechanical SLMs

Deformable mirror SLMs can have surprisingly fast response times. One example is the optically-addressed DMD [32] which is based on a VLSI circuit containing input photo-transistors which control, by electrostatic attraction, a flexible membrane stretched over holes etched into the silicon substrate. A 128 x 128 array was demonstrated with a frame-time of ~ 30 μs. A similar EASLM with analogue address circuitry and 50 μm square pixels has also been constructed [33].

By suitably processing a silicon wafer it is possible to include microscopic overhanging (cantilevered) features of SiO$_2$ or metal [34]. This can act as a deflecting element, again controlled by the fields from the Si circuit. Response times, for single-pixel address of ~ 10 μs can be achieved. Further mechanical SLMs, along with examples of many of the other types of SLM are included in the review article by Fisher [35].

8.3.4 Optical Discs as SLMs

Optical disc technology has had a major impact in both the IT and entertainment markets (e.g. CDs, CD-ROMs etc.). The optical disc can be regarded as a two-dimensional SLM with a scanned optical address. Of course, in normal use they are read sequentially but there is currently considerable interest in parallel interrogation.

Table 8.1 Mechanisms and materials for spatial light modulation. The names of various example SLMs are given in parenthesis.

Refractive	Absorptive	Mechanical
Electro-Optic	**Electro-absorption**	**Deformation**
Liquid crystals (LCLV)	GaAs (Franz-Keldysh)	Oil film (Eidophor)
LiNbO$_3$ (MSLM, EBSLM)	GaAs MQW (SEED arrays)	Thermoplastic (TP-PC)
KDP (TITUS)	InGaAs MQW (mod-arrays)	Elastomer (Ruticon)
BSO (PROM)		Polymer (MLM)
PLZT (PLZT-SLM)		
Nonlinear refraction	**Nonlinear absorption**	**Deflection**
GaAs (OB arrays)	InGaAs SQW (SEL arrays)	Silicon/silica (micro-mechanical)
Photorefractive	**Photochromic**	
BSO (PICOC)	Fulgides Bacterio-rhodopsin	
Thermo-optic	**Thermo-absorption**	
ZnSe (ETIF)	GaAs, InSb ZnSe, CdS	
Magneto-optic		
YIG (LIGHT-MOD)		

A variety of mechanisms have been exploited to record information on such discs - most involving the heating effect of a scanned laser spot. In the case of the "write-once read-many" discs the modulation is in the form of changes in absorption/reflection or surface deformation, while the re-writable devices exploit magneto-optic polarisation rotation effects or absorption changes due to induced phase transitions. It should be noted that current recording techniques do not appear to be compatible with highly parallel writing because of the milliwatt power levels needed for each pixel (i.e. a 100 x 100 array would require tens of watts).

"Write-once" discs have been used as the basis of optical interconnect experiments. Both binary array patterns, suitable for vector-matrix cross-bar interconnects, and computer generated holograms, for optical neural network connections, can be implemented. Disc rotation can provide a large number of different interconnect configurations, despite the fixed nature of the recorded pattern. An optical character recognition system has been implemented [36] using an optical disc both to store the required interconnect configuration and to implement it optically.

Using re-writable discs more complex reconfigurable interconnects could be envisaged, e.g. with a new connection pattern being defined on one part of the disc while a previous configuration is still being used.

8.4 Examples of SLM-Based Switching Networks

Although all the mechanisms and SLM concepts discussed in sections 8.2 and 8.3 (see Table 8.1) could, in principle, be used to develop optical interconnect systems, only a limited subset have actually been tested in prototype interconnect switching systems. This section presents four examples.

8.4.1 Vector-Matrix Processor Interconnect (MILORD)

The cross-bar architecture shown in figure 8.1 permits, at any instant, each output channel to receive signals from any of the transmitting sources. This high-bandwidth broadcast mode of interconnect is very flexible and is of interest in multi-processor computer systems.

A group at CERT, Toulouse, France have been developing such a system - dubbed MILORD - aimed at providing arbitrary interconnections between multiple transputer processors [37]. This uses an optically-addressed Hughes liquid-crystal light valve (LCLV) configured as a digital switch array. The state of each element in the array (high or low reflectivity) is controlled with a CRT, coupled via a fibre-optic faceplate directly to the write stage of the LCLV. This makes it similar to an e-beam addressed device.

The system was configured to interconnect up to 34 input/output channels (i.e. 34 processors) and the SLM therefore acted as a 34 x 34 array of switch elements. Note that the LCLV OASLM is a non-pixellated device and hence the number of independent channels that could be switched is limited, at first sight, only by the system spatial resolution (both SLM and associated optics) and the size of the active area. However, it is essential that the switch contrast is significantly larger

188

than the number of channels being routed. It was concluded that the limit in this sytem was about 100 channels.

Although the data transmission bandwidth of "transparent" switches, such as this, are very high, reconfiguration rates can be severely limited by the SLM-system frame time. In this example, switch reconfiguration times of ~ 200 ms were obtained. This is clearly a severe problem when one considers the amount of data that would accumulate while the switch is being reconfigured (e.g. 4 Mbits for a 20 Mbit/s data rate). Much faster SLMs are required for such a system.

8.4.2 Matrix-Matrix Processor Interconnect (OCPM)

An SLM-based multi-processor system aimed at overcoming some of the problems encountered by the MILORD group is currently under development in the UK under the name: optically connected parallel machine (OCPM) [38]. The system uses a somewhat different architecture, described as a matrix-matrix geometry [39]. This approach, shown schematically in figure 8.4, takes a two-dimensional input array and uses Fourier-plane diffractive optics, to fan-out the incoming signals over the SLM as multiple replicas of the input pattern.

Lens arrays, on the output side of the SLM then collect the light so that (as with the vector-matrix approach) each output channel receives a component of each input. Again, by switching particular elements in the SLM array any one of these inputs can be selected. As is the case with the alternative geometry, for N input/outputs the SLM has to handle N^2 channels. However there are considerable advantages in the associated optics, particularly for the output half of the system, in that anamorphic elements are avoided and much more efficient use is made of lens field sizes.

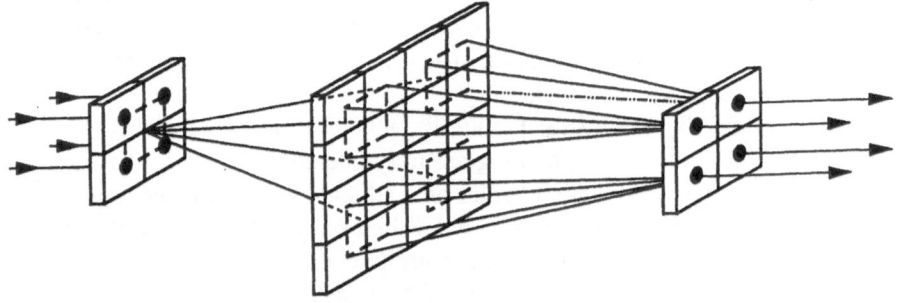

Fig. 8.4 Schematic of the matrix-matrix interconnect architecture (cf. Fig. 8.1). A 2-D array of input channels is fanned out across the SLM and collected into a similar output array. Again, arbitrary connects can be realised by switching the relevant pixel in the SLM array.

The SLM in the OCPM system is a silicon-backplane liquid crystal modulator. This is a directly addressed EASLM which can exploit the faster-responding ferroelectric liquid crystals [40] and hence achieve frametimes < 1 ms [8]. Present work, using ferroelectric or electroclinic liquid crystals is aimed at switch reconfiguration times of < 10 μs. The target switch size in the OCPM project is 64 x 64, while data transmission rates in the range 10 Mbits/s to 1 Gbit/s are envisaged - corresponding to a multi-THz throughput capacity. Current estimates indicate that switch systems of significantly larger size should ultimately be realisable.

8.4.3 Spatial Switching with Nonlinear F-P Cavity Devices

Interferometric spatial switching of optical signals, of the type more commonly realised using electro-optic $LiNbO_3$ directional couplers, can be implemented with nonlinear Fabry-Perot devices. Instead of the induced refractive index changes causing a mismatch between coupled waveguides (and hence changing the route taken by the signal), refractive tuning of a Fabry-Perot cavity can be used to switch between reflecting and transmitting states [41].

Nonlinear-refraction SLMs could be used to provide arrays of such 1x2 or 2x2 spatial switches which, when suitably interconnected, could act as high capacity parallel switching networks (see figure 8.5). Optothermal nonlinear interference filters have been used to construct an all-optical 1x4 switch network, with the routing determined by header pulses, preceding the optical data-stream [42]. The network comprised three bistable elements acting as data-transparent, latching 1x2 nodes. This type of device is compatible with reconfiguration times of around 100 μs, but has contrast levels that are limited (mainly by manufacturing tolerances) to ~ 10 : 1.

An electrically controlled version of this type of switch could be based on the ETIF SLM [20, 21], either by using it as the routing switch directly, or indirectly, by having it supply the necessary control/clock signals to a similar optically-addressed SLM.

8.4.4 S-SEED Photonic Switching Fabrics

The above examples all correspond to data-transparent switch schemes in which high-bandwidth optical data is routed directly through the switch with no signal regeneration. This approach is not directly scalable - requiring more demanding specifications of the SLMs (and other components) as the switch size is increased. The alternative approach is to have individual elements in a (digital) SLM array switching in sympathy with the transmitted information, i.e. at the data rate. This ensures full signal regeneration at every stage, and is the approach exploited in electronic switching fabrics. It can be reproduced in the optical domain by using optically-triggered latching modulators, read by independent sets of (standardised) power beams.

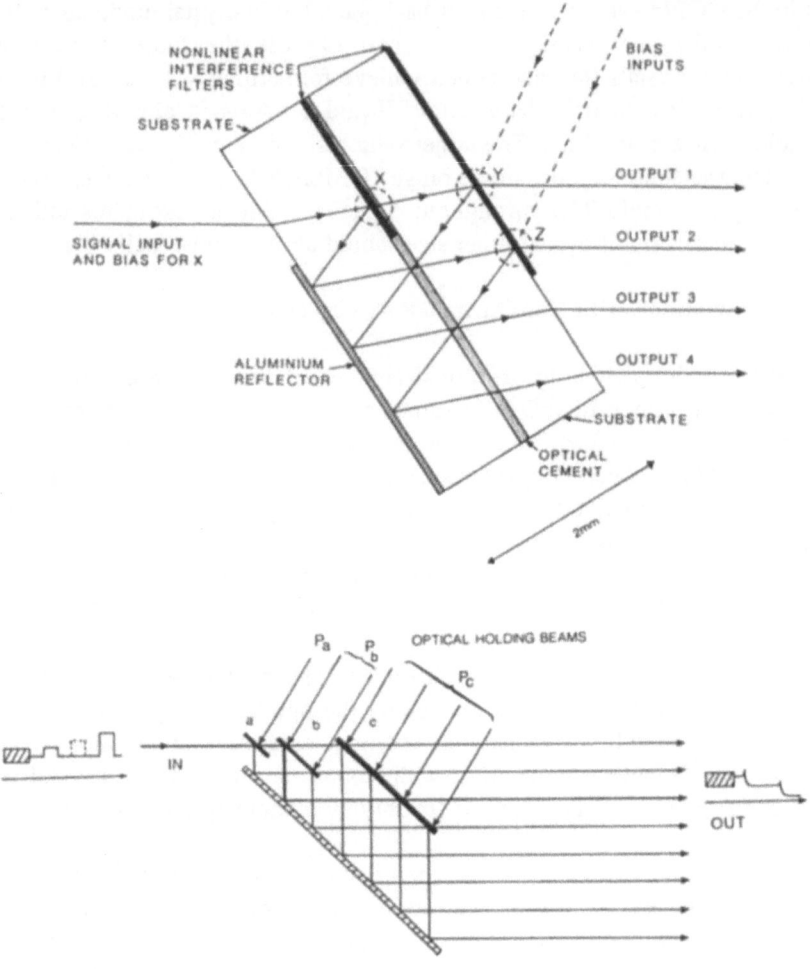

Fig. 8.5 Top: A 1 x 4 optical spatial switch, based on nonlinear refractive elements, capable of acting as a data-transparent signal routing network. Bottom: A schematic of a similar 1 x 8 switch showing how header signals could set a route for a subsequent data stream.

A series of such "photonic switching fabrics" have been constructed, for demonstration purposes, at AT&T Bell Laboratories (Naperville, USA). These have been based on two-dimensional arrays of symmetric-SEEDs (S-SEEDs) with various free-space interconnect patterns implemented between each node-plane (e.g. cross-over, Banyan). By fanning out signals between each plane to two or more nodes, and controlling which nodes are active (i.e. ready to switch state in the presence of a "high" data bit), specific paths through the network can be set up. Figure 8.6 shows a schematic of such a multi-stage interconnection network.

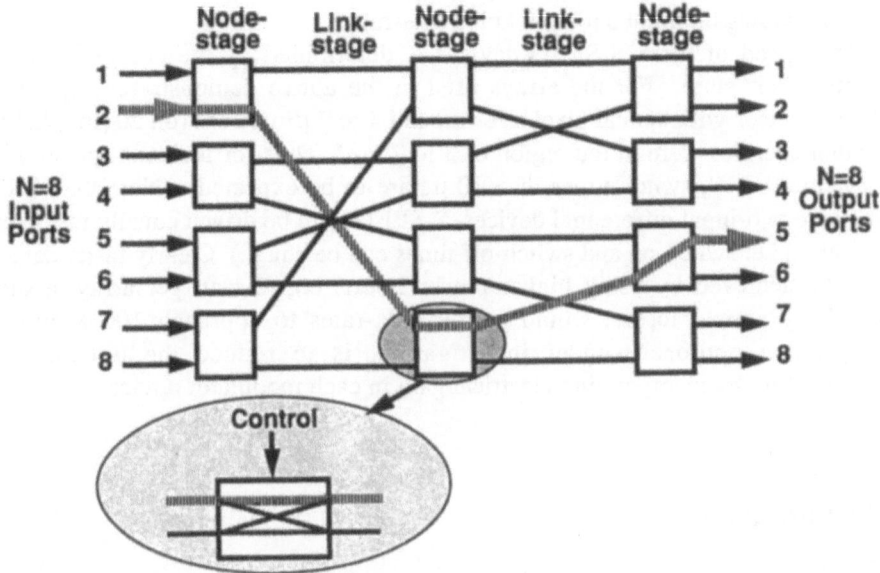

Fig. 8.6 A multi-stage interconnect network of the type implemented using (data-rate switching) S-SEED arrays. The nodes actively route input signals according to control instructions, while the links provide passive connections between consecutive node stages.

Two demonstrator experiments have been recently reported. The first [43] consisted of a series of three 16 x 8 arrays of S-SEEDs with cross-over interconnects between each stage. The system handled 64 input/output optical data channels and required another 64 optical inputs for control purposes. The S-SEED arrays operated in a time-sequential-gain mode, in which each device pair (forming a single array element) was switched according to the ratio of the two inputs and, after latching, was then read by higher power clock pulses. These outputs then became the inputs to the next stage. Each array was clocked by a single 850 nm laser diode, with the few milliwatt output split by a binary phase grating into the 256 beams needed to power all the elements. The system was operated at ~ 30 kbit/s, limited by the low optical powers reaching the S-SEED devices - < 1 μW per pixel.

The second demonstrator switching fabric [44] had a larger capacity - over a thousand channels - and included more stages (to reduce the blocking probability). Importantly, its overall size was significantly smaller, covering an area of ~ 20 x 30 cm. It comprised 6 stages of 32 x 32 S-SEED arrays, each customised for the particular application. One significant difference from the earlier demonstration was the use of electrical control lines, connected to each element in the array, to set up the required routes through the network. Note that, the devices remained optically addressed but additionally had an electronic enable/disable facility. The system was again limited to ~ 30 kbit/s data-rate.

The speed of these S-SEED devices is determined by the energy required to switch their state. For the arrays used in the above demonstrator experiments, using devices with optical pixel sizes around 5 to 7 μm across (on 20 μm pitch), the switch energies were in the region of 3 to 7.5 pJ. Thus for incident optical powers of below 1 μW, switch times of > 10 μs are to be expected. (Note that, because they are two-input differential devices, S-SEEDs can be driven equally fast in either direction i.e. switch-on and switch-off times can be equal.) Clearly faster data-rates can be achieved by using higher power lasers: e.g. 1 watt per array, assuming similarly optical losses, would permit data-rates to approach 100 Mbit/s. The alternative approach, under investigation, is to reduce the optical energy requirement by incorporating electrical gain in each modulator device.

8.5 Summary

A broad range of physical phenomena have been used as the basis of many different types of OASLMs and EASLMs. A lot of these devices have progressed no further than initial laboratory demonstrations. Others which have been exploited to develop experimental optical interconnect systems, are still far from ideal for this application. Important parameters that need to be improved include frame-rates (particularly for EASLMs), contrast, and power consumption. And studies aimed at using EASLMs as real-time computer-generated holograms [45] - a capability that would see many applications in the context of reconfigurable optical interconnects - are placing increasing demands on space-bandwidth product and optical quality.

Despite the many difficulties, significant progress continues to be made. At present, technical advances are commonly driven by other, perhaps more immediate, applications, such as in the area of displays, or developments in related fields, e.g. semiconductor optoelectronics. It is clearly the case, however, that given the necessary breakthrough in terms of required performance for specific approaches, there exists a large range of SLM-based optical interconnect schemes that would be expected to rapidly advance to commercial significance.

Acknowledgements

The author thanks Rebecca Wilson for her help in preparing this review, along with other colleagues at Heriot-Watt University.

References

1. Grinberg, J., Jacobsen, A., Bleha, W.P., Miller, L., Fraas, L., Boswell, D. and Myer, B.: Optical Engineering 14, 1975, p.217.
2. Auborg, P., Huignard, J.P., Hareng, M. and Mullen, R.A.: Applied Optics 21, 1982, p. 3706.
3. Baillie, W.L., Openshaw, P.M., Hart, A.D. and Makh, S.S.: IEE Proceedings 133, Parts 1 and 2, 1986, pp. 60-69.
4. Efron, U., Braatz, P.O., Little, M.J., Schwartz, R.N. and Grinberg, J.: Optical Engineering 22, 1983, p. 682.
5. Welkowsky, M.S., Efron, U., Byles, W. and Goodwin, N.W.: Optical Engineering 26, 1987, pp. 414-417.
6. McKnight, D.J., Vass, D.G. and Sillitto, R.M.: Applied Optics 23, 1989, pp. 4757-4762.
7. Underwood, I., Vass, D.G., Sillitto, R.M., Bradford, G., Fancey, N.E. and Al-Chalabi, A.O.: Proc. SPIE Vol. 1562. 1991.
8. Crossland, W.A., Birch, M.J., Davey, A.B. and Vass, D.G.: Proc. of Colloquium: "Two Dimensional Optoelectronic Device Arrays" (IEE, London) Digest No. 1991/158, paper-7, 1991.
9. Donjon, J., Dumont, F., Grenot, M., Hazan, J-P., Marie, G. and Pergrale, J. IEEE Trans. Elec. Dev. ED-20, 1973, p. 1037.
10. Sprague, R.A. and Nisenson, P.: Optical Engineering 17, 1978, pp. 256-266.
11. Warde, C., Weiss, A.M., Fisher, A.D. and Thackara, J.I.: Applied Optics 20, 1981, p. 2066.
12. Lee, S.H., Esener, S.C., Title, M.A. and Drabik, T.J.: Optical Engineering 25, 1986, 250-260.
13. Groh, G. and Marie, G.: Optics Communications 2, 1970, p. 133.
14. Casasent, D.: Optical Engineering 16, 1977, pp. 295-301.
15. Shinoda, K. and Suzuki, Y.: SPIE Vol. 613 "Nonlinear Optics and Applications", 1986, pp. 158-164.
16. Schwartz, A., Wang, X-Y. and Warde, C.: Optical Engineering 24, 1985, pp. 119-123.
17. Pulliam, G.R., Ross, W.E., MacNeal, B.E. and Bailey, R.F. J.: Appl. Phys. 53, 1982, p. 2754.

18. Lee, Y.H., Warren, M., Olbright, G.R., Gibbs, H.M., Peyghambarian, N., Venkatesan, T., Smith, J.S., and Yariv, A.: Appl. Phys. Lett. 48, 1986, pp. 754-756.
19. McKnight, D.J., Redmond, I.R., Walker, A.C., Taghizadeh, M.R., Buller, G.S., Mathew, J.G.H. and Smith, S.D.: Optical Computing and Processing 1, 1991, pp. 137-144.
20. Walker, A.C., Smith, S.D., Campbell, R.J. and Mathew, J.G.H.: Optics Letters 13, 1988, pp. 345-347.
21. Walker, A.C., Smith, S.D., Wilson, R.A., Campbell, R.J. and Mathew, J.G.H. Proc. Int. Topical Meeting on Optical Computing (SPIE), 1990, p. 399.
22. Marrakchi, A., Tanguay, A.R., Yu, J. and Psaltis, D.: Optical Engineering 24, 1985, p. 124.
23. Nichols, K.B., Burke, B.E., Aull, B.F., Goodhue, W.D., Gramstorff, B.F., Hoyt, C.D. and Vera, A.: Appl. Phys. Lett. , 1988, pp. 1116-1118.
24. Rejman-Greene, M.A.Z., Scott, E.G., Webb, R.P. and Healey, P.: OSA Proceedings on Photonic Switching (OSA, Washington DC) Vol. 8, 1991, pp. 210-212.
25. Barnes, N., Healey, P., Rejman-Greene, M.A.Z., Scott, E.G., Webb, R.P. and Wood, D.: Electronics Lett. 26, 1990, p. 1110.
26. Ryvkin, B.S.: Sov. Phys. Semicond. 15, 1981, p. 796.
27. Chirovsky, L.M.F., Focht, M.W., Freund, J.M., Guth, G.D., Leibenguth, R.E., Przybylek, G.J. and Smith, L.E.: OSA Proceedings on Photonic Switching (OSA, Washington DC) Vol. 8, 1991, pp. 56-59.
28. Jewell, J.L., Harbison, J.P., Scherer, A., Lee, Y.H. and Florez, L.T. IEEE J. Quantum Electron. 27, 1991, pp. 1332-1346.
29. Jewell, J.L., Lee, Y.H., Scherer, A., McCall, S.L., Olsson, N.A., Harbison, J.P. and Florez, L.T.: Optical Engineering 29, 1990, pp. 210-214.
30. Kirkby, C.J.G. and Bennion, I.: IEE Proceedings 133, 1986, pp. 98-104.
31. Thoma, R., Hampp, N., Brauchle, C. and Oesterhelt, D.: Optics Lett. 16, 1991 651-653.
32. Pape, D.R.: Optical Engineering 24, 1985, pp. 107-110.
33. Hornbeck, L.J.: Proc. International Display Research Conf., 1982, p. 76.
34. Brooks, R.E.: Optical Engineering 24, 1985, pp. 101-106.
35. Fisher, A.D. and Lee, J.N.: SPIE Vol. 634 "Optical and Hybrid Computing", 1986, pp. 352-371.
36. Yamamura, A.A., Neifeld, M.A., Kobayashi, S. and Psaltis, D.: Optical Computing and Processing 1, 1991, pp. 3-12.
37. Comte, D., Siron, P., Thibault, X., Fraces, M. and Bouzinac, J.P. Proc. Int. Symp. Optics in Computing (Toulouse 1989), pp. 113-121.
38. OCPM started in 1991 as a project supported by the UK Department of Trade and Industry and the Science and Engineering Research Council under the LINK scheme. Partners are: British Aerospace, BNR Europe, Meiko Ltd., Heriot-Watt University and the University of Bath.

39. Kirk, A.G., Crossland, W.A. and Hall, T.J.: SPIE Vol. 1574 (1991); and Proc. Third Int. Conf. Holographic Systems Components and Applications (IEE, London), 1991, 137-141.

40. Collings, N., Crossland, W.A., Ayliffe, P.J., Vass, D.G. and Underwood, I. Applied Optics 28, 1989, pp. 4740-4747.

41. Paton, C.R., Smith, S.D. and Walker, A.C.: "Photonic Switching" Electronics and Photonics Series (Springer-Verlag) Vol. 25, 1988, pp. 111-114.

42. Buller, G.S., Paton, C.R., Smith, S.D. and Walker, A.C.: Appl. Phys. Letts. 53, 1988, pp. 2465-2467.

43. McCormick, F.B., Tooley, F.A.P., Cloonan, T.J., Brubaker, J.L., Lentine, A.L., Morrison, R.L., Hinterlong, S.J., Herron, M.J., Walker, S.L. and Sasian, J.M.: OSA Proceedings on Photonic Switching (OSA, Washington DC) Vol. 8, 1991, pp. 48-55.

44. McCormick, F.B., Tooley, F.A.P., Brubaker, J.L., Sasian, J.M. , Cloonan, T.J., Lentine, A.L., Morrison, R.L., Criscri, R.J., Walker, S., Hinterlong, S.J. and Herron, M.J.: "Optical Applied Science and Engineering" SPIE Vol. 1533, 1991, paper 12.

45. Barnes, T.H., Eiju, T., Matsuda, K., Ichikawa, H., Taghizadeh, M.R. and Turunen, J.: Proc. Third Int. Conf. Holographic Systems Components and Applications (IEE, London), 1991, pp. 142-146.Figure Captions

9 Reprogrammable Components: Photorefractive Materials

D. Bize
ONERA - CERT, Toulouse.

9.1 Introduction

Photorefractive materials such as lithium niobate (LiNbO$_3$), potassium niobate (KNbO$_3$), barium titanate (BaTiO$_3$), strontium barium niobate (SBN) and bismuth silicon oxide (Bi$_{12}$SiO$_{20}$) are attractive candidates for real-time optical connections (OC).

Photorefractive materials have a wide range of applications such as :
- volume holographic storage [1, 2, 3, 4, 5]
- convolution, correlation [6, 7]
- logic operations [8, 9]
- phase conjugation [10, 11, 12]
- edge enhancement [13, 14]
- non linear image filtering [15, 16, 17]
- image amplifiers [18, 19].

The promise of photorefractive materials in OC is substantial. A large number of parallel operations can be processed in a single crystal. Two wave mixing with high gains have been observed [18]. Operations on a nanosecond time scale have been demonstrated [20]. Finally, the requirements on write and erase energy density in some photorefractive materials (like BSO, [13]) are comparable to the high resolution photographic plates (50 to 100 μJ/cm^2).

As a result, many researchers without previous experience with photorefractives require information on the different materials and their use. So, with the digest of several excellent overviews of photorefractive processes and materials [21 to 25] we intend to provide basic information for the effective selection and use of photorefractive materials for optical information processing applications.

Section 1 begins with a description of the photorefractive effects. We first give a description of the charge transport model of a grating formation and erasure. The internal photorefractive processes of storage, fixing, gain and fan-out are described. In section 2 we divide photorefractive applications in two areas, those requiring only holographic recording (such as data storage and beam steering) and those requiring gain (such as coherent amplification and self pumped conjugation). We present the optimum material characteristics for each area, then review the best available material, listing their advantages, disadvantages, and experimentally determined operating parameters.

9.2 The photorefractive effect

The first observation of the photorefractive effect was in 1965, when it was found that a high intensity laser produced local changes in the refractive index in LiNbO$_3$, BaTiO$_3$ and KTN [25, 26] that were first considered as a local damage. It was later discovered that these variations could be erased by uniform high intensity illumination or by briefly heating the crystal to 200°C. This reversible effect was clearly different from catastrophic damage occuring at much higher power densities. Researchers were quick to see the applications to reversible volume holographic storage. Since then, the photorefractive effect has been observed in a number of materials and the basic mechanisms of the effect have been uncovered and analysed.

Excellent books and mainly reference 21 can provide the reader with basic information on the photorefractive effect. So, we give here a short description of the main phenomena in the photorefractive materials.

9.2.1. Grating formation and storage

Energy transfer between beams crossing the crystals was observed, leading to a variety of real time processing possibilities. The process can be qualitatively described as follows.

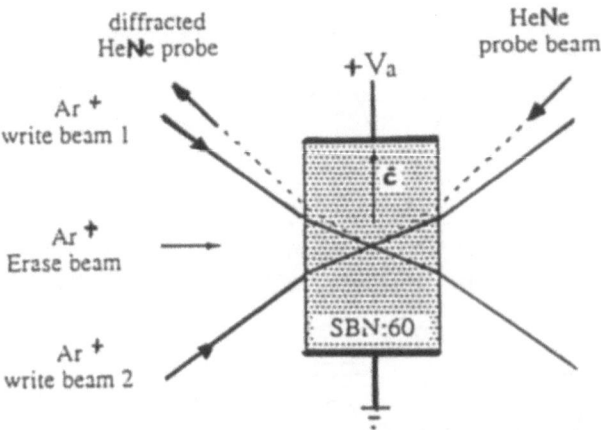

Fig. 9.1. Index gratings written and erased by 514.5 nm argon laser light and continuously monitored by a weak Bragg-matched He-Ne (632.8) while an external voltage is applied along the crystal's optic axis C. θ is the internal half angle between recording beams, and β is the angle between the grating wavevector k_g and the crystal optic axis.

Consider two plane waves crossing in a photoconductive and electro-optic material, as shown in Fig. 9.1. They produce a sinusoïdal fringe pattern (Fig. 9.2). In the regions of high intensity, photo absorption excites charge carriers to the conductive band, allowing them to drift or diffuse towards darker areas, where they are trapped. The resulting carrier distribution (2b) generates a space charge electric field in accordance with the Poisson's equation (2c). The field modulates the index of refraction through the electro optic effect (2d), creating a sinusoïdal phase grating. If the writing beams are switched off, the phase grating remains stored in the materials until thermal or optical excitation redistributes the trapped charge carriers. The grating diffracts each writing beam into the direction of the other. If the grating coincides spatially with the fringe pattern, there is no net energy transfer. However, a phase difference between the index grating and the fringes causes constructive interference with one of the writing beam, coherently amplifying that beam at the expense of the other.

9.2.2 Main parameters

Response time. The photorefractive effect is generally described as slow (on the order of milliseconds or longer). However, light intensities comparable to those used for conventional χ^3 materials are applied to photorefractives, very fast

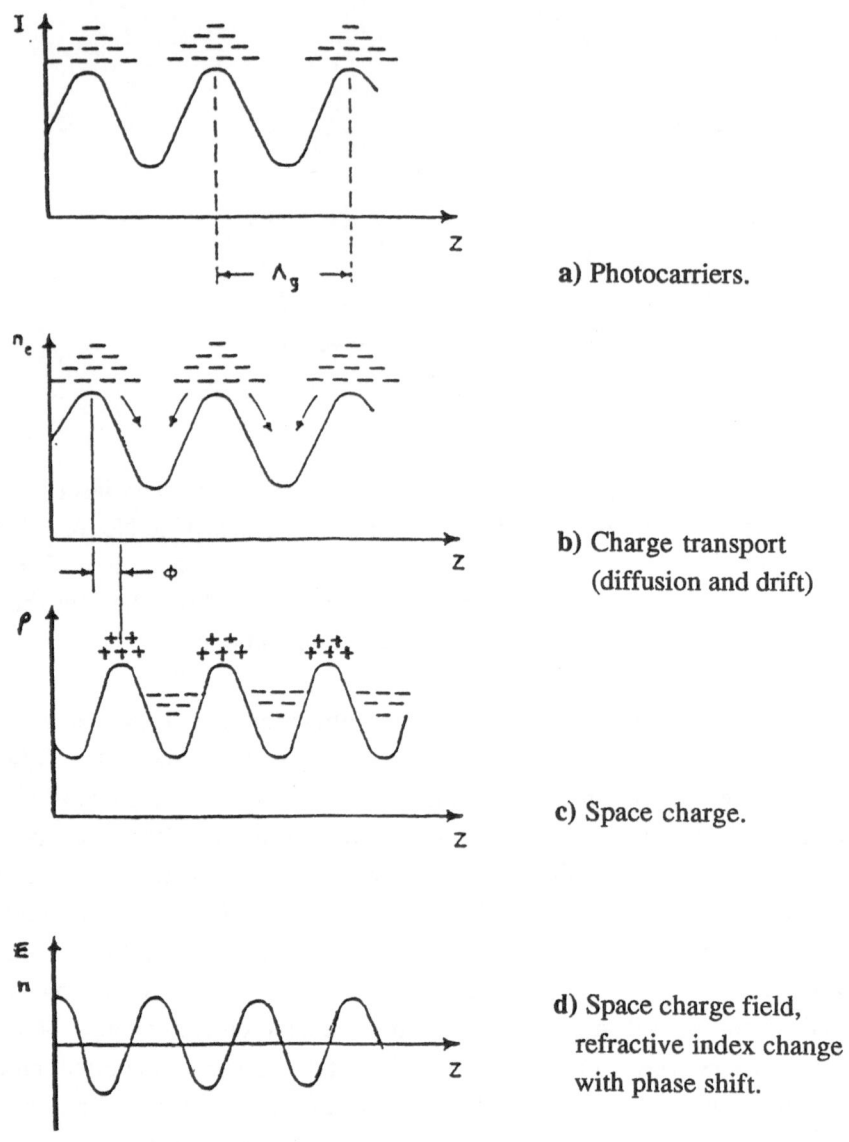

a) Photocarriers.

b) Charge transport
(diffusion and drift)

c) Space charge.

d) Space charge field,
refractive index change
with phase shift.

Fig. 9.2. After reference 22.

responses are seen. There is of course a limit to the speed obtained by increased
intensity. The limit can be set by the charge carrier generation rate or by the
charge carrier transport time [27]. Experimentally, high intensity nanoseconde

pulses have been used for photographic storage in $LiNbO_3$ [28], $BaTiO_3$ [29], BSO [96] and GaAS,InP [97, 99] and for four wave mixing and holography in BSO [30].

30 to 40 picosecond-long pulses have been used to store index gratings in $BaTiO_3$ [31] and GaAs [32]. If the intensity is hundreds of MW/cm^2, many electrons are brought simultaneously to the conduction band so that the electron plasma will itself produce a large index change, that vanishes when the electrons recombine in their new locations, leaving behind them the normal photorefractive index change.

Energy to write a grating with 1% diffraction efficiency. The energy per unit area required to write a grating of 1% diffraction efficiency in a crystal 1 mm thick is a useful figure of merit in photorefractive applications when a large efficiency is not required.

This parameter allows comparison between fast materials with small values of electro optic coefficient and slow materials with large value of electro optic coefficient. However this figure of merit has the disadvantage that some materials never reach a 1% diffraction efficiency in a 1 mm length.

Photorefractive sensitivity. The photorefractive sensitivity S is defined as the index amplitude modulation per unit absorbed energy per unit volume.

The photorefractive sensitivity is an other useful figure of merit because it tells how well a material uses a given amount of optical energy. It allows comparison of materials with different absorption coefficients on an equal basis.

Gain. The gain G experienced by a beam (1 or 2) is the ratio of its output intensity in the presence of the other beam (2 or 1) to its output intensity in the absence of the other beam, then [19, 21, 43, 75]

$$G_1 = \frac{I_{1out}(I_{2in \neq 0})}{I_{1out}(I_{2in=0})}$$

$$G_2 = \frac{I_{2out}(I_{1in \neq 0})}{I_{2out}(I_{1in=0})}$$

Multiple hologram storages. The storage capacity of a volume holographic recording media at wavelengh λ is fundamentally bound to one information per λ^3 [33]. For $\lambda = 1\mu m$, this number is of order $10^{12}/cm^3$. Because this far exceeds the information which can be transmitted through the volume's aperture by a single optical wavefront, the problem becomes one of information storage and access. One way to solve this problem is to arrange the information into pages, which are stored and retrieved sequentially in time. In this approach, data are recorded in a set of superimposed holograms, each coded with a unique reference wavefront.

When holograms are superimposed in photorefractive materials, each successive recording partially erases the existing holograms. To achieve equal final diffraction efficiency for all holograms, the optical energy used for each recording must vary according to an exposure schedule. A recording schedule which takes into account different write and erase sensitivity has already been described [34, 35]. More details are given in chapter 4 and architectures exploiting multiple hologram storages are given in chapter 13.

Fixing. Under the right circumstances, the photorefractive index grating can be semi-permanently stored, or "fixed", so that the hologram can be read out for extended periods without significant erasure.

Fixing has been accomplished by two photons absorption in $LiNbO_3$, KTN and PLZT [37] but the most effective methods use ionic charge compensation. Fixing by charge compensation has been demonstrated in $LiNbO_3$, $BaTiO_3$ and BSO as well as in SBN [38, 39].

The process of fixing by charge compensation consists of several steps. An electronic distribution is created and maintained by illuminating the crystal with the recording beams. An external control mechanism, such as applying a strong external field or heating the material, is then used to allow ionic transport. The free ions, which are not subject to photoexcitation, are attracted to the electron dense regions.

When the index modulation reaches its minimum, the external control is turned off, fixing the ions in place. Finally the latent grating is revealed by illuminating the crystal uniformly, which redistributes the electrons and exposes the ionic charge distribution.

Fanning. Noise is inevitable in any optical system. In photorefractive materials, coherent input light scatters from crystal imperfections and impurities, producing wide angular bandwidth temporally coherent noise.

While the initial noise intensity is negligible, noise propagating in a gain media can be amplified to significant intensity levels. This process called "fanning" becomes particularly important when the material gain coefficient is large or when there is no other signal competing for the gain available [23].

Bandwidth. The total amount of information which can be sent through an optical system is determined by its dynamic range and its space bandwidth product (SBP), which is the product of the usable input area (of the entrance profile) and the spatial frequency bandwidth.

Whether the response at a given spatial frequency is acceptable or not depends on the application requirements, but normally the bandwidth is defined by the maximum number of line pairs per millimetre the system can process keeping the response uniform to within a factor of two. For example, a memory must uniformly reproduce all features of a stored image which are within its bandwidth. A coherent amplifier must amplify uniformly all features of the input image within its band limit.

9.3 Photorefractive materials

In order to present the available photorefractive materials in as useful a fashion as possible, we choose to separate their applications into two categories. Applications which require the volume storage of an index pattern corresponding to an intensity distribution will be called *passive*. Examples are holographic image /data storage, diffractive beam steering and incoherent-to-coherent conversions. In each of these cases, any coupling between the interfering beams generally tends to degrade performance through fanout and spatial and temporal variation in contrast ratio. At the other extreme are *active* applications, which demand coupling in order to operate. Self-pumped phase conjugation, cavity oscillation, edge enhancement, novelty filtering and of course coherent image amplification are all active. There may be some overlap between the two areas such as an application where a weak signal must be amplified and stored, but the division helps to evaluate a wide range of materials. A given material may be useful for either or both areas, depending on the doping and the user-determined operating parameters such as input beam angles, externally applied fields, temperature and input wavelength.

It is important to be aware of the strong effect that doping has on the photorefractive performance of a particular crystal sample. The impurities and concentrations involved are not completely understood. As with many crystal growth issues, the results sometimes seem to be as much due to magic as technology. The effects of dopant concentrations and oxidation / reduction state vary in each material, and have been thoroughly investigated in only certain materials. Where possible, we will cite references to such studies.

Each of the following two sections will describe the desirable performance characteristics for one area, then present several of the best available materials. We will give references to some previous applications of each material, point out its strengths and weaknesses and summarize its experimentally determined operation parameters.

9.3.1 Passive applications

Passive applications require accurate storage of an intensity distribution in the induced index modulation. Any dynamic interactions between the storage beams tends to reduce contrast and the depth of the modulation. The photorefractive incoherent to coherent optical converter or PICOC [41] demonstrated by Marrakchi et al is a good example of a passive application. The PICOC configuration, an application of photorefractive materials to spatial light modulation, is treated in chapter 8. Two plane waves store a uniform sinusoïdal grating in a photorefractive material (Fig. 9.3). A weak signal beam counterpropagating to beam two is diffracted by the grating into the direction of beam one. Information enters the PICOC via an intense incoherent image, which selectively erases the stored grating, so that the coherent output signal is modulated.

What material characteristics are important for photorefractive applications ? We would like materials with high sensitivity and fast response and which accept reasonable operating conditions (range of beam input angles, wavelengths, required applied field, etc.). Ideally, crystals of large size with high optical quality should be readily available at low cost. These characteristics apply to all photorefractive applications. Passive applications demand, in particular, a high peak diffraction efficiency. They may also require fixation for long-term holographic storage. Unfortunately, no single material fulfilling all these requirements has yet been developed. We will restrict ourselves to materials that have been examined in some detail.

Lithium niobate (LiNbO₃). Lithium niobate's used as an electro-optic modulator led to the first observation of the photorefractive effect in the form of modulator damage [26]. Photorefraction in LiNbO₃ has been thoroughly investigated first by researchers trying to eliminate the problem and later by those trying to enhance it. LiNbO3 has proven an excellent material for holographic storage, providing high diffraction efficiency and long term stability, although its photorefractive sensitivity is comparatively low. The demand for LiNbO3-based electro-optic modulators has resulted in wide availability of large high quality crystals, making it probably the most easily-obtained photorefractive material. LiNbO3 has been primarily used for holographic storage [36,39] although it has also been used for coherent amplification [42] and four-wave mixing [43].

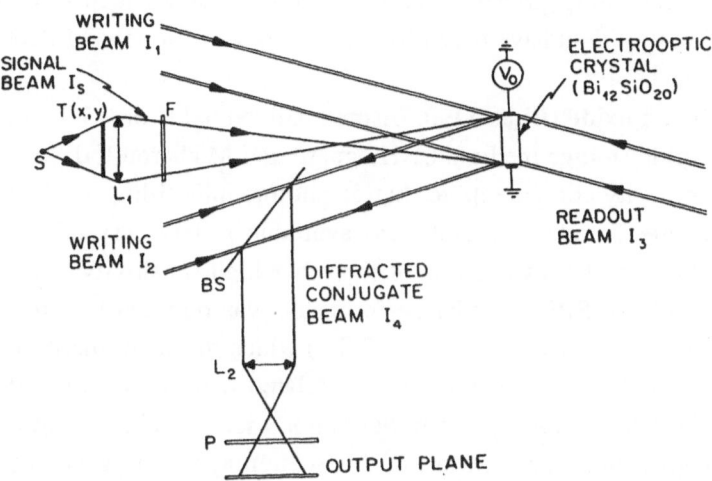

Fig. 9.3. Experimental configuration for photorefractive incoherent to coherent optical conversion in simultaneous erasure/writting mode. After [41].

Lithium niobate like Potassium niobate and Barium titanate is a uniaxial ferroelectric oxide. Its photorefractive sensitivity is strongly enhanced by Fe doping [44]. The Fe atoms provide both the donor (Fe^{2+}) and the acceptor (Fe^{3+}) sites. Unannealed crystals with a 0.05 % Fe concentration, 25 % in the Fe^{2+} state, have been shown [45] to provide the best recording sensitivity. Absorption is primarily from the Fe impurities. For this dopant concentration, the absorption is ~ 10/cm at $\lambda = 0.5$ μm.

LiNbO3 is photovoltaic, rather than photoconductive with ionic transport dominated by short-distance electronic diffusion. Application of external electric fields enhances the sensitivity somewhat but does not affect the gain [46].

$LiNbO_3$ is an outstanding material for multiple hologram storage [47]. Using thermally-controlled fixation Staebler et al. could store 500 angularly-multiplexed holograms, each with more than 2.5% diffraction efficiency. Selective erasure of particular sub-holograms has been demonstrated in $LiNbO_3$ by superposing an identical π phase shifted hologram [43]. Completely non-selective holographic erasure can be achieved by raising the crystal's temperature to 300 C.

The sensitivity of $LiNbO_3$ has been shown to increase for higher intensity recording light. Holograms recorded with MW/cm^2 intensities required only one tenth as much energy as those recorded at approximately one W/cm^2 [48]. Multiphoton recording processes involving both near infrared (1.06 μm) and visible (0.532 μm) light have been observed in Fe- and Cu-doped $LiNbO_3$ [49].

Bismuth silicon oxide ($Bi_{12}SiO_{20}$). Bismuth silicon oxide or BSO was first used as a holographic storage medium by Huignard and Micheron [50] in 1976. They reported sensitivity comparable to that of photographic film, much higher than Lithium niobate. In addition, BSO was available in large crystals of excellent optical quality which was not true of the other high-sensitivity photorefractive materials KTN and SBN. Similar performance was reported for an isomorphic compound, bismuth germanium oxide (BGO) which differs primarily in its lower electro-optic coefficient. Another closely related material is Bismuth titanium oxide (BTO) which shows promise but is not widely available and has not yet been thoroughly investigated. Most of this section applies in general to all three materials. Because BSO may or may not provide photorefractive coupling depending on the operating parameters, it has also been included in the next section, on active materials. In this section, we will deal with BSO's use for passive applications.

As a high quality, high sensitivity material, BSO has been attractive for the full range of passive photorefractive applications. It has been used for holographic storage, edge enhancement, convolution and cross-correlation, time-average interferometry, phase conjugation, beam steering and incoherent to coherent conversion [50 to 63 and 98, 100].

The linear electro-optic coefficient of BSO (r_{41} : 4.5 pm/V) is lower than that of many other photorefractive materials. To compensate for this, an external electric field, can be applied to the crystal. This increases the index modulation.

The applied field also alters $\emptyset$, the grating phase shift angle which in turn affects photorefractive coupling. A moving grating may be used to improve coupling but this is not necessary for most passive applications.

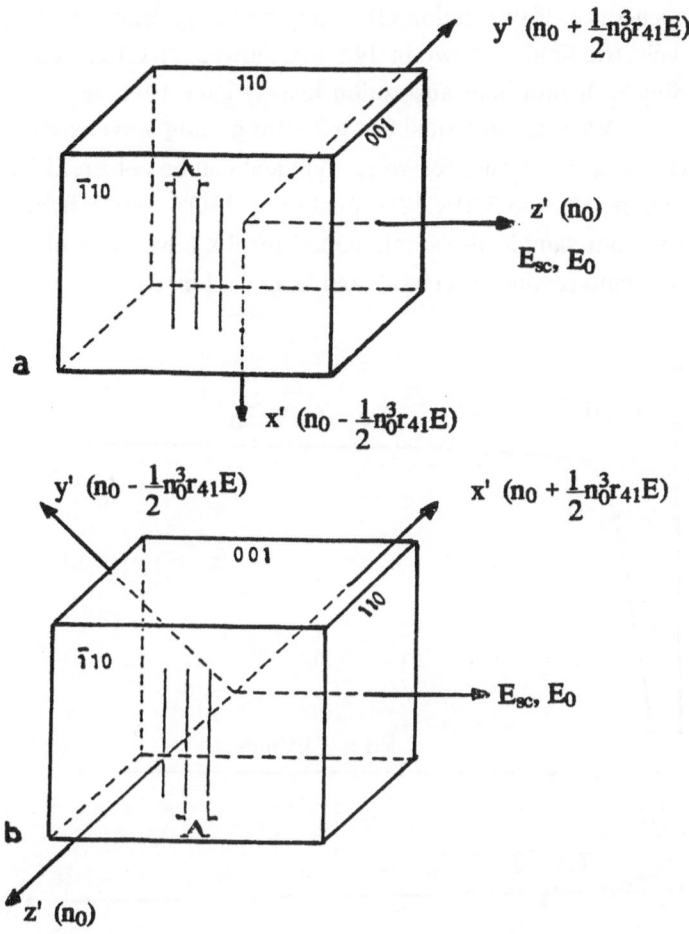

Fig.9.4. a) and **b)** Crystallographic orientation required to optimize either **a)** the energy transfer $E_o \perp 001$ or **b)** the diffraction efficiency $E_o \perp 110$ in cubis BSO crystals. 'x,y',z' are the principal axes of birefringence. After [19].

The crystal orientation for both storage and amplification in BSO is shown in Fig. 9.4. For maximum diffraction efficiency, the electric field is applied in the (110) direction and the optical input enters the $(\bar{1}10)$ face symmetrically ($\beta = 0°$).

The dependence of diffraction efficiency on the grating wavelength and on the applied electric field is shown in Fig. 9.5. Diffraction efficiencies as high as 95% (regarding reflection and absorption losses) have been achieved in BGO [58] using a 15 kV/cm applied field and a 20 μm grating wavelength. The sensitivity of BSO is near maximum for green light and can be enhanced by increasing the crystal temperature to 300°C. The quality of BSO's photorefractive performance can vary from sample to sample. Good results have been obtained both from nominally pure (undoped) crystals and from Cr doped ones.

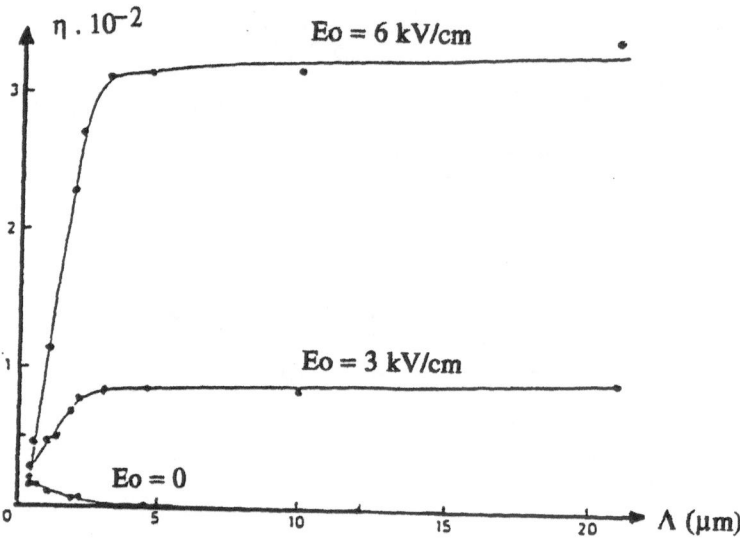

Fig. 9.5. Diffraction efficiency η versus fringe spacing Λ for different applied electric field. After [19].

BSO like BGO and BTO, is optically active ; the direction of polarization of linear polarized light is rotated when it propagates through the crystal. The rate of rotation is substantial (38.6 °/mm at 514.5 nm) and is important in determining optimum input light polarization angles [101]. Its effect on photorefractive performance was analyzed in detail by Marrakchi, Johnson and Tanguay [53]. One reason BTO is interesting is that its optical activity for green light is about one third that of BSO.

Because electrons are the dominant charge carrying species in BSO, hologram fixing via charge compensation is possible and has been demonstrated using an applied electric field on crystals at room temperature. The write-read response time asymmetry measured is small compared to that found in $LiNbO_3$, but the effect is still under investigation and improvements in performance are probable.

Strontium Barium Niobate ($Sr_x Ba_{1-x}Nb_2O_6$). We include SBN as a representative of the tungsten-bronze crystal family, which also includes BSKNN [91]. Both materials have demonstrated great potential for photorefractive applications and are the subject of an extensive development program at Rockwell International Science Center. SBN-60 in particular provides both high gain and excellent holographic storage performance under the control of an external electric field.

Holographic storage in SBN was first reported by Thaxter in 1969 using SBN-75 ($Sr_{75}Ba_{25}Nb_2O_6$) and an applied bias electric field to enhance photorefractive sensitivity. Later investigations showed how the electric field could be used to control both recording and reconstruction of holograms. Other SBN compositions from SBN-25 to SBN-75 were also grown and tested. Although SBN-75 has the highest electro-optic coefficient ($r33 = 1400$ pm/V), SBN-60 is the most interesting because it has better photorefractive properties and because it is most easily grown in large (25 mm cube) high quality crystals. The photorefractive sensitivity of SBN to visible light is increased by Ce doping [64 to 71].

Because of its extremely high gain, recent work on SBN has concentrated more on its characterization and use for active than for passive applications. SBN can be used for storage applications where gain is allowed. SBN is highly sensitive, and the use of electrical control is potentially an advantage. Application of an electrical field to SBN while recording not only increases the performances, it introduces a strong asymmetry in the erase time, indicating that the hologram is simultaneously recorded and fixed.

9.3.2 Active applications

The distinguishing characteristic of active photorefractive applications is their use of photorefractive gain. For example, consider self-pumped phase conjugation, an example of degenerate four wave mixing. When a single input signal is sent into a strong gain material, a pair of counterpropagating beams can spontaneously arise out of the amplified scattered light, and begin oscillating as a laser cavity with the crystal/air boundaries as mirrors. An index grating is written by the interaction of these beams with the input. The fourth wavefront, the phase conjugate output, is produced by the diffraction of the oscillation from this grating. Without photorefractive gain, this complex process cannot occur.

What are the ideal material characteristics for active applications ? As for any photorefractive applications, we would like to have high sensitivity, fast response and reasonable operating requirements, all in a material in large, high optical quality crystals. In addition, the material must have a high and uniform gain coefficient.

Barium Titanate (BaTiO$_3$). Barium titanate has by far the highest measured gain coefficient of any known photorefractive material. It was briefly mentioned in the first paper on optical damage [26] in 1966, and reported in more detail in 1970 by Townsend and LaMacchia [72], but it was not until 1979 that barium titanate's amplifying properties were investigated [73]. Most of the possible active applications have been demonstrated in BaTiO$_3$. Coherent amplification [74, 75], edge enhancement [14], contrast reversal, optical limiting [15], novelty filtering [17], and phase conjugation using both conventional and self-pumped [76, 77] geometries, have all been demonstrated.

Barium titanate is a uniaxial ferroelectric crystal with point group symmetry 4mm. The impurity species primarily responsible for charge transport is iron, with the charge transport species either electrons or, normally holes, depending on the reduction ratio [78, 79]. In its as-grown state, the crystal has multiple allowed domain orientations. In order to define a unique optic axis and maximise the strength of the photorefractive effect, the material must be electrically or mechanically poled. Electrical poling involves applying a strong field (2-5 kV/cm) along the optic axis while holding the crystal just below the Curie temperature (121°C). Once poled, the crystal is stable unless subjected to extreme temperature or electric field.

The usable spectrum ranges from visible to near-infrared, with low (~1cm^{-1}) absorption. The moderate sensitivity implies a corresponding moderate response time. Because the spontaneous charge diffusion rate in barium titanate is high, stored holograms will decay within hours of writing, even without erasure. As a result, it is a poor long-term holographic storage material. Barium titanate is comparatively expensive, especially for the largest samples (1 cm^3), in part due to the difficult poling process. At present, there is only few commercial vendors, although several researchers are now growing their own crystals.

The most important operational characteristics for active applications is G, the exponential gain coefficient [21] Barium titanate produces an extremely high G without using an applied electric field or moving grating. G is a strongly varying function of the beam input angles θ and β (θ angle between the beams, β : angle between the grating vector and C axis according to Fi. 9.1)).

Another photorefractive material closely related to BaTiO$_3$ is Potassium Niobate (KNbO$_3$) which is similar in performance to BaTiO$_3$, with slightly better sensitivity and response time but lower gain. It has been used for coherent amplification, holographic storage, spatial light modulation, and phase conjugation. Since KNbO$_3$ is not yet commercially available and has not seen widespread application, we will forego further discussion of it [83 to 86].

Bismuth Silicon Oxide (Bi$_{12}$SiO$_{20}$). Because BSO crystals are readily available with high quality, they are is useful both as an active and passive material, . No impurity doping is required. BSO has high photorefractive sensitivity and fast response, but a relatively small electro-optic coefficient, only r$_{41}$= 4.5 pm/V compared to barium titanate's r$_{51}$= 1640 pm/V. Application of a strong electric field increases the index modulation, but also reduces the gain by changing the phase $\emptyset$ from its optimum value. Because of this, BSO was initially thought of primarily as a storage medium. In fact, because the grating in any photorefractive material takes some time to form, the phase shift can be manipulated by moving the intensity pattern at a constant velocity relative to the photorefractive media. Huignard and al. [87 to 90] generated a moving grating in BSO by introducing a small frequency shift between the two interfering beams, resulting in gains as high as 1000. The crystal and field orientation is diagrammed in Fig. 9.4. The optical input enters the ($\bar{1}$10) crystal face, with the field along the (001) direction (Fig. 9.6).

Strontium Barium Niobate (Sr$_x$Ba$_{1-x}$Nb$_2$O$_6$). Of the various possible compositions of SBN, SBN-60 (Sr$_6$Ba$_4$Nb$_2$O$_6$) is the most useful for photorefractive applications. It is the most sensitive, has the highest gain, and can now be grown in large (25mm cube), high quality crystals [68]. Doping with cerium [66] increases both sensitivity and gain. The optimum configuration for two wave mixing in SBN is with the grating wave vector approximately aligned with the crystal optic axis, and with the input light linearly polarized in the plane of incidence. Application of an external electric field should, theoretically, increase the gain substantially, but material deformation due to induced stress tends to reduce image quality [81, 82].

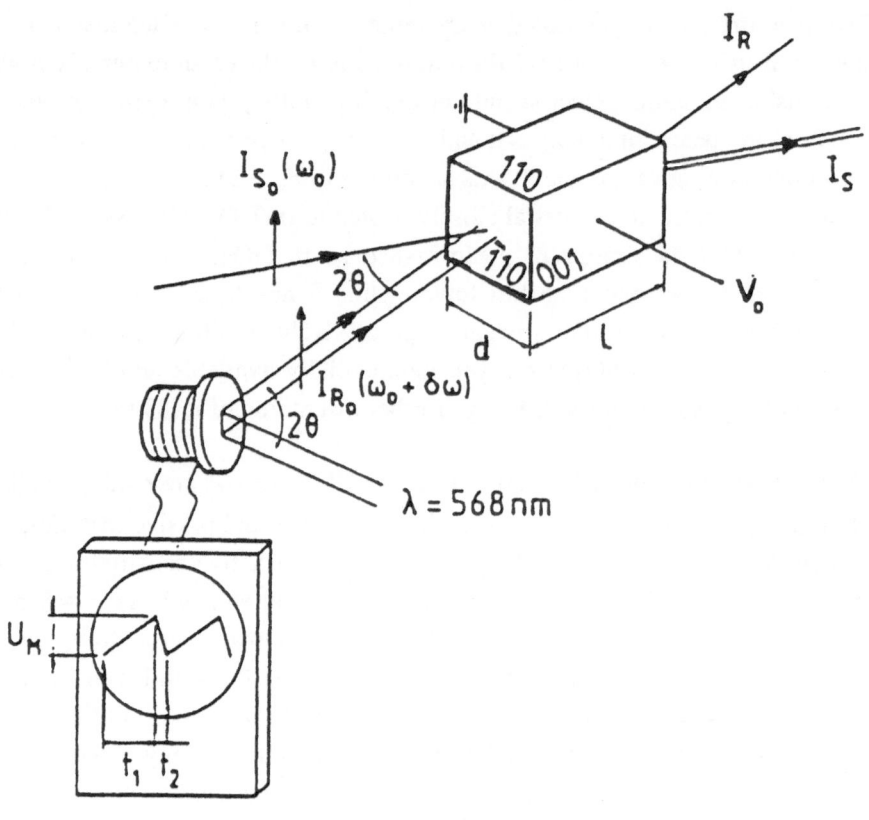

Fig. 9.6. Two wave mixing in photorefractive BSO with moving interference fringes.

Gallium Arsenide (GaAs). Many information processing and communications applications require use of infrared light. The band gap in several electro-optic semiconductors allows photoexcitation by low energy photons, and while the relatively high dark conductivity makes long-term storage in such materials impossible, photorefractive amplification and wave mixing have been observed in InP and CdTe, as well as chromium-doped and undoped GaAs. GaAs is particularly interesting because it has the potential for direct integration with existing GaAs electronic and opto-electronic technology. Its industrial applications for electronics and optical detection also result in good material availability [92, 93].

Gallium arsenide is a face-centered cubic material with similar operation requirements to BSO. GaAs is opaque to visible light, and strongly absorptive (a=1.2/cm) at normal (1μm) operating wavelengths. The high electron mobility and strong absorption result in high sensitivity and fast response times; although the peak index modulation is low. In order to provide amplification, the photorefractive coupling coefficient G must be high enough to overcompensate for absorption. Recently, gain coefficients of 6-7/cm have been measured in Cr-doped GaAs using both a moving grating and an applied field [94], similar to arrangements for BSO [58].

The motion of the grating relative to the material can be caused by actually translating the medium, or by inducing a small frequency shift in one of the interacting beams [87, 90]. As in BSO, the optimum value of ß is zero (normal average beam incidence), with input light linearly polarized parallel to the applied field. Unlike BSO, GaAs is not optically active. The picosecond response characteristics and theoretical behaviour of GaAs were recently investigated [94, 95].

9.4 Conclusion

The optical computing expansion depends especially on the research results on the photorefractive materials. A lot of research laboratories are working on them to achieve a best understanding of the physical mechanisms which occur in the photorefractive effect. Now it is sometimes possible to synthesize crystals with photorefractive properties decided in advance.

We are just at the beginning of the applications of the photorefractive materials in optical and optoelectronic devices.

References

1. F.S Chen, J.T La Macchia and D.B Fraser: Holographic storage in lithium niobate. Appl. Phys. Lett. 13, 1968, pp 223-225

2. D.W Vahey: A non linear coupled wave theory of holographic storage in ferromaterials. J. Appl. Phys. 46, 1975, pp 3510-3515

3. D.Von der Linde and A.M Glass: Photorefractive effects for reversible holographic storage of information. Appl. Phys. 8, 1975, pp 85-100

4. D.L Staebler and J.J Annodei. J. Appl. Phys. 43, 1972, p 1042

5. D.L Staebler and W. Phillips. Appl. Opt. 13, 1974, p 788

6. J.O White and A. Yariv: Real time image processing via four-wave mixing in a photorefractive material. Appl. Phys. Lett. 37, 1980, pp 5-7

7. L. Pichon and J.P Huignard. Opt. Comm 36, 1981, p 277

8. Y. Fainman, C.C Guest and S.H Lee: Optical digital operation by two beam coupling in photorefractive material. Appl. Opt 25, 1986, pp 1598-1603

9. H. Rajbenbach: Digital optical processing with photorefractive materials : Applications of a parallel half-adder circuit to algorithmic state machines. J. Appl. Phys. 62, 1987, pp 4675-4681

10. J.P. Huignard, J.P. Herriau, P. Aubourg and E.Spitz: Phase conjugate wave front generation via real-time holography in BSO crystal. Opt. Lett. 4, 1979, pp 21-23

11. J. Feinberg: Phase conjugating mirror with continuous wave gain. Opt. Lett. 5, 1980, pp 519-521

12. M.D Levenson, K.M Johnson, V.C Hanchett, K. Chiang. J. Opt. Soc. Am 71 1981, p 737

13. J.P. Huignard, J.P. Herriau: Real time coherent edge reconstruction with BSO crystals. Appl. Opt 17, 1978, pp 2671-2672

14. J. Feinberg: Real time edge enhancement using the photorefractive effect. Opt. Lett. 5, 1980, pp 330-332

15. M. Cronin-Golomb and A. Yariv: Optical limiters using photorefractive nonlinearities. J. Appl. Phys. 57, 1985, pp 4906-4910.

16. A. Ford, Y. Fainman and S.H. Lee: Single beam interferometry using photorefractive fanout. Opt . Lett, 1988.

17. D.Z. Anderson, D.M. Lininger and J. Feinberg: Optical tracking novelty filter. Opt. Lett. 12, 1987, pp 123-125.

18. N.V. Kukhtarev, V.B. Markov and S.G. Odulov. Opt. Comm. 23, 1977, p 338.

19. A. Marrakchi and J.P. Huignard. Appl. Phys. 24, 1981, pp 24-131.

20. L.K. Lam, T.Y. Chang, J. Feinberg and R.W. Hellwarth. Opt. Lett. 6, 1981, p 475

21. P. Günter, J.P. Huignard: Photorefractive materials and their applications. Vol I and II Springer Verlag, 1981

22. G.C. Valley and M.B. Klein: Optimal properties of photorefractive materials for optical data processing. Opt. Eng. 22, 1983, pp 704-711

23. P.Gunter: Holography, coherent light amplification and optical phase conjugation with photorefractive materials. Phys. Rep. 93, 1982, pp 199-299

24. P. Gunter: Electro-Optic and Photorefractive Materials. Springer-Verlag 1987

25. G.C. Valley, M.B. Klein, R.A. Mullen, D. Rytz and B. Wechsler: Photorefractive materials. Ann. Rev. Mater. Sci 18, 1988

26. A. Ashkin, G.D Boyd, J.M Dziedzic, R.D Smith: Optically-induced refractive index inhomogeneities in LiNbO3 and LiTaO3. Appl. Phys. Lett .9, 1966, pp 72-74,

27. P.Yeh: Fundamental limit of the speed of photorefractive effect and its impact on device applications and material research: author's reply to comment. Appl. Opt. 26, 1987, pp 3190-3191

28. C.T Chen, D.M Kim and D.Von Linde: Efficient pulsed photorefractive process in LiNbO3: Fe for optical storage and deflection. IEEE J. Quant. Elec. 16, 1980, pp 126-129

29. L.K Lam, T.Y Chang, J.Feinberg and R.W Hellwarth: Photorefractive-index gratings formed by nanosecond optical pulses in BaTiO3. Opt. Lett. 6, 1981, pp 475-477

30. P Herman, J.P Herriau and J.P Huignard, Nanosecond four-wave mixing and holography in BSO crystals. Appl. Opt. 20, 1981, pp 2173-2175

31. A.L Smirl, G.C. Valley, R.A. Mullen, K. Bonhert, C.D MIre and T.F Boggess: Picosecond photorefractive effect in BaTiO3. Opt. Lett. 12, 1987, pp 501-503

32. G.C. Valley, A.L. Smirl, M.B. Klein, K. Bonhert and T.F. Boggess: Picosecond photorefractive beam coupling in GaAs. Opt. Lett. 11, 1986, pp 647-649

33. P.J. Van Heerden: Theory of optical information storage in solides. Appl. Opt. 2, 1963, pp 393-400

34. D. Psaltis, D. Brady, K. Wagner: Adaptative optical networks using photorefractive crystals. Appl. Opt. 27, 1988, pp 1752-1759

35. Y. Fainman, H. Rajbenbach, S.H. Lee: Application of photorefractive crystals as basic computational modules for optical computing. J. Opt. Soc. Am. B3, 1986, p 16

36. J.P. Huignard, J.P. Herriau and F. Micheron: Selective erasure and processing in volume holograms superimposed in photosensitive ferroelectrics. Ferroelectrics 11, 1976, pp 393-396

37. D. Van der Linde, A.M. Glass, K.F. Rodgers, Appl. Phys. Lett. 26, 1975, p 22

38. J.P Herriau, J.P Huignard: Hologram fixing process at room temperature in photorefractive BSO. Appl. Phys. Lett. 49, 1986, pp 1140-1142

39. L. Staebler, W.J. Burke, W. Phillips, J.J. Amodei: Multiple storage and erasure of fixed holograms in Fe-dopted LiNbO3. Appl. Phys. Lett. 26, 1975, pp182-184.

40. M. Cronin-Golomb, B. Fischer, J.O White and A. Yariv: Theory and applications of four-wave mixing in photorefractive media. IEE J. Quant. El. 20, 1984, pp 12-29.

41. A. Marrakchi, A.R. Tanguay, Jr, J. Yu and D. Psaltis: Physical characterization of the photorefractive incoherent to coherent optical converter. Opt. Eng. 24, 1985, pp 124-131

42. N.V. Kukhtarev, V.B. Markov, S.B. Odulov, M.S. Soskin and V.L. Vinestskii: Holographic storage in electro-optic crystals.I.steady state, and II. beam coupling light amplification. Ferroelectrics 22, 1979, pp 949-960 and pp 961-964

43. N.V. Kukhtarev and S. Odulov, Opt. Comm. 32, 1980, p 183

44. D.L. Staebler and W. Phillips: Fe-doped LiNbO3 for read-write applications. Appl. Opt. 13, 1974, pp 788-794

45. R.R. Shah, D.M. Kim, T.A. Rabson and F.K. Tittel: Characterization of iron-doped lithium niobate for holographic storage applications. J. Appl. Phys. 47, 1976, pp 5421-5431

46. D.L. Staebler and J.J. Amodei: Coupled-wave analysis of holographic storage in LiNbO3. J. Appl. Phys. 43, 1972, pp 1042-1049

47 H. Kurz: Photorefractive recording dynamics and multiple storage of volume holograms in photorefractive LiNbO3. Optica Acta 24, 1977, pp 463-473

48. C.T. Chen, D.M. Kim and D. Von Der Linde: Efficient pulsed photorefractive process in LiNbO3 : Fe for optical storage and deflection. IEEE , J. Quant, El.16, 1980, pp 126-1290

49. D. Von Der Linde, A.M. Glass and K.F. Rogers: Multiphoton photorefractive processes for optical storage in LiNbO3. Appl. Phys. Lett. 25, 1974, pp 155-157

50. P. Huignard and F. Micheron: High sensitivity read-write volume holographic storage in $Bi_{12}SiO_{20}$ and $Bi_{12}GeO_{20}$ crystals. J. Appl. Phys. 48, 1977, pp 3686-3690

51. M. Peltier and F. Micheron: Volume hologram recording and charge transfer process in $Bi_{12}SiO_{20}$ and $Bi_{12}GeO_{20}$. J. Appl. Phys. 48, 1977, pp 3683-3690

52. J.P. Hermann, J.P. Herriau and J.P. Huignard: Nanosecond four-wave mixing in BSO crystals. Appl. Opt. 20, 1981, pp 2173-2175

53. A. Marrakchi, R.V. Johnson and A.R. Tanguay. Jr: Polarization properties of photorefractive diffraction in electrooptic and optically active sillenide crystals (Bragg conditions). J. Opt. Soc. Am B3, 1986, pp 321-336

54. M.A. Powell and C.R. Petts: Temperature enhancement of the photorefractive sensitivity of BSO and BGO. Opt. Lett. 11, 1986, pp 36-38

55. R.A. Mullen and R.W. Hellwarth, Optical measurement of the photorefractive parameters of $Bi_{12}SiO_{20}$ crystals. J. Appl. Phys. 58, 1985, pp 40-44

56. P. Herriau and J.P. Huignard: Hologram fixing process at room temperature in photorefractive $Bi_{12}SiO_{20}$ crystals. Appl. Phys. Lett. 49, 1986, pp 1140-42

57. J.P. Huignard, J.P. Herriau and T. Valentin, Time-average holographic interferometry with photoconductive electrooptic $Bi_{12}SiO_{20}$ crystals. Appl. Opt. 16, 1977, pp 2796-2798

58. J.P. Herriau, D. Rojas, J.P. Huignard, J.M. Bassat and J.C. Launay: Highly efficient diffraction in photorefractive BSO-BGO crystals at large applied fields, Ferroelectrics 66, 1986

59. J.P. Huignard and J.P. Herriau: Real time coherent object edge reconstruction with $Bi_{12}SiO_{20}$ crystals. Appl. Opt. 17, 1978, pp 2671-2672

60. J.O. White and A. Yariv: Real time image processing via four-wave mixing in a photorefractive medium. Appl. Phys. Lett.37, 1980, pp 5-7

61. L. Pichon and J.P. Huignard: Dynamic joint-Fourier-transform correlator by Bragg diffraction in photorefractive $Bi_{12}SiO_{20}$ crystals. Opt. Comm. 36, 1981, pp 277-280

62. J.P. Huignard, J.P. Herriau, P. Aubourg and E. Spitz: Phase-conjugate wavefront generation via real-time holography in $Bi_{12}SiO_{20}$ crystals. Opt. Lett. 4, 1979, pp 21-23

63. G. Pauliat, J.P. Herriau, A. Delboulbe, G. Roosen and J.P. Huignard: Dynamic beam deflection using photorefractive gratings in $Bi_{12}SiO_{20}$ crystals. J. Opt. Soc. Am B3, 1986, pp 306-314

64. J.B. Thaxter: Electrical control of holographic storage in strontium-barium niobate. Appl. Phys. Lett. 15, 1969, pp 210-212

65. J.B. Thaxter and M. Kestigian: Unique properties on SBN and their use in a layered optical memory. App. Opt. 13, 1974, pp 913-924

66. K. Megumi, H. Kosuka, M. Kobayashi and Y. Furuhata: High sensitivity holographic storage in Ce-doped SBN. Appl. Phys. Lett.3, 1977, pp 631-633

67. G. Salamo, M.J. Miller, W.W. Clark III, G.L. Wood and E. Sharp: Strontium barium niobate as a self-pumped phase conjugator. Opt. Comm. 59, 1986, pp 417-422

68. R.R. Neurgaonkar, W.K. Cory, J.R. Oliver, M.D. Ewbank and E.F. Hall: Development and modification of photorefractive properties in the tungsten bronze family crystals. Opt. Eng 26, 1987, pp 392-405

69. M.D. Ewbank, R. Neurgaonkar and W.K. Cory: Photorefractive properties of strontium-barium niobate. J. Appl. Phys. 62, 1987,pp 374-380

70. G.L. Wood, W.W. Clark III, M.J. Miller, E.J. Sharp, G.P. Salamo and R.R Neurgaonkar: Broadband photorefractive properties and self-pumped phase conjugation in Ce-SBN : 60. IEEE J. Quant. El.23, 1987, pp 2126-2134

71. F. Micheron, C. Mayeux and J.C. Trotier: Electrical control in photoferroelectric materials for optical storage. Appl. Opt. 13, 1974, pp 784-787

72. R.L. Towsend and J.T. LaMacchia: Optically induced changes in BaTiO3

73. J.Feinberg, D. Heiman. A.R Tanguay, Jr and R.W Hellwarth. J. Appl. Phys. 51, 1980, p1297

74. Y.Fainman, E.Klancnik and S.H Lee: Optimal coherent image amplification by two wave coupling in photorefractive BaTiO3. Opt. Eng. 25, 1986, pp 228-234

75. F. Laeri, T. Tshudi and J. Abers: Coherent CW image amplifier and oscillator using two-wave interaction in a BaTiO3 crystal. Opt. Comm. 48, 1983, pp 247

76. J. Feinberg: Self-pumped, continuous-wave, phase conjugator using internal reflection. Opt. Lett. 7, 1982, pp 486-488

77. M. Cronin-Golomb, B. Fischer, J.O White and A. Yariv: Passive (self-pumped) phase conjugate mirror: Theoretical and experimental investigation. Appl. Phys. Lett. 41, 1982, pp 689-691

78. Ducharme and J. Feinberg: Altering the photorefractive properties of BaTiO3 by reduction and oxidation at 650°C. J. Opt. .Soc. Am .B3, 1986, pp 283-292

79. B. Klein: Physics of the photorefractive effect in BaTiO3,Electro-Optic and Photorefractive Materials. Ed.P.Gunter, Springer-Verlag, 1987, pp 266-282

81. D. Bize,J.E. Ford, T.Y. Taketomi, S.H. Lee: Effects of applied voltage on holographic storage in SBN : 60. SPIE Conf - San Diego - August 1988

82. J.E. Ford, Y. Taketomi, D. Bize and all: Multiplex holography in SBN: 60 with applied field. Sumitted to JOSA - A . special issue on Progress in Holography 9 /16 / 91

83. A. Krumins and P. Gunter: Diffraction efficiency and energytransfer during hologram formation in reduced KNbO3. Appl. Phys. 19, 1979, pp 153-163

84. P. Gunter and A. Krumins: High-sensitivity read-write volume holographic storage in reduced KNbO3 crystals. Appl. Phys. 23, 1980, pp 199-207

85. E.Volt and P. Gunter: Photorefractive spatial light modulation by anisotropic self-diffraction in KNbO3 crystals. Opt. Lett. 12, 1987, pp 769-771

86. P. Gunter: Electric-field dependence of phase-conjugate wave-front reflectivity in reduced KNbO3 and B12GeO20. Opt. Lett. 7, 1982, pp 10-12

87. J.P. Huignard, A. Marrakchi. Opt Com - 38 , 1981, p 249

88. G.C. Valley: Two-wave mixing with an applied field and a moving grating. J. Opt. Soc. Am B1, 1984, pp 868-873

89. H. Rajbenbach, J.P. Huignard, and B. Loiseaux. Opt. Comm. 48, 1983, p 247

90. Ph. Refregier, L. Solymar, H. Rajbenbach and J.P. Huignard: Two-beam coupling in photorefractive Bi12SiO20 crystals with moving grating: Theory and experiments. J. Appl. Phys. 58 ,1985, pp 45-57

91. J.Rodriguez, A.Siamakoun, M.J Miller, W.W.ClarkIII, G.L Wood, E.J. Sharp and R.R Neurgaonkar: BSKNN as a self-pumped phase conjugator. Appl. Opt. 26, 1987, pp 1732-1736

92. A.M. Glass, A.M. Johnson, D.H. Olson, W. Simpson and A.A. Ballman: Four-wave mixing in semi-insulating InP and GaAs using the photorefractive effect. Appl. Phys. Lett. 44, 1984, pp 948-950

93. G.C. Valley, A.L. Smirl, M.B. Klein, K. Bonhert and T.F. Boggess: Picosecond photorefractive beam coupling in GaAs. Opt. Lett.1 1, 1986, pp 647-649

94. A.L. Smirl, G.C. Valley, K. Bohnert and T.F. Boggess: Picosecond photorefractive and free-carrier transient energy transfer in GaAs at 1 µm. IEEE .J. Quant. Elec., Janv.1988

95. G.C. Valley and A.L. Smirl: Theory of transient energy transfer in gallium arsenide. IEEE.J. Quant. Elec., Janv 1988

96. J.M. Cohen-Jonathan, Ph. Rossignol, G. Roosen: Photorefractive grating build-up by a 28 ps light pulse in BSO. Journ. Phys., Suppl n° 6 Tome 49, 1988, p C2-267

97. J.C. Fabre, J.M. Cohen-Jonathan, G. Roosen: Photorefractive beam coupling in GaAs and InP generated by nanosecond light pulses. J. OptSoc. Am. B Vol. 5, 1988, p 1730

98. G. Pauliat, C. Besson, G. Roosen: Polarisation properties of two wave mixing under an alternating electric field in BSO crystals. IEEEJ. of Quantuum Elect., vol QE - 25 n° 7, 1989,p 1736

99. G. Pauliat, A. Villing, J.C. Launay , G. Roosen: Optical measurement of charge carrier mobilities in photorefractive sillenite crystals. Journ. Opt. Soc. Am. B, vol 7, 1990, p 1481

100. C. Besson, J.M.C. Jonathan, A. Villing, G. Pauliat, G. Roosen: Influence of alternating field frequency on enhanced photorefractive gain in two beam coupling. Opt. Lett., vol 14, n° 24, 1989, p 1359

101. G. Pauliat, G. Roosen: Theoretical and experimental study of diffraction in optically active and linearly birefringent sillenite crystals. Ferroelectrics, vol 75, 1987, p 281

10 Acousto-optic devices

Jari Turunen
Heriot-Watt University, Edinburgh, UK

Early theoretical and experimental investigations on the interactions between light and sound by Brillouin [1], Debye and Sears [2], Lucas and Biquard [3], Raman and Nath [4], and others [5], showed that it is possible to deflect and modulate beams of light using sound-induced permittivity variations in dielectric media. These control capabilities of light by sound proved valuable after the invention of the laser; during the 1960's, acousto-optics experienced a rapid technological expansion, becoming a well established field with a wide range of applications in modern optical engineering [6-8].

Acousto-optics offers the means to achieve various types of real-time reconfigurable optical interconnections, as will be discussed below. The emphasis will be on the use of volume acoustic waves (and bulk devices), relevant in free-space optical interconnection schemes. However, surface acousto-optics will also be covered briefly, because of its recent applications to guided-wave optical switching, and the potential of using the reflection of light from surface acoustic waves [9] in connection with free-space integrated optics [10].

10.1 Fundamentals of acousto-optics

In an idealized acousto-optic device illustrated in Fig. 10.1, a piezoelectric transducer converts a periodic electronic signal $V(t) = V(t+T)$ into a plane pressure wave that propagates in an isotropic, dielectric, and elasto-optic material at the speed of sound V, generating a progressive refractive index modulation structure

$$n(x, z, t) = n(Kx - \Omega t), \ 0 \le z \le L, \tag{10.1}$$

where $K = 2\pi/\Lambda$, $\Omega = 2\pi\nu = 2\pi V/\Lambda$ and $\Lambda = VT$. A unit amplitude optical plane wave with wavenumber $k = 2\pi/\lambda$ and wavelength λ in the interaction medium (refractive index n_0) is incident on the index-modulated region at an angle θ (angles inside the medium will be used throughout this chapter; angles outside are obtained using Snell's law). The acousto-optic cell then acts like a moving holographic phase grating: owing to the periodicity of $n(x,t)$ the incident beam is split into a set of diffraction orders propagating in the directions given by the grating equation

$$\sin\theta_m = \sin\theta + m\lambda/\Lambda. \qquad (10.2)$$

Since the optical frequency $\omega \gg \Omega$, the diffraction efficiencies I_m of the diffracted orders can be calculated using the theory of stationary index-modulated diffraction gratings (see chapter 4). The motion of the grating implies, however, that the frequencies ω_m of the diffracted orders are Doppler-shifted:

$$\omega_m = \omega + m\Omega. \qquad (10.3)$$

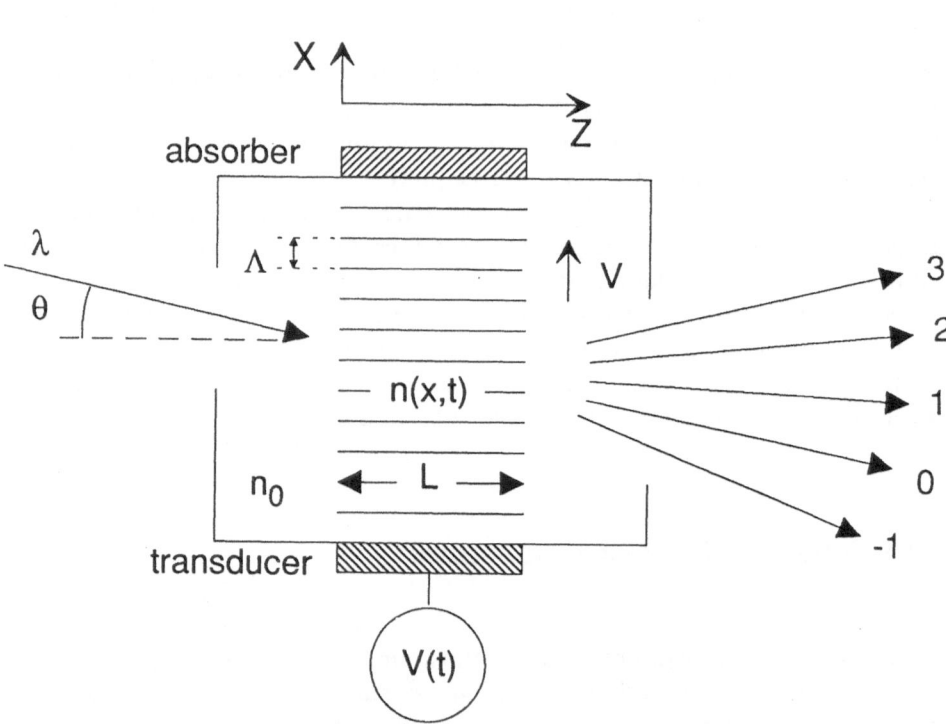

Fig. 10.1. Diffraction of a plane optical wave by a plane acoustic wave.

The diffraction orders are therefore mutually uncorrelated in the sense of a time average over at least a few periods T, and the output field is partially coherent (see [11] and the references cited therein).

The efficiencies I_m of the diffracted orders depend in a complicated manner on parameters such as the thickness L of the interaction region, the optical and acoustic wavelengths, the angle of incidence θ and the relative strengths of the harmonics of the index modulation profile. A wide variety of rigorous and approximate methods are available to solve the diffraction problem. These include normal mode, coupled wave, and thin grating decomposition techniques [8,12,13]; see also chapter 4. In general, numerical methods must be applied.

In case of a sinusoidal refractive index modulation of the form

$$n(x,t) = n_0 + \Delta n \sin(Kx - \Omega t) \tag{10.4}$$

it is customary to introduce two parameters, i.e. the Klein-Cook parameter $Q = K^2 L/k$ and the Raman-Nath parameter $v = kL\Delta n/n_0$, to characterize two fundamentally different interaction regions, where approximate analytic solutions are possible. Under conditions Q<<1 and Q/v<<1 (for more precise limits, see reference [14]), the grating may be considered optically thin, thus inflicting a pure phase change upon the incident wave: for small θ (up to $\approx 10°$),

$$I_m = |J_m(v)|^2, \tag{10.5}$$

where J_m is the m:th order Bessel function of the first kind. This is the so-called Raman-Nath regime. If Q>>1, Q/v>>1 and the angle of incidence θ is close to the Bragg angle $\theta_B = \arcsin(\lambda/2\Lambda)$, the grating is optically thick and only two orders (first and zeroth) have significant intensities: this is the Bragg regime. The well-known two-wave coupled wave theory gives, for θ and θ_B up to $\approx 10°$,

$$I_1 = 1 - I_0 = (v/v')^2 \sin^2(v'/2), \tag{10.6}$$

where $v'^2 = v^2 + (KL\Delta\theta)^2$ and $\Delta\theta = \theta - \theta_B$. For exact Bragg incidence $\theta = \theta_B$, equation (10.6) reduces to $I_1 = \sin^2(v/2)$, i.e. a first order diffraction efficiency of 100% is achieved with $v = \pi$.

10.2 Acousto-optic Bragg deflectors

Acousto-optic devices working in the Bragg domain are of greatest technological importance. These include the Bragg cell frequency shifter that up- or downshifts the frequency of the incident laser beam according to equation

(10.3), and the acousto-optic modulator that allows electronic control of the first-order intensity $I_1 = \sin^2(v/2)$ by modulation of the Raman-Nath parameter v through the index-modulation amplitude Δn.

For optical interconnect applications, the most interesting device is the Bragg deflector. This is a component that facilitates, for a fixed input beam angle θ, the control of the first-order output beam direction by varying the grating period Λ. The deflector is typically operated either in the linear frequency modulation mode or in the random access mode, i.e. the acoustic frequency v is either swept linearly over a given bandwidth Δv, or certain output directions within this bandwidth are addressed sequentially. In the random-access mode, the time required to reconfigure the output beam direction (the access time) is

$$\tau = D/V. \tag{10.7}$$

Here D is the effective diameter of the optical beam, e.g. the width of the Bragg cell aperture in case of uniform illumination, or the $1/e^2$ full width of a non-truncated Gaussian incident beam.

The number of resolvable output beam directions for a given angular range $\Delta\Phi = \Delta v\lambda/V$ and optical beam divergence $\Delta\phi = \gamma\lambda/D$ is

$$R = \Delta\Phi/\Delta\phi = \Delta vD/V\gamma = \tau\Delta v/\gamma. \tag{10.8}$$

The number $\tau\Delta v$ is commonly called the time-bandwidth product of the deflector. The value of γ depends on the incident field distribution and the required resolution criterion: $\gamma = 1$ for a uniformly illuminated rectangular aperture if the Rayleigh resolution criterion is used (appropriate e.g. in scanning applications). For a discussion of resolution with truncated Gaussian beams, see reference [15]. In the linear frequency modulation mode commonly used in acousto-optic scanning of laser beams, the resolution is actually somewhat lower than indicated by equation (10.8) and cylindrical lensing effects appear [15]. This mode of operation is, however, of limited interest in the present context.

Evidently, to maximize R, one can maximize either the bandwidth Δv or the deflector aperture D, or use a material with a slow acoustic speed V. Both D and V are limited by considerations such as acoustic attenuation, material properties, and the dimensions of the final system, so Δv is the parameter to be optimized (see [6,16] for a more detailed discussion of deflector design). The angular bandwidth $\Delta\Phi$ of an acousto-optic Bragg cell (isotropic material) is proportional to the diffractive spread of the acoustic beam: at any acoustic frequency v, only one of the acoustic plane wave components in the angular spectrum of the sound field satisfies the Bragg condition exactly. The bandwidth Δv is then of the order of v_C/Q, where v_C is the center frequency [7,8].

The fractional bandwidth $\Delta v/v_C$ may be increased by transducer design. An obvious approach is to use an acoustic wave field with a wide angular spectrum of plane waves [17]. This, however, wastes acoustic power since the optical field is diffracted by only a small portion of the acoustic plane waves. More efficient schemes may be devised using phased array transducers that steer the slant angle of the acoustic wavefront as a function of Λ [17-19]. Now the Bragg condition is satisfied over an increased range Δv, and a significant portion of the acoustic energy is available for deflection at any frequency v inside the bandwidth.

In an optically anisotropic medium, the refractive indices for the incident and the diffracted wave may differ as light is coupled between two optical modes in the scattering process. Therefore, significant modifications to the Bragg condition can occur and a wide bandwidth can be achieved without the use of beam steering. Assume that both the acoustic and optical beams propagate in a plane normal to the optic axis of a uniaxial crystal, and that the refractive indices associated with the ordinary (diffracted) and the extraordinary (incident) wave are n_O and n_e, respectively. Then, using the wave vector diagram of Fig. 11.2a, the conservation of momentum condition $\mathbf{k} + \mathbf{k_1} = \mathbf{K}$ gives the expressions

$$\sin\theta = \frac{\lambda_o}{2n_e\Lambda}\left[1 + (n_e^2 - n_o^2)\Lambda^2 / \lambda_o^2\right] \qquad (10.9)$$

and

$$\sin\theta_1 = \frac{\lambda_o}{2n_o\Lambda}\left[1 - (n_e^2 - n_o^2)\Lambda^2 / \lambda_o^2\right] \qquad (10.10)$$

for the angles θ and θ_1 of the incident and the first-order diffracted beams, respectively [20]. In equations (10.9) and (10.10), λ_o is the optical wavelength if vacuum. These expressions may be viewed as an extension of the conventional Bragg law, to which they reduce if $n_e = n_o$.

The optimal center frequency v_C for wideband scanning applications is that for which the variation of the angle of incidence as a function of the acoustic frequency is minimized [21]:

$$v_C = (n_e^2 - n_o^2)^{1/2} V / \lambda_o. \qquad (10.11)$$

As shown in Fig. 10.2b, the diffraction angle θ_1 is zero. and the wave vectors of the acoustic wave and the diffracted optical wave are perpendicular at v_C. As a result, an increased range of diffraction angles are accessible with a fixed input beam before excessive Bragg mismatch takes place (compare with Fig. 10.2c, which describes isotropic diffraction). In fact, the fractional bandwidth $\Delta v / v_C$ is now of the order of $Q^{-1/2}$, i.e. an increase of $\sqrt{Q}$ is obtained compared to the isotropic case.

The first deflection experiments using anisotropic materials were performed using birefringence in quartz and sapphire [20,22]. The best material for relatively low-frequency deflection is, however, TeO_2 operated in the slow shear mode with $V = 617$ m/s. The relatively small index difference in TeO_2 that arises from optical activity implies that the optimal center frequency given by equation (10.11) is conveniently below 100 MHz for visible and near-infrared wavelengths. Large TeO_2 crystals (up to approximately 50 mm aperture) with good optical quality and acceptable acoustic absorption characteristics can be grown, and deflectors with time-bandwidth products well in excess of 1000 and above 50% Bragg efficiency are commercially available for visible and near-infrared wavelengths. The access time of these devices is typically on the range 1-50 μm.

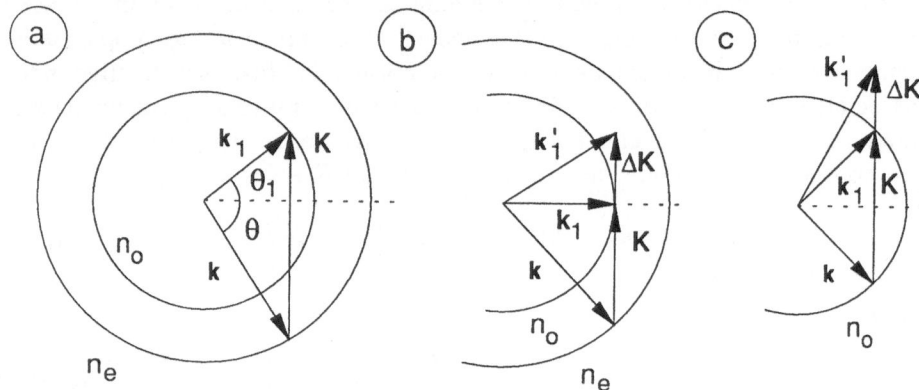

Fig. 10.2. (a) Wave vector diagram illustrating the generalized Bragg condition in an anisotropic medium. A deviation ΔK of an acoustic wave number from a center value K causes a smaller mismatch $\Delta k_1 = k'_1 - k_1$ from the Bragg condition in the optimized anisotropic case (b) than in the isotropic case (c).

Other commonly used acousto-optic materials are gallium phosphide (for 100-1000 MHz frequency range), lithium niobate (for GHz frequencies), and lead molybdate. See references [6,7,23,24] for more detailed discussion on the characteristics and the selection of acousto-optic materials.

10.3 Interconnect capabilities of bulk devices

A Bragg deflector operating in the random access mode can obviously act as a reconfigurable optical interconnect (or switch) between one input and one output selected from M candidates. If these M potential receivers are arranged in a linear array with a compression ratio

$$C = \frac{\text{size of the receiver window}}{\text{separation between two adjacent receivers}} \qquad (10.12)$$

a deflector with resolution R can switch between $M = R/C$ output channels. In interconnect applications, the resolution criteria are tighter than in scanning devices: for a uniformly illuminated aperture, it is appropriate to choose $\gamma = 2$ in equation (10.8) (and to increase γ similarly for truncated Gaussian beams) for a meaningful definition of a compression ratio of unity.

The standard deflector can broadcast the information from a single input to several selected receivers in the output array (see Fig. 10.3) if the frequencies corresponding to these output directions are simultaneously present in a Bragg cell with sufficient bandwidth. This multifrequency mode of operation, also encountered in acousto-optic spectrum analysis [6], is fundamentally different from the usual frequency (and amplitude) modulated scanning mode, which can produce apparently similar patterns. However, in the scanning mode, the desired outputs are visited sequentially, and continuous broadcasting is therefore not possible. The reconfiguration time in the multifrequency mode is the same as the access time in the random-access mode, i.e. given by equation (10.7).

If the M potential receivers are arranged in a regular array, the drive signal $V(t)$ in the multifrequency Bragg mode must be of the form

$$V(t) = A(t) \cos[2\pi v_C t + \phi(t)], \qquad (10.13)$$

where the functions $A(t)$ and $\phi(t)$ are periodic, with period T_0. Then the deflector behaves operationally as a hybrid hologram [25], i.e. the high-frequency carrier deflects the wavefront like a Bragg grating, while the low-frequency part inflicts amplitude and phase modulation upon the diffracted wavefront. It therefore acts like a thin (Raman-Nath regime) hologram with an amplitude transmission function

$$t(x) = A(x)\exp[i\phi(x)] = \sum_{p=-\infty}^{\infty} U_p \exp(i2\pi px / \Lambda_0), \qquad (10.14)$$

where the spatial period $\Lambda_0 = vT_0$, and the coefficients

$$U_p = \Lambda_o^{-1} \int_0^{\Lambda_o} t(x)\exp(-i2\pi px / \Lambda_o)dx \qquad (10.15)$$

represent the amplitudes of the diffraction orders of the thin grating. To achieve a compression ratio C, one must choose $\Lambda_0 < D/C$, where D is the effective input beam width.

The broadcast mode can be achieved simply by driving the deflector with a synthetic signal that is a superposition of the desired frequencies. However, cross correlations and intermodulation orders appear [26], which can be minimized by synthesizing the Raman-Nath hologram using one of the optimization techniques of computer-generated holograms covered in chapter 5. This approach was first used to demonstrate a programmable array illuminator with a binary drive signal [27]. Arbitrary weighted patterns have been generated with multilevel drive signals [28], and the method has been further extended using continuous drive signals [29,30]. Fan-out of two or three can be achieved in the Raman-Nath regime using the simple sinusoidal index modulation of equation (10.4), as discussed in reference [31]. In general, the hybrid technique that can utilize wideband anisotropic Bragg diffraction is far superior to the pure Raman-Nath regime approach: the latter is highly sensitive to acoustic diffraction and absorption, and requires more acoustic power to achieve the full phase modulation range $0 \le \phi \le 2\pi$.

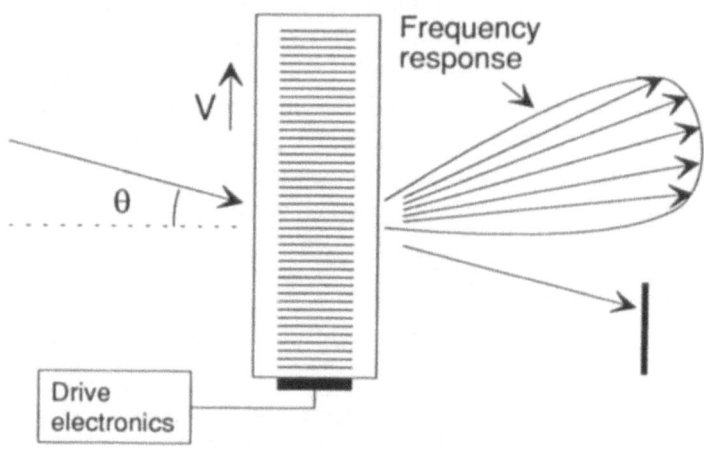

Fig. 10.3. Use of a wideband acousto-optic deflector as a space-invariant programmable fan-out element.

The reconfigurable acousto-optic on to M arbitrary spot array generators discussed above can also be used as programmable Fourier plane filters to perform space-invariant interconnections between two linear arrays of sources and detectors as illustrated in Fig. 10.4: light from each source is broadcast onto a specified set of the neighbors of the image receiver. One of the simplest operations is to broadcast from each of the M_1 sources in the input array to all of the M_2 receivers (M_1 to M_2 star coupling, see [32]). Alternatively, the two nearest neighbors of the image receiver, or a predefined central block of such neighbors, may be addressed. Need for the latter arises e.g. in morphological image processing, where these synthetic acousto-optics holograms have been used [30]. Any of the operations described above could, of course, be performed using a stationary synthetic hologram, but the acousto-optic approach facilitates the reconfigurability of the pattern within the access time τ.

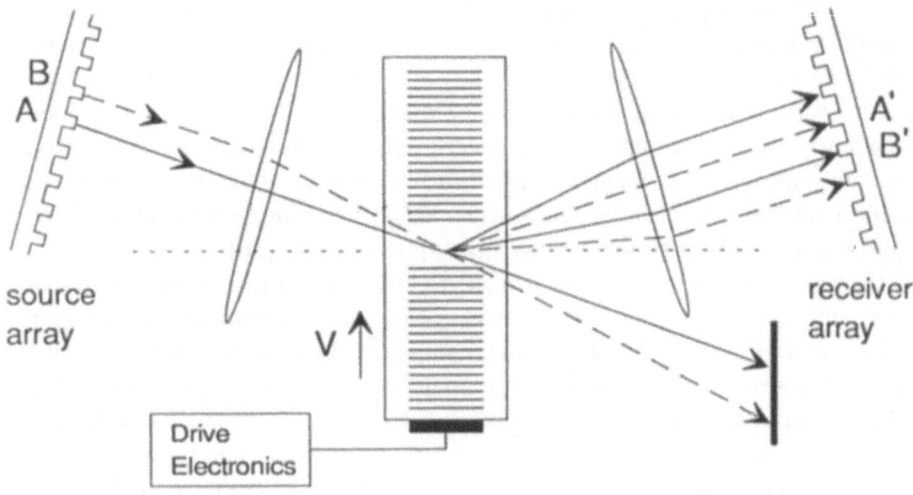

Fig. 10.4. Use of a wideband acousto-optic deflector as a space-invariant Fourier-plane optical interconnect: connection of transmitters A and B to the nearest neighbors of the image reveivers A' and B' are shown.

A rather unique feature of the acousto-optic realization of synthetic holograms is that transmission functions containing both amplitude and phase variations can be generated by an appropriate signal V(t); the amplitude freedom can be utilized to remove the undesired higher diffraction orders of a phase-only hologram. This has been demonstrated [29,30], and it can be significant in M to M -type space-invariant operations because the cross-talk due to higher orders is removed. The required grating profiles are most conveniently designed using the technique of reference [33]: the relative phases of the diffracted beams are optimized, and thereby the amplitude fluctuations in the grating profile are minimized, leading to high diffraction efficiency.

In space-invariant optical interconnect applications of the type described above the sources are often mutually coherent, being produced originally from a single source using an optical array illuminator [34]). In such circumstances the interference that occurs when beams from two or more sources are recombined may cause problems. The frequency degeneracy and therefore also the mutual coherence of these beams is lifted by the acousto-optic device [see equation

230

(10.3)], and the interference effects disappear if the receivers time-average over at least a few periods T_0 of the signal grating.

The capability of an acousto-optic deflector to perform space-invariant many to many interconnections is limited by the dependence of the bandwidth on the angle of incidence: the standard wideband deflectors are not designed for wide input angle range and optimal geometries for this purpose remain to be found. The tight input angle acceptance may, however, be utilized to realize space-variant operations in a Bragg cell. Two or more input beams can be scanned independently over reasonably wide (overlapping) angular ranges if acoustic waves with appropriate propagation directions are simultaneously present in the interaction medium. Such acoustic waves can be generated either using tilted transducer arrays [35] or suitable beam steering techniques [36].

So far, the discussion has been limited to linear one-dimensional arrays of sources and receivers. Two-dimensional devices capable of performing operations that are separable in two dimensions can be constructed by crossing two one-dimensional devices. In isotropic materials, this only requires mounting transducers on two facets of the interaction medium. In case of anisotropic devices, it is necessary to cross two Bragg cells physically. The operations that such devices can perform include connecting one source to a selected receiver in a two-dimensional array, and generation of separable spot arrays, e.g. two by two nearest neighbor interconnects and MxM structuring elements for morphological processing [30].

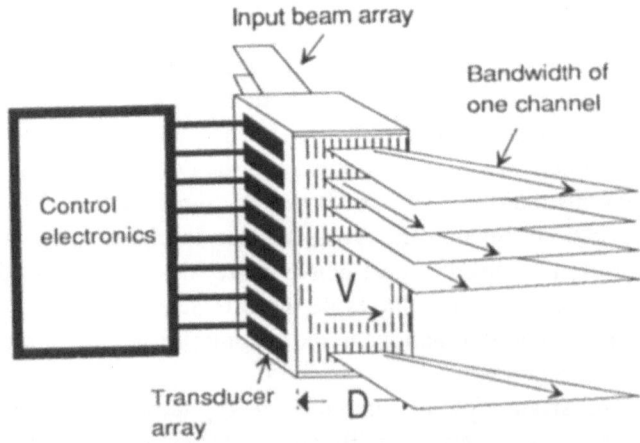

Fig. 10.5. A multichannel acousto-optic deflector.

More arbitrary interconnect operations are achieved using multichannel acousto-optic deflectors illustrated in Fig. 10.5. An array of transducers is mounted on one side of the interaction medium to launch several acoustic waves inside the cell. Each channel can, at least in principle, perform any of the one-dimensional operations described above. Various types of optical systems can be used in conjunction with multichannel devices to realize space-invariant, space-variant, and mixed interconnection systems. For example, fully arbitrary, reconfigurable, two-dimensional array generation is possible [27]. Acousto-optic crossbar switches have also been suggested and demonstrated [37-41]. These devices utilize the capability of each channel to connect one source in an input array into one (or more if broadcasting is required) receiver in the output array. The same task can be performed using the space-variant interconnect capabilities of a single Bragg cell [35,36]. This suggests that, with multichannel devices, many such switches could be operated in parallel.

Multichannel acousto-optic deflectors with up to 64 channels have been demonstrated [42,43]. Slow shear mode and longitudinal mode TeO_2 deflectors, and GaP deflectors with 32 channels, time-bandwidth products on the range 100-1000, efficiencies of 20-70%, and center to center channel spacings of 0.5-2.5 mm are commercially available.

10.4 Surface acoustic wave devices

With the development of guided-wave integrated optics, the role of acousto-optic devices based on surface acoustic waves has grown in importance. Surface acoustic waves propagate in a thin surface layer of the substrate, causing both surface ripples and an exponentially decaying index modulation structure [44]. Such waves can be efficiently launched on piezoelectric materials such as quartz or lithium niobate using interdigital transducers [45,46] as illustrated in Fig. 10.6.

Many of the acousto-optic interconnection geometries discussed above, being two-dimensional in nature, may be directly applied to guided-wave optics. The realization of complicated transducer geometries and multiple transducer arrays required in space-variant operations is greatly simplified as the transducers can be patterned directly on the substrate using standard microlithography. In fact, some of the examples on interconnection capabilities of acousto-optic devices were first demonstrated in guided-wave optics [35].

Surface acoustic waves can also be used to deflect unguided beams that pass through or are reflected by them. Owing to the thin interaction region, the surface acoustic waves act as Raman-Nath regime gratings [44]. Their use for the realization of wide bandwidth deflectors was investigated in the 1970's [9,47], and the most effective interaction geometry was found to be the one in which light incident from inside the substrate and reflected back by the acoustic waves as illustrated in Fig. 10.6b. The reason why this mode of operation of surface acoustic wave devices is mentioned here is its obvious compatibility with the free-space integrated optics scheme of Jahns and Huang [10], in which the light bounces back and forth inside a dielectric substrate, being controlled by optical components patterned on one side of the substrate. First-order diffraction efficiencies up to a few per cent have been demonstrated using this method. These values are substantially lower than those achieved using Bragg devices, which is why the interest in this type of acousto-optic scanner has diminished. However, the efficiencies compare favorably with those of most other types of reconfigurable interconnect devices, and this scheme could prove useful especially if the interconnect architecture allows one to utilize elsewhere the undiffracted portion of the input beam.

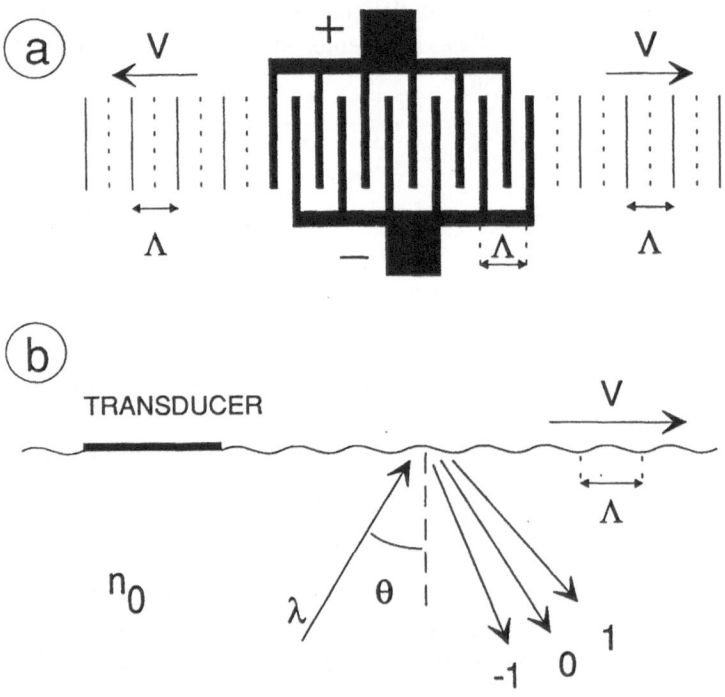

Fig. 10.6. (a) A surface acoustic wave generated by an interdigital transducer. **(b)** The geometry for most efficient deflection of a free-space optical wave by a surface acoustic wave.

10.5 Conclusions

Acousto-optics has much to offer in the field of reconfigurable optical interconnection and photonic switching, not least because it is a mature technology and the basic components are commercially available. However, the needs in optical interconnection do not always coincide with the requirements of scanning devices: in particular, a wide input angle acceptance required in space-invariant (Fourier-plane) interconnection is not a feature usually considered in the deflector design. On the other hand, some schemes, such as the one given in reference [47]), which are not particularly attractive in scanning applications may prove to be so in optical interconnection.

References

1. L. Brillouin: Ann. Phys. 17, 88 (1922).
2. P. Debye, F. W. Sears: Proc. Nat. Acad. Sci. 18, 409 (1932).
3. R. Lucas, P. Biquard: J. Phys. 71, 464 (1932).
4. C. V. Raman, N. S. N. Nath: Proc. Ind. Acad. Sci. 2, 406 (1935); ibid. 2, 413 (1935); 3, 75 (1936); 3, 119 (1936); 3, 459 (1936).
5. A. Korpel, Ed.: Selected papers on Acousto-Optics SPIE Milestone Series MS 16 (1990).
6. N. J. Berg, J. N. Lee, Eds.: Acousto-Optic Signal Processing (Marcel Dekker, New York, 1983).
7. M. Gottlieb, C. L. M. Ireland, J. M. Ley: Electro-Optic and Acousto-Optic Scanning and Deflection (Marcel Dekker, New York, 1983).
8. A. Korpel: Acousto-Optics (Marcel Dekker, New York, 1988).
9. A. Alippi, A. Palma, L. Palmieri, G. Socino, E. Verona: Opt. Acta 27, 1061 (1980).
10. J. Jahns, A. Huang: Appl. Opt. 28, 1602 (1989).
11. Y. Ohtsuka: J. Opt. Soc. Am. A 3, 1247 (1986).
12. M. Born, E. Wolf: Principles of Optics, Sixth Edition (Pergamon, London, 1980), Chapter XII.
13. T. K. Gaylord, M. G. Moharam: Proc. IEEE 73, 894 (1985).
14. W. R. Klein, B. D. Cook: IEEE Trans. Son. Ultras. SU-14, 123 (1967).
15. L. Dickson: Appl. Opt. 11, 2196 (1972).
16. E. K. Sittig: in Progress in Optics X, E. Wolf, Ed. (North-Holland, Amsterdam, 1972), p. 229.
17. E. I. Gordon: Proc. IEEE 54, 1391 (1966).
18. A. Korpel, R. Adler, P. Desmares, W. Watson: Proc. IEEE, 54, 1429 (1966).

19. G. A. Coquin, J. P. Griffin, L. K. Anderson: Proc. Trans. Son. Ultras. SU-17, 34 (1970).
20. R. W. Dixon: IEEE J. Quant. Electr. QE-3, 85 (1967).
21. A. W. Warner, D. L. White, W. A. Bonner: J. Appl. Phys. 43, 4489 (1972).
22. E. G. H. Lean, C. F. Quate, H. J. Shaw: Appl. Phys. Lett. 10, 48 (1967).
23. N. Uchida, N. Niizeki: Proc. IEEE 61, 1073 (1973).
24. I. C. Chang: Opt. Eng. 24, 132 (1985).
25. H. Bartelt, S. K. Case: Appl. Opt. 21, 2886 (1982).
26. D. L. Hecht: IEEE Trans. Son. Ultrason. SU-24, 7 (1977).
27. J. Turunen, E. Tervonen, A. T. Friberg: J. Appl. Phys. 67, 49 (1990).
28. E. Tervonen, J. Turunen, A. T. Friberg, J. Westerholm, M. R. Taghizadeh: Opt. Lett. 16, 1274 (1991).
29. D. W. Prather, J. N. Mait: Opt. Lett. 16, 1720 (1991).
30. R. A. Athale, J. N. Mait, D. W. Prather: Opt. Commun. 89, 99 (1992).
31. Y. Ohtsuka, Y. Arima, Y. Imai: Appl. Opt. 24, 2813 (1985).
32. U. Killat, G. Rabe, W. Rave: Fib. Integr. Opt. 4, 159 (1982).
33. H. P. Herzig, D. Prongue, R. Dandliker: Jpn. J. Appl. Phys. 29, L1307 (1990).
34. N. Streibl: J. Mod. Opt. 36, 1559 (1989).
35. C. S. Tsai, P. Le: in Photonic Switching (Optical Society of America, Washington, DC, 1991), p. 67.
36. K. Wagner, R. Weverka, A. Mickelson, K. Wu, C. Garvin: in Photonic Switching (Optical Society of America, Washington, DC, 1991), p. 67.
37. A. VanderLugt: Proc. SPIE 651, 51 (1986).
38. P. C. Huang, W. E. Stephens, T. C. Banwell, L. A. Reith: Electr. Lett. 25, 252 (1989).
39. W. E. Stephens, P. C. Huang, T. C. Banwell, L. A. Reith, S. S. Chang: Opt. Eng. 29, 183 (1990).
40. D. O. Harris, A. VanderLugt: Opt. Lett. 14, 1177 (1989).
41. D. O. Harris: Appl. Opt. 30, 4245 (1991).
42. M. Amano, E. Roos: Proc. SPIE 753, 37 (1987).
43. D. R. Pape: in Optical Computing, (Optical Society of America, Washington, DC, 1991) p. 185.
44. E. G. Lean: in Progress in Optics XI, E. Wolf, ed. (North-Holland, Amsterdam, 1973) p. 123.
45. S. Datta: Surface Acoustic Wave Devices (Prentice-Hall, Englewood Cliffs, 1986)
46. C. S. Tsai, Ed.: Guided-Wave Acousto-Optics. Interactions, Devices and Applications (Springer, Berlin, 1990).
47. A. Alippi, A. Palma, L. Palmieri, G. Socino: Appl. Opt. 15, 2400 (1976).

Second section: Interconnection schemes and systems

Part 2.1 Parallel schemes

11 Density of parallel optical interconnects

Norbert Streibl
Physikalisches Institut der Universität Erlangen-Nürnberg

The fundamental limits for the spatial density of parallel optoelectronic interconnections will be identified and discussed: In the beginning (Sect. 11.1) some basic concepts are introduced. Interconnections can be classified according to their *topology* (Sect. 11.2) as point-to-point- or multipoint-interconnections. Optoelectronic interconnections will support different degrees of *parallelism* (Sect. 11.3) depending on the question whether they are two- or threedimensional interconnections. For transporting the optical signals between the electronic terminals either waveguides, such as fibre bundles, or, alternatively, free space optics, based on imaging systems (Sect. 11.4) may be employed. This choice determines largely the possible interconnection density because it depends on the *randomness* or the regularity of the required "wiring". To calculate the achievable spatial density of optoelectronic interconnections the number of degrees of freedom that is proportional to the number of independent information channels of the optical systems has to be investigated. It is limited by *crosstalk* (Sect. 11.5) between adjacent channels that can be tolerated. Regular interconnects (as opposed to random interconnects) allow space sharing (Sect. 11.6) as is also pointed out in chapter 2, that is, a high packing density becomes possible if several information channels are transported by a common optical system. In Chap. 13 a different approach is discussed, namely thick elements, which may lead to different constraints. Another important property of an interconnect is *fanout*, i. e. beam splitting, and *fanin*, i. e. beam combination, for multipoint connections: it allows broadcasting functions, such as clock distribution, and the implementation of bus systems. On the other hand it costs energy and/or spatial density (Sect. 11.7). Practical issues include cooling and *heat dissipation* (Sect. 11.8) and the *alignment tolerances* (Sect. 11.9) required for the packaging of the optical and optoelectronical components. At the end, in the summary (Sect. 11.10), it will be attempted to put all issues together and to come up with realistic estimations of spatial interconnect densities for different system concepts.

11.1 Introduction

Fibre optical interconnects over long distances are today in wide use, for example in the field of telecommunications: fibre optics can transmit data over longer distances without a repeater than electronic communications using copper wires. Glass fibres are cheaper in comparison and they offer virtually unlimited temporal bandwidth. The overall data rate in fibre communications is either limited by the bandwidth of the optoelectronic devices (lasers and detectors) - currently in excess of 10 GBit/s - or by practical limitations in wavelength multiplexing. This bandwidth is much larger than in electronic long distance communications using copper cables. The extremely high optical carrier frequency of some 10^{14} Hz still allows for further increase in bandwidth of fibre communications in theory. In addition to the increased bandwidth fibre links are also very immune to electromagnetic interference because the optical signals can be well confined; crosstalk between adjacent lines is no problem. Finally, the transmitter and receiver are "optically isolated" from each other, the links are separating their electrical potentials. Hence, long distance optical communications has already shown its superiority in bandwith and in cost. Long range interconnects have been optimized with respect to losses (repeater distance), material dispersion and modal dispersion (for example: dispersion shifted single mode fibres).

Is there any motivation for using optics also for communications over short distances? Shortrange optoelectronic interconnections not only between but moreover within data processing systems are attractive for a variety of reasons [8,10,16,1,25]. Optics handles high data rates, eliminates ground loop problems between modules and there is no crosstalk or electromagnetic interference along the line - albeit there might be some amount of it at the optoelectronic terminals. One major advantage is the potential for a large number of parallel channels, that can be handled by optical systems. If twodimensional arrays of light-transmitters and receivers are employed, optical communication through the threedimensional space *above* the arrays is possible; this is topologically very different from electronical connections suffering from quasi-planar constraints in twodimensional space and on multilayer boards. Modern pin-grid chip packages offer several hundred electronic connections; limitations include topology (bonding pads are situated only along the edge of the chip with conventional technology), area requirements (size of the bonding pad and the necessary driver electronics) and heat dissipated for communication off-chip (large currents in ECL-technology). These figures can be improved by employing parallel one- or twodimensional arrays of optoelectronic devices on-chip ("optical pins") or very near to a chip in a hybrid package. Optoelectronic devices can be fabricated small and therefore with high impedances. Thus, they allow communication with a lower energy dissipation than low impedance all-electronical systems for basic quantum mechanical reasons [21]. The required optoelectronic devices with low heat dissipation are in some cases commercially available or at least in different states of research and development (lasers with low threshold, quantum well light modulators, detectors and monolithically integrated arrays of such devices with driver and processing electronics, so called "smart pixels").

For short range interconnections completely different design issues arise than in long distance communications: extinction losses in the waveguide material become quite irrelevant because connections are short. Rather, there is the demand to implement highly parallel connections. This requires high spatial density leading to coupling and packaging problems. Therefore, much effort is needed regarding the design and development of monolithic optoelectronic ICs, optoelectronic hybrid setups and the packaging of such devices together with optics. In particular, the heat dissipation of optoelectronic elements and their cooling will become a major consideration. Hence, for short range optical interconnects the main cost factors will be the optoelectronic terminals, their packaging and the achievement of high interconnection densities. In contrast, in long range communications the main cost is in transporting the optical signals over the distance.

Different areas of applications require different degrees of parallelism: A bus system in a computer is a typical example for a board-to-board-connection and it contains on the order of 100 parallel lines. For chip-to-chip-connections either few channels (with extremely high data rates each) or communication with a high degree of parallelism on the order of 1.000 or more channels are required to compete with electronics. Applications involving only few parallel channels will not be discussed in this investigation, since they can well be handled by using individual optical interconnections with discrete fibres, fibre connectors, etc.

Highly parallel systems require novel optoelectronical and microoptical components: It is assumed, that one- or twodimensional arrays of optoelectronical devices are to be interconnected. In an array constructed with hybrid technology there might be a pitch (i.e. distance between adjacent optoelectronical terminals) on the order of 1 mm or less. In an array that is monolithically integrated on one single chip the pitch may be on the order of 100 μm or less. Hence, the optical system must provide communications with an interconnect-density between 10^2 - 10^4 cm^{-2} and with several hundred or even thousand parallel channels. Given high data rates per channel, low crosstalk, galvanic isolation and a rather low waste of energy per transmission line optoelectronic interconnections with such densities are very attractive for high performace systems: telephone switches [4,11,19,20] or high performance computers, such as signal processors or multiprocessors. The high spatial density of optics may be helpful in processing architectures that are especially well suited for optical implementations, such as cellular automata, neural nets, multichannel correlators and certain digital array processors [14,3,12,29,26]. Indeed, a lot of fundamental research on optoelectronic and all-optical computing was motivated by the anticipation of very high interconnection densities made available by optics. This area is currently impeded, however, by slow progress on low-energy all-optical logic devices.

11.2 Topology

Data links can be classified into different categories: a single telephone line, for example, provides a point-to-point connection between one talker and one listener. In parallel systems often there are bundles of point-to-point connections that can be realized by employing bundles of fibres or waveguides.

A somewhat more complicated class of point-to-point interconnections are permutation elements: there is still one link between pairs of two terminals, but in the parallel system the geometrical arrangement of the channels at the input and at the output end is permuted. This "data shuffling" introduced by such interconnection patterns finds applications [4,17] in parallel sorting and searching algorithms and in rapid transformations such as the Fast Fourier Transformation. Consequently these techniques are often implemented in hardware at the heart of switching and routing networks in telecommunications or within multiprocessors.

Often several listeners have to be provided with the same message: broadcast systems (one talker and many listeners) are a class of multipoint connections. They rely on "fanout", that is splitting information into several channels at the talker. The simplest application of this kind is the distribution of clock signals.

In computer systems there is also often a need, that many potential talkers communicate with just one listener. For example, this is the case for an interupt line, where peripherals can tell the CPU that some special event needs to be handled. Such events happen seldom and it is therefore economic to share one single channel between many participants. Hence, one needs a "fanin" capability, that is joining of information from several channels at one listener.

The bus interconnection provides the combination of fanout and fanin: many potential talkers and listeners are sharing the same communication line. Either special hardware or a software protocol are used for arbitration, that is for avoiding collisions of messages. Typical applications are the bus between computer peripherals and the CPU. A bus system relies on time multiplexing of an interconnection. It is most attractive in case ressources (number of communication links) are limited and if communication requests occur only from time to time.

Finally, reconfigurable interconnects are of interest. An example is the telephone system - the biggest machine that man has ever built. In many practical cases reconfigurable interconnects between many participants are constructed as multistage switches [4,11,19,20 and Chapt. 14]; they contain small "atomic" information routing switches (such as crossbars) and fixed interconnects, such as various types of permutation elements [5], bus systems etc. inbetween. Hence, the analysis of interconnect density for the optical implementation of reconfigurable interconnects can be traced back to the analysis for the above mentioned fixed interconnects.

Another important issue is whether a line is unidirectional or bidirectional. In optics the communication channels are bidirectional (the propagation of light can be reversed), but the optoelectronic terminals are usually unidirectional (detectors do not emit light). Therefore multiplexing and coupling schemes are required for bidirectional optoelectronic interconnections, or, alternatively, the number of lines is simply doubled. Table 11.1 summarizes the classification of interconnects according to topology.

Table 11.1 Comparison of different topologies for interconnect systems.

Type	number of talkers	number of listeners	typical example	schematic topology
point-to-point	one	one	wires	
permutation	one	one	routing network	
broadcasting	one	many	clock distribution	
interupt line	many	one	interupt	
bus line	many	many	bus system	
reconfigurable	many	many	telephone system	

11.3 Parallelism

One single information channel, i. e. an individual connection between two terminals, is a onedimensional channel. Examples are wires or fibres. Connections between onedimensional arrays of terminals, such as a ribbon cable, a printed circuit board or integrated optics, have essentially twodimensional topology. Connections between twodimensional arrays of terminals, such as performed by an imaging system, require the threedimensional space above the arrays to support the communications between them. They are therefore called threedimensional interconnects. Fig. 11.1 illustrates this difference schematically.

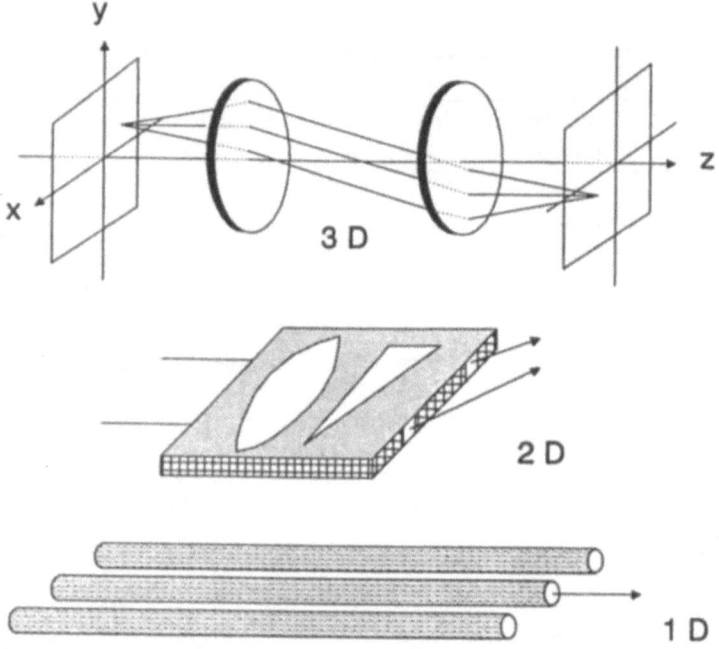

Fig. 11.1 Schematical view of one- two- and threedimensional interconnects [29].

One-, two- and threedimensional interconnects are vastly different from each other in terms of scaling: Onedimensional interconnections have only one terminal at each end. If the pitch, that is the distance between adjacent terminals, is p and the size of the array along one edge is d, then the twodimensional connection supports a linear array of $N = d/p$ parallel channels, whereas the threedimensional connection supports a matrix of $N^2 = (d/p)^2$ parallel channels. With a pitch of $p = 300$ μm and an edge of $d = 1$ cm twodimensional communications may provide a density of ρ_{2D} = 32 channels/cm, whereas threedimensional communications may provide ρ_{3D} = 32 x 32 $\approx$ 1.000 channels/cm^2. The density of interconnects is thus given by

$$\rho_{1D} = 1$$
$$\rho_{2D} = p^{-1}$$
$$\rho_{3D} = p^{-2} \tag{11.1}$$

for one-, two- or threedimensional communications between zero-, one- or twodimensional arrays of terminals respectively.

A very important measure for the performance of an interconnect is its connectivity C: it is the number of "potential communications patterns" supported by the system. A point-to-point connection has connectivity $C = 1$. A broadcasting system or an interupt line with M participants has a connectivity C = M. A bus line with M participants supports C = M(M-1) $\approx$ M^2 possible patterns of dialogue. The overall connectivity of a parallel communications system is given by the product of the number of parallel lines (N or N^2 for two- or threedimensional communications, respectively) and fanin and fanout.

$$C_{1D} = 1$$
$$C_{2D} = N \cdot \text{fanin} \cdot \text{fanout}$$
$$C_{3D} = N^2 \cdot \text{fanin} \cdot \text{fanout}. \tag{11.2}$$

11.4 Randomness

The connectivity counts the number of potential communication patterns provided by an interconnection system and is a topological measure. However, also geometrical considerations are necessary: a signal has not only to be transported a certain distance, it must also emerge from the connector at the right place. Two types of point-to-point interconnections have been identified: (i) regular connections with N parallel lines situated side by side or (ii) permutation elements with N lines that might cross each other in an arbitrary manner. Global interconnections of this type are needed for sorting [17], and find their applications in various areas that require the switching and the routing of information [4,11,19,20,5,26 and Chapt.15]. Fig. 11.2 illustrates regular and random interconnects. It also shows some special *global interconnects*, such as the perfect shuffle, the butterfly and the crossover, that are frequently used in sorting.

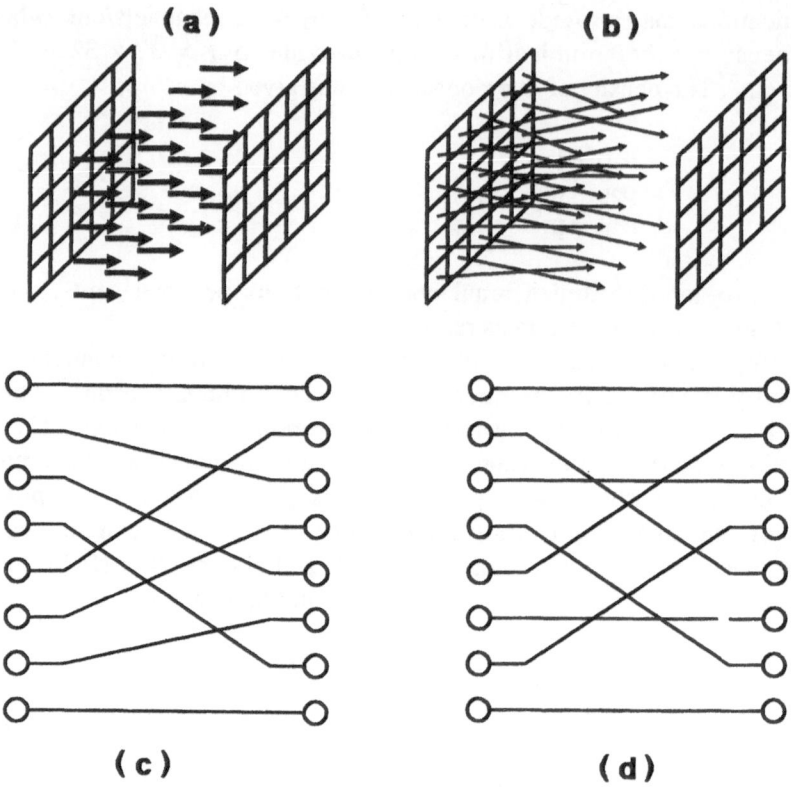

Fig. 11.2 a) regular point-to-point interconnection, **b)** random interconnection, **c)** perfect shuffle interconnection **d)** butterfly interconnection.

A fibre bundle or a good lens for one-to-one imaging provides a regular interconnection between twodimensional arrays. In principle each pixel that is imaged could serve as a separate information link. Fibre links have the advantage that they allow freedom in the geometrical arrangement of the participants: they are flexible and can be bent, the positioning of the transmitter and receiver with respect to each other is uncritical (of course alignment of optoelectronic terminals and fibres with respect to each other is necessary!). Thus fibres help to minimise alignment.

On the other hand, imaging systems allow to implement regular fanout patterns, i. e. they allow to implement multipoint interconnections: An imaging system is characterized by its point spread function (PSF) $p(x,y)$. The intensity distribution in the image of an incoherently radiating object is given by the convolution integral of the object intensity and the PSF [6]. From a discrete array of light sources as object with pitch p_x and p_y and with emitted intensities

$o_{k,l} = o(kp_x, lp_y)$ and with a discrete PSF $p_{m,n} = p(mp_x, np_y)$, we obtain as image at a discrete array of receivers the intensity values

$$i_{m,n} = \sum_k \sum_l o_{k,l} p_{k-m,l-n} \tag{11.3}$$

Each emitter channel (k,l) features the same "footprint" $p_{m,n}$ in the receiver plane. This type of interconnection is called shift-invariant or space-invariant. For the "footprint" $p_{m,n}$ one has a lot of freedom in the optical design, especially if holographic optical elements or kinoforms are used as spatial filters.

Regular point-to-point interconnections without fanout are the most trivial case of a space-invariant interconnection $p_{m,n} = \delta_{m,n}$. Space-invariant interconnect patterns with fanout are needed in several applications: certain systems architectures are inherently space-invariant, for example systolic array computers, cellular automata and symbolic substitution [22,12,13,3,29,26].

Interconnections that are not space-invariant are called space-variant. In this (much larger) class of interconnections each object-channel $o_{k,l}$ has ist own "footprint" or point spread function $p_{k,l,m,n}$ which can be completely different from adjacent channels. A permutation element featuring irregularly distributed point-to-point connections is the simplest example without fanout. With fanout very complicated interconnection patterns are possible.

$$i_{m,n} = \sum_k \sum_l o_{k,l} p_{k,l,m,n} \tag{11.4}$$

How much more complicated a space-variant interconnection is than a space-invariant one, can be seen by counting the number of degrees of freedom needed to specify the interconnection pattern in both cases: For a onedimensional array of emitters and receivers with N elements the point spread function p_m will have N different entries in the case of a space-invariant system; however, there can be as many as N^2 different entries for the general space-variant system. In threedimensional interconnections one has to specify N^2 or N^4 entries of the point spread function for space-invariant and space-variant interconnections. As a consequence of this enormous number of variables in a general space-variant system complicated (and therefore large) optical systems will be needed to implement the interconnections. How to embed such a large number of degrees of freedom in the optical system will be discussed lateron. The spatial density of space-variant interconnections turns out to be lower than in the space-invariant case.

Inbetween the two extreme cases of a completely space-invariant and a completely space-variant system, there are important intermediate cases. For example, the perfect shuffle interconnection or the butterfly, shown in fig. 11.2, can be decomposed into two or three space-invariant operations [18]. Consequently, two or three space-invariant optical systems can be used to implement such an operation. The degree of space-variance σ [23] is the minimum number of space-invariant systems that is needed to build a given interconnect pattern. For example, for a space-invariant interconnect $\sigma = 1$, for the perfect shuffle $\sigma = 2$, the butterfly $\sigma = 3$, for the most general random interconnect between M participants $\sigma = M$. Intermediate systems with a low degree of space-variance can be built by pupil division [18] or by splitting the operation into several space-invariant systems. Due to beam splitting an combination unavoidable losses in intensity or in resolution are encountered (Sect. 11.7). Hence, there is a tradeoff between energy efficiency of a space-variant optical interconnect and its degree of space-variance σ. Table 11.2 summarizes the numbers of degrees of freedom encountered in several interconnect types.

Table 11.2 Number of degrees of freedom that must be provided for different patterns of interconnects between arrays of N elements in each dimension. The factor σ depends in the 3-D case on the chosen 3-D generalization of the patterns. Also, for example the crossover requires different "granularity" within a multistage network, which increases σ compared to the here given values.

	2-D interconnection	3-D interconnection
space-invariant	N	N^2
completely space-variant	N^2	N^4
partially space-variant	σN	σN^2
perfect shuffle	2N	$2N^2$
crossover	2N	$2N^2$
butterfly	3N	$3N^2$

Whether or not completely space variant interconnections are necessary is an open question architecturely. It was shown, that in principle every general purpose computation could be done in a space invariant system [26], although this might not be efficient. Efficiency can be increased a lot by employing interconnect patterns such as the perfect shuffle or the butterfly. However, the number of components is in these cases probably still higher than with random wiring.

11.5 Crosstalk

Each mode of the electromagnetic field is an orthogonal component of the overall wavefield and thus the smallest and most basic information channel in optics. In telecommunications single mode fibres are used for optical long-distance interconnects, that is, the information is transported by a channel with minimum crossection. Multimode fibres are sometimes used for short distance connections. They may carry the same information in several modes at the same time. Their main disadvantage is, that different modes propagate at different speeds, and this modal dispersion blurs the transmitted signals over a long distance. On the other hand, the alignment tolerances are much less critical, because light can be coupled to anyone or several of the available modes; in case of a singlemode fibre it must exactly be matched to the only propagating mode. As a rule of thumb one can estimate the numerical aperture of a fibre by

$$N.A. \approx \sqrt{n_{core}^2 - n_{cladding}^2} \qquad (11.5)$$

and its diameter by several times $\delta x \approx \lambda/N.A.$, with λ denoting the wavelength in vacuo.

A lens system with numerical aperture N.A. focuses a beam of light to a focal spot with minimum diameter $\delta x \approx 1.22\,\lambda/N.A.$ (whereby the first zero of the Airy diffraction disc of a circular pupil was taken for this estimation). Hence, for fibres as well as for a diffraction limited imaging lens the minimum cross section of a mode is determined by the aperture and yields an absolute theoretical limit for packing densities as its consequence

$$\begin{aligned} \rho_{2D} &\approx N.A./\lambda \\ \rho_{3D} &\approx (N.A./\lambda)^2 \end{aligned} \qquad (11.6)$$

For the wavelength of a laser diode on the order of $\lambda = 1\ \mu m$ this translates into orders of magnitude $\rho_{2D} \approx 10^4$ cm^{-1} and $\rho_{3D} \approx 10^8$ cm^{-2}. These numbers are however quite unrealistic in practical applications: namely, there is crosstalk between adjacent channels caused by cross coupling, if the packing density approaches the theoretical limit. If the cladding of a fibre is not sufficiently thick light will leak out of the cladding instead of being totally internally reflected. Light leaking out of one fibre into adjacent ones contributes to crosstalk. If there is not enough distance between detectors, the diffraction pattern of one channel will spill over onto the detector of an adjacent channel. This diffractive crosstalk shall be estimated now. The "encircled energy" contained in a circle with radius R around a diffraction limited focus is according to Rayleigh [2] given by

$$E(R) = 1 - J_0^2(\pi R \cdot N.A./\lambda) - J_1^2(\pi R \cdot N.A./\lambda) \qquad (11.7)$$

whereby the overall energy in the focus is normalized to 1 and J_0 and J_1 denote Bessel-functions. Within a ring with radius R and thickness d there will be an energy of

$$E(R+d) - E(R) \approx d\frac{\delta E}{\delta R} = 2d/R\ J_1^2(\pi R \cdot N.A./\lambda) \qquad (11.8)$$

Now we assume a regular cartesian matrix of detectors with a pitch p and a detector area of d^2. Then one detector receives from an adjacent one an amount of energy given by the ratio of its area and the area of the ring with radius p $d^2/(2\pi pd)[E(p+d) - E(p)]$. If we assume that the pitch p is large compared to the size of the diffraction disc δx then we may use an asymptotic expansion for the Bessel function and also average over its residual oscillations. This yields an estimation for the diffractive crosstalk c from an adjacent channel:

$$c \approx \frac{d^2\lambda}{\pi^3 p^3 N.A.} \approx \frac{1}{1.22\pi^3} \frac{d^3}{p^3} \frac{\delta x}{d} \tag{11.9}$$

which translates into a simple rule of thumb

$$c[\text{in dB}] \approx -15\text{dB} - 10\log_{10}(d/\delta x) - 30\log_{10}(p/d). \tag{11.10}$$

Table 11.3 Crosstalk between adjacent channels in an optical interconnect for different size detectors and for different pitch of the channels p (measured in terms of the diameter d of the detector).

detector size	fill factor p/d				
d/δx	1	2	5	10	100
1	-15 dB	-24 dB	-36 dB	-45 dB	-75 dB
10	-25 dB	-34 dB	-46 dB	-55 dB	-85 dB

These figures are of course dependent on the beam shape which was not considered in the estimation above: Gaussian beams or apodisation lead to some improvements. In any case, from such an estimation it becomes clear, that for a diffraction limited detector $d \approx \delta x$ and a "filling factor" p/d = 10 diffraction crosstalk on the order of -45 dB can be expected. Table 11.3 gives a few more numerical values. We conclude, that large filling factors p/d are required for low diffraction crosstalk. Only "dilute" arrays of light sources and detectors are useful for optoelectronic interconnections. If we try to put different messages onto modes that are too near to each other in phase space, communication will suffer from crosstalk. The bit error rates in intracomputer communications usually require exceptionally low bit error rates which correspond to low values of crosstalk.

In practice our estimation might still be unrealistic in many cases: apart from diffraction there are many other reasons for crosstalk, for example aberrations of a lens that blur the point spread function or straylight caused by surface reflections and scattering within the optical system. However, equations similar to eq. 11.9 and 11.10 but with other numerical constants will hold approximately. The crosstalk must depend on the ratio of the size of the spot δx and the detector diameter d. The asymptotic decrease of intensity of the Airy-function, the shape of a diffraction limited focus, $\left|J_1(\zeta)/\zeta\right|^2 \approx \zeta^{-3}$ must result in a dependence on the filling-ratio $(p/d)^3$ as long as aberrations are not extremely large. From this estimation of the crosstalk limits of the interconnect density can be deduced:

$$\rho_{2D} = \frac{1}{p} \approx \frac{1}{d}\sqrt[3]{1.22\pi^3 cd/\delta x} \qquad (11.11)$$

Hence, if we require $c < 10^{-3}$ and $d \approx \delta x$, for example, then the pitch must be at least $p \approx 3.d$, and for $c < 10^{-4}$ one must require $p \approx 6d$. Therefore, under reasonable assumptions

$$\rho_{2D} \approx \frac{1}{5d} \quad \text{and} \quad \rho_{3D} \approx \frac{1}{25d^2}$$

This has important consequences: a (not too expensive) lens may have 1.000 resolvable picture elements in one dimension (or 10^6 in two dimensions). If - for the sake of avoiding crosstalk through spillover of intensity - only every third or every tenth picture element in a row is occupied with a channel for communications, this means that this single lens supports only 100 - 300 channels in one dimension and 10^4 - 10^5 for twodimensional arrays of receivers. For a larger number of parallel channels a single lens is insufficient. If larger numbers of channels are required lenslet arrays matched to the arrays of terminals have to be used.

11.6 Space sharing

In the preceding section the number of mutually independent channels was related to the number of resolvable picture elements of a lens, the so called *space-bandwidth-product* SBP. The filling ratio d/p was determined to be on the order of 0.2 (in one dimension). For a space-invariant system this gives directly the number of available parallel channels $P_{INVARIANT}$:

$$P_{INVARIANT} = (d/p)^2 \cdot SBP \qquad (11.13)$$

If no energy is to be wasted by beam combination (Sect. 11.7), each type of channel in a space-variant interconnect needs its own optical system or at least an individual portion of the aperture; thus, for each type of channel only $1/\sigma$ of the SBP is available. Therefore the overall number of available channels $P_{VARIANT}$ scales with the degree of space-variance σ:

$$P_{VARIANT} = (d/p)^2 \, SBP \, / \, \sigma \qquad (11.14)$$

For a completely random interconnect the degree of space-variance equals the number of channels $\sigma = P_{\text{VARIANT}}$, which results in

$$P_{\text{RANDOM}} = d/p \sqrt{SBP} \qquad (11.15)$$

Specifically we find

$$P_{\text{RANDOM}} = [P_{\text{INVARIANT}}]^{1/2} \qquad (11.16)$$

Table 11.4 Number of parallel optical channels in a space-invariant imaging system and for a completely random interconnect. A overall space-bandwidth-product of SBP $\approx 10^6$ and filling ratios d/p = 0.1 and 0.3 are assumed.

SBP $\approx 10\,6$	d/p ≈ 0.1	d/p ≈ 0.3
space-invariant connect	10.000	100.000
random interconnect	100	300

Table 11.4 gives some numerical examples, showing that the number of parallel optical channels is not as large as is sometimes assumed. For "optically thin" components a SBP $\approx 10^6$ may be a limiting value for most practical systems. Optically "thick" elements may pose different constraints [23, Chapt. 13]. Lenses with larger SBP than 10^6 can of course be built and used; however, practical issues such as the allowable volume and the cost become a big burden in this case.

Optical random interconnections are often implemented by using the matrix multiplier architecture shown in fig. 11.3 [7]. Albeit only onedimensional arrays of $P_{\text{RANDOM}} \approx N$ terminals are interconnected, the interconnection is threedimensional: the interconnection pattern is determined by a matrix with N^2 entries, that is illuminated from the emitters by using cylindrical lenses. The penalty for the complete space-variance of the system is payed by the fact that only one-dimensional arrays of terminals with spatial density $\rho_{2D} \approx 1/p$ are available.

A second class of random interconnects are multichannel systems: thereby no optical element is shared by several or all channels. Rather, each channel has its own separate optical system (fibre or lenslet). A point-to-point connection with a fibre bundle is an example. The density of interconnections is directly related to the density of fibres. Therefore the arrays will be fairly "dilute" with low fill factor d/p. For complicated topologies involving permutations, fanout and fanin, fibre bundles are not well suited: elaborate technologies for "weaving" such bundles are missing.

Multichannel systems can also be built out of fly's eye lenses and arrays of other free space optical components. Usually a first array of optical elements behind an array of light emitters is used for collimation and deflection of the light towards the detector array and there, in a distance L, a second array of elements collects the light on the detector array (Fig. 11.4). Holographic optical elements are well suited for this kind of array optics [24]. One channel may have an diameter p, which is directly related to the density of connections $\rho_{2D} \approx 1/p$ and $\rho_{3D} \approx 1/p^2$. Between the two arrays of optical components the light beams will diverge due to diffraction at the apertures with diameter p. If no significant divergence can be tolerated, which will be necessary to avoid straylight and crosstalk, then the width of the channels and the interconnect distance L are related by

$$p \approx \sqrt{8\lambda L} \tag{11.17}$$

whereby it was assumed the second component catches the Fraunhofer diffraction pattern of the first array including the +/- fourth minimum.

Crossbar

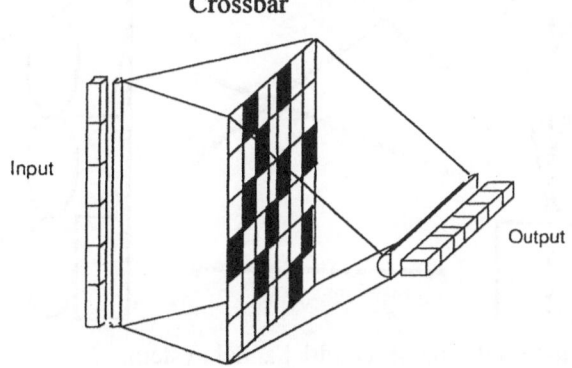

Fig. 11.3 Schematic rendering of an optical matrix multiplier.

In order to arrive at some numbers, we estimate - for practical reasons - that the lateral size of the array with N channels side by side should be on the same order as its length $L \approx N\,p$. In this case the overall number of parallel channels is given by

$$N_{1D} \approx L/p \approx p/8\lambda \approx \sqrt{L/8\lambda} \tag{11.18}$$

for onedimensional arrays and

$$N_{2D} \approx L/(8\lambda) \tag{11.19}$$

for twodimensional arrays, respectively. For $L \approx 1$ cm and a wavelength of 1 μm this translates to $N_{1D} \approx 30$ and $N_{2D} \approx 1,000$. The result of our very rough estimation is, that the number of parallel channels has the same order of magnitude as discussed before in eq. (11.15). The interconnection density is on the order of $\rho_{2D} \approx 30 \ cm^{-1}$ for linear arrays and $\rho_{3D} \approx 1,000 \ cm^{-2}$ for twodimensional arrays. An important advantage of multichannel systems is, that they can grow in the number of parallel channels simply by adding on more channels on its side; they are not limited in growth by the overall space bandwidth product of a lens that is shared by all channels.

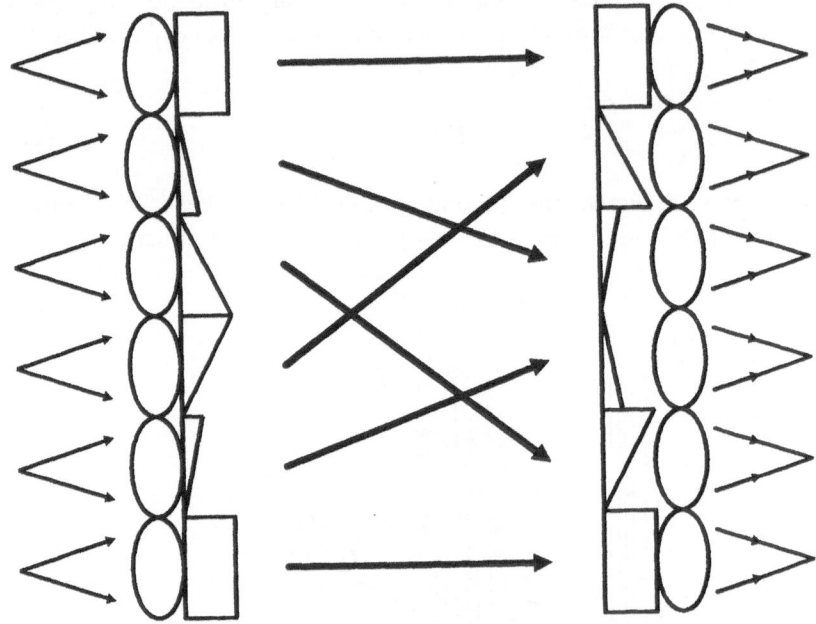

Fig. 11.4 Schematic rendering of a multichannel system.

11.7 Beam combination

Fanout of an incoming signal into a multitude of outgoing channels is an operation that can be performed by multiple beam splitters [27,28] with energy efficiencies over 90 %. Fanin, that is the combination of beams from different sources, is much different: The second law of thermodynamics forbids lossless combination; there is a penalty to pay, as was first noticed in the context of optical interconnects by Goodman [9]. There are two basic ways for beam combination:

(a) In the case of multimode-fanin both beams are combined into different modes of the electromagnetic field. For example, a polarizing beamsplitter can combine two beams of different polarization losslessly. Or, if the detector is large enough, different beams might be focused down on different areas of the detector by using separate optical systems. Or, different beams can share different areas of the aperture ("pupil division") and hit the detector from different directions. All implementations of multimode fanin rely on some kind of multiplexing (in space, direction, time, wavelength or polarization). It is common to all these approaches, that the detector needs to absorb at least as many modes of the electromagnetic field as there are beams to be combined. Unless polarization or wavelength is used for beam combination, the penalty is a decrease of the available space-bandwidth by the factor of fanin, SBP/fanin, and therefore a decrease in packing density.

(b) In the case of singlemode-fanin all incoming beams from different sources are combined into the same mode of the electromagnetic field, thus avoiding the loss in space-bandwidth incurred with multimode-fanin. The problem with this approach is illustrated best by considering the combination of two mutually incoherent beams of light with a half-silvered mirror: Light from both sources will reach the detector in the same mode. However, half of the light will leave the beamsplitter "the wrong way", that is not in the direction of the detector. In general, with an N-fold beamsplitter only 1/N of the incoming intensity will be coupled into the desired mode, (N-1)/N of the incoming light intensity will leave the beam splitter into undesired directions. These N-1 parasitic beams contribute to light loss, in the best case, and to straylight and crosstalk, in the worst case. Singlemode-fanin is a waste of light in many cases.

An example where singlemode fanin makes sense, is if the N-1 parasitic beams can be reused elsewhere in the system. This is the case for interconnections with bus topology, where besides a fanin of N at the same time a fanout of N is required. Appropriate systems architectures for optical bus interconnections with low light loss have been proposed in [15]. There the fanout component serves at the same time as fanin component. We summarize: In singlemode interconnection systems, where the parasitic beams cannot be reused, there is a choice: either one looses a factor of 1/fanin in power because much of the light leaves the beam combiner "the wrong way"; or one looses a factor of 1/fanin in space-bandwidth by using some sort of multiplexing such as pupil division or large area detectors, which amounts essentially to retreating to multimode fanin.

11.8 Heat

The temporal bandwidth (or the datarate) of a shortrange optical interconnection is limited typically neither by material absorption or dispersion, as it is the case in long distance communication through singlemode fibres, nor by modal dispersion, as it is the case in long distance communications through multimode fibres. Rather, the most stringent limit is nowadays the datarate of the optoelectronic devices, such as lasers or detectors. It has for a single device on a well packaged chip the order of magnitude of some 10^{10} Bit/s. For arrays of devices the bandwidth is mainly limited by packaging and cooling. A forced cooling system can remove heat energy at a rate of $R_{HEAT} \approx 1 - 10$ W/cm^2 without too much expense. Hero experiments, where more than 1000 W/cm^2 have been removed from a substrate by use of forced liquid cooling through microchannels have been reported in the literature, but practical systems have not yet reached these numbers.

A light transmitter and/or receiver dissipates a certain amount of power P_{DISSIP}. (for example, a semiconductor laser with low threshold dissipates 1 - 10 mW, the driver electronics may dissipate a similar amount of power). This means that the number of devices that can run in parallel on a chip is limited. The dissipation limit to spatial density is determined by

$$\rho_{3D} \approx R_{HEAT} / P_{DISSIP} \tag{11.20}$$

For devices with $P_{DISSIP} = 1$ mW and a cooling system with $R_{HEAT} = 1$ W/cm^2 the density is limited to $\rho = 1.000$ channels/cm^2. This leads - for a twodimensional array - to a pitch of about $p \approx 300$ μm or a 32 x 32 array on the area of 1 cm^2.

Surface emitting lasers with vertical cavity as well as quantum well modulators, such as the SEED [26], have a power dissipation of this order of magnitude. Unfortunately, an additional amount of heat is generated by driver electronics, etc. Onedimensional arrays are not as critical to run as twodimensional arrays because there is more space available for cooling. It must be concluded from the figures above, that a reduction of the power dissipation of the optoelectronic devices is very important task for the near future or one should consider fancy cooling systems, which may render dense optoelectronic interconnections impractical for many applications. Table 11.5 gives a few more numbers.

Table 11.5 Interconnect density for different power dissipations per communication channel and for different cooling rates.

power diss. per channel	rate of cooling			
	100 mW/cm^2	1 W/cm^2	10 W/cm^2	1 kW/cm^2
10 μW	10^4 cm^{-2}	10^5 cm^{-2}	10^6 cm^{-2}	10^8 cm^{-2}
1 mW	10^2 cm^{-2}	10^3 cm^{-2}	10^4 cm^{-2}	10^6 cm^{-2}
100 mW	1 cm^{-2}	10 cm^{-2}	100 cm^{-2}	10^4 cm^{-2}

With the kind of heat dissipation of today's devices crosstalk due to diffraction is not a practical problem, because the devices must be packed in a very "dilute" array. Otherwise cooling is insufficient if they all run in parallel.

Up until recently most of the development effort in semiconductor lasers went to high power lasers, which is the opposite of what is needed for parallel optical interconnections. If we assume that several thousand photons are needed for communicating 1 Bit of information (corresponding to an energy of about 1 fJ) and we assume a data rate of 1 GBit/s/channel, a laser with a power in the order of magnitude of some microwatts would be sufficient. From the point of view of laser physics semiconductor lasers with sufficiently low threshold power are theoretically possible [30].

11.9 Tolerances

Various components are necessary for an optoelectronic interconnect, all of which have to be specified and positioned within a certain tolerance range. First of all, the images of the transmitter array and the detector array must be aligned with respect to lateral shifts Δx, that is displacements within the receiver plane. For detectors with diameter d a reasonable tolerance will be, let's say, $\Delta x \approx d/4$. In a diffraction limited system where the detector size is matched to the diameter of the Airy-disc, the lateral shift tolerance is estimated by $\Delta x \approx \lambda/N.A.$, or actually a value somewhat smaller than this. Small detectors with 1 μm diameter will require submicron alignment accuracy. Even detectors with 10 μm diameter pose serious problems in alignment.

Secondly, there is a tolerance against longitudinal shifts Δz along the optical axis. It is basically determined by the depth of focus, which can be estimated by $\Delta z \approx d^2/4\lambda$ or in the diffraction limited case $\Delta z \approx \lambda/N.A.^2$, or a value smaller than this.

Third, there may be rotation around the optical axis $\Delta\phi$. If the lateral shift of the outmost detector in an array of N detectors with pitch p must be less than the lateral shifting tolerance, one has to require that $\Delta\phi \approx d/(4Np)$.

Fourth, there may be tilt with respect to the optical axis $\Delta\iota$ of the receiver plane. The longitudinal shift of the outmost detector must be within the depth of focus, leading to a tolerance of $\Delta\iota \approx d^2/(4\lambda Np)$.

Fifth, one very critical error is magnification mismatch Δm. It occurs in a space-invariant imaging system, if a lens is mounted in a wrong position or if it has an error in focal length. Again one must require that the lateral shift of the outmost detector stays within tolerances, that is $\Delta m \approx d/(4Np)$. In a multichannel system consisting of an array of transmitters and a matched array of lenslets, a magnification error occurs if the pitches of the transmitters and the lenslet array are slightly different ("Moiré magnification"). If the diameter of the channel equals the pitch, then the mismatch in pitch $\Delta p/p$ must be less than $\Delta p/p \approx 1/(4N)$, otherwise the outermost channel looses its alignment completely.

Another problem that directly translates into magnification mismatch is thermal expansion: Obviously, different materials with different expansion coefficients α will be used for optoelectronic transmitters, receivers and the optical components. The difference in expansion coefficients $\Delta\alpha$ of the materials used times the temperature range ΔT, in which the interconnection system is to be used, gives a thermal magnification mismatch. For packaging and mounting either special materials with matched thermal expansion coefficients are needed or a tight limit is put on the temperature range of highly parallel systems $\Delta\alpha\,\Delta T \approx d/(4Np)$. Table 11.6 is a summary of these tolerance considerations and gives some numerical values.

Table 11.6 Tolerances with respect to various misalignments; d denotes the detector diameter, $p \approx 5d$ the pitch of the arrays, $\lambda \approx 1$ mm the wavelength and N the linear dimension of the array, i. e. the number of channels along one edge.

Error	formula	$d = 1\,\mu m$	$d = 10\,\mu m$	$d = 100\,\mu m$
lateral shift	$\Delta x \approx d/4$	$0.25\,\mu m$	$2.5\,\mu m$	$25\,\mu m$
longit. shift	$\Delta z \approx d^2/(4\lambda)$	$0.25\,\mu m$	$25\,\mu m$	$2.5\,mm$
rotation	$\Delta\phi \approx d/(4Np)$	$1/(20\,N)$	$1/(20\,N)$	$1/(20\,N)$
tilt	$\Delta\iota \approx d^2/(4\lambda Np)$	$1/(20\,N)$	$1/(2\,N)$	$50/N$
magnification	$\Delta m \approx d/(4Np)$	$1/(20\,N)$	$1/(20\,N)$	$1/(20\,N)$
error	formula	N = 10	N = 30	N = 100
rotation	$\Delta\phi \approx 1/(20.N)$	17'	6'	1.7'
magnification	$\Delta m \approx 1/(20.N)$	$5\ 10^{-3}$	10^{-3}	$5\ 10^{-4}$

11.10 Conclusions

The spatial density of optical interconnections depends strongly on the type of the interconnections. Optical crosstalk and limited space bandwidth product limit space-invariant systems to a number of on the order of 10^4 parallel channels with densities of $\rho_{2D} \approx 100$ channels/cm and $\rho_{3D} \approx 10^4$ channels/cm^2 for two- and threedimensional interconnections respectively. Random interconnects have smaller densities: The matrix multiplier architecture exhibits $\rho_{MM} \approx 100$ channels/cm and multichannel systems (with lenslet arrays) have $\rho_{2D} \approx 30$ channels/cm or $\rho_{3D} \approx 1.000$ channels/cm^2 in two or three dimensions. These figures have been obtained by allowing virtually no crosstalk between adjacent channels. If some amount can be tolerated or if elaborated (and therefore voluminous and expensive) lens systems are employed, these numbers can be increased by at least one order of magnitude.

If optical interconnections are to be mass produced the mechanical and optical tolerances of the components will pose serious problems if high packing densities are chosen. For example, magnification has to be controlled to within less than 0.1 %. Off the shelf lenses, however, have much larger tolerances in focal length. Similarly, other mechanical tolerances become difficult to maintain for spatial densities $\rho_{2D} > 30$ channels/cm and $\rho_{3D} > 1.000$ channels/cm^2. Again there is not so much a fundamental limit, but rather a tradeoff between spatial density and cost.

Finally it should be noted, that by now (1992) the spatial density of high throughput optoelectronical interconnections is mainly limited by the heat dissipation of the optoelectronic emitters, receivers and driver electronics. With practical cooling systems (1-10 W/cm^2) and with lasers with threshold powers on the order of 1 mW spatial densities of $\rho_{3D} = 10^3 - 10^4$ channels/cm^2 can be achieved. These are comparable with the diffraction limits of spatial density in spacevariant interconnections. If space invariant interconnects with their increased spatial density are to be used, either devices with lower heat dissipation or better cooling are necessary.

Acknowledgement

Partial support from different sources helped for the research leading to this review and is gratefully acknowledged: it came from the German Bundesministerium für Forschung und Technologie (BMFT) under contract No. TK 584/4, from the Deutsche Forschungsgemeinschaft (DFG) under the contract SFB 182, and from the European Community under the ESPRIT Basic Research Action. The responsibility for the contents of this report, however, lies solely with the author.

References

1. Bergmann L. A., Wu W. H., Johnston A. R., Nixon R., Esener S. C., Guest C. C. Drabik T. J., Feldman M., Lee S. H. (1986): "Holographic optical interconnects for VLSI", Opt. Engin. **25**, 1109
2. Born M. and Wolf E. (1980): "Principles of optics", section 8.5.2, sixth edition, Pergamon Press, Oxford.
3. Chavel P. and Taboury J. (1991): "On alleged and real advantages of optical interconnects", Annales de Physique, coll. 1, suppl. 1, vol 16, pp. 153.
4. Feng T. (1981): "A survey of interconnection networks" IEEE Computer December 1981, 12-27

5. Giglmayr J. (1989): "Spatial extension of multistage interconnection networks", OSA Proc. Photonic Switching vol. 3, eds: Midwinter J. E., Hinton H. S.
6. Goodman J. W. (1968): "Introduction to Fourier optics" McGraw Hill.
7. Goodman J. W., Dias A. R., Woody L. M. (1978): "Fully parallel high speed incoherent optical method performing discrete Fourier transformations" Opt. Lett. **2**, 1
8. Goodman J. W., Leonberger F. J., Kung S. Y., Athale R. A. (1984): "Optical interconnections for VLSI-systems", Proc. IEEE **72**, 850.
9. Goodman J. W. (1985): "Fanin and fanout with optical interconnections" Opt. Acta **32**, 1489.
10. Hase K. R. (1984): "Ein Beitrag zur Realisierung rechnerinterner optischer Bussysteme mit planaren Lichtleitern" Dissertation University Duisburg, Germany.
11. Huang A. and Knauer S. (1984a): "STARLITE - a wideband digital switch" Proc. Globecom 1984, 141-148.
12. Huang A. (1984b): "Architectural considerations involved in the design of a optical digital computer" Proc. IEEE **72**, 780.
13. Hwang K. and Briggs F. A. (1985): "Computer architecture anf parallel processing", McGraw Hill, New York.
14. Ishihara S. (ed.) 1990: "Optical Computing in Japan", Nova Science Publ., New York
15. Krackhardt U., Sauer F., Stork W., Streibl N. (1992): "Concept for an optical bus-type interconnection network" Appl. Opt. **31**, 1730-1734.
16. Kostuk R., Goodman J. W., Hesselink L. (1985): "Optical imaging applied to microelectronic chip to chip interconnections", Appl. Opt. **24**, 2851.
17. Knuth D. (1973): "The art of computer programming, vol. 3, sorting and searching", Addison-Wesley, Reading Massachusetts, USA.
18. Lohmann A. W. (1986): "What classical optics can do for digital optical computing" Appl. Opt. **25**, 1543.
19. Midwinter J. E. (1985): "Light electronics - myth or reality", Proc. IEE **132** 371.
20. Midwinter J. E. (1987): "Novel approach to the design of optically activated switching matrices" Proc. IEE **134**, 261.
21. Miller D. A. B. (1989): "Optics for low energy communication inside digital processors: quantum detectors, sources and modulators as efficient impedance converters", Opt. Lett. **14**, 146.
22. Preston K., Duff M. J. B., Levialdi S., Norgren P. E., Toriwaki J.-I. (1979): "Basics of cellular logic with some applications to medical image processing", Proc. IEEE **67**, 826.
23. Prise M. E., Streibl N., Downs M. M. (1988): "Optical considerations in the design of digital optical computers", Opt. and Quantum Electron. **20**, 49-77.

24. Schwider J., Stork W., Streibl N., Völkel R: (1991): "Possibilities and limitations of space-variant holographic optical elements for switching networks and general interconnections", Appl. Opt. accepted for publication.

25. Stork W. (1987): "Optical crossbar" Optik **76**, 173.

26. Streibl N., Brenner K.-H., Huang A., Jahns J., Jewell J., Lohmann A. W., Miller D. A. B., Murdocca M., Prise M. E., Sizer T. (1989): "Digital Optics" Proc. IEEE **77**, 1954.

27. Streibl N. (1989): "Beam shaping with optical array generators" J. Mod. Opt. **36**, 1559.

28. Streibl N. (1992): "Multiple beam splitters", in "Optical Computing Hardware", vol. 2, eds: S. Lee, J. Jahns, Academic Press (1992).

29. Stucke G. (1989): "Digitaler optischer Computer", Reihe Informatik, Bd. 69, BI Wissenschaftsverlag, Mannheim/Wien/Zürich.

30. Yariv, A (1988): "Scaling laws and minimum threshold currents for quantum-confined semiconductor lasers", Appl. Phys. Lett. **53**, 1033-1035.

12 Interconnects with optically thin elements

A. Kirk and T. Hall
Department of Physics, King's College London.

12.1 Introduction

Electronic architectures are typified by their flatness. Wires and active devices are laid out on boards and interconnection occurs in two dimensions only. It is difficult and expensive to manufacture integrated circuits which have more than a few simple `crossovers' - points at which one wire passes over another. In contrast to this, the ability of light to propagate through space allows optically interconnected systems to make full use of all the space available. Rather than using the planar design inherited from electronics, many of these architectures envisage the use of the `parallel optics concept'. In such schemes two dimensional arrays of electronic and optoelectronic logic and switching devices are interconnected through free-space optics. These designs are typified by a much greater parallelism and a higher degree of interconnection than in electronic circuits. A more detailed discussion of these architectures will be made later in this chapter, but initially we will concentrate upon the devices used to define the optical interconnections.

In these parallel architectures devices are required to direct light from a source - which may be a laser diode, an LED or a modulated beam of light directed from elsewhere - through free space to one or more receivers or modulating devices. This should be achieved with as little crosstalk as possible between channels and with the loss of as little light as possible. In many architectures the most suitable and flexible candidates for this task are diffractive optical elements. In this chapter we will discuss the application of thin diffractive optical elements to parallel optical interconnections. In section 12.2 we will show the way in which Fourier plane array generators can be used to perform fanout and multiple imaging operations. In section 12.3 techniques by which fixed-weight shift-variant interconnections can be obtained will be described. Shift-invariant interconnections are also of great interest for optoelectronic information processing applications and these are considered in section 12.4. One of the most important applications for diffractive elements is in creating and multiply imaging arrays of beamlets which can then be switched by spatial light modulators. This allows reconfigurable interconnections to be obtained and is discussed in section 12.5. In section 12.6 we consider the ability of diffractive elements to allow very compact optical interconnects to be defined. Finally we will discuss some of the developments which may occur in this area in the near future.

12.2 Fourier plane array generators

In 1971 H.Dammann and K.Görtler introduced a technique for the design of binary phase diffraction gratings to split a single beam of light into several equally intense diffraction orders [1]. Such a grating is shown in Fig.12.1

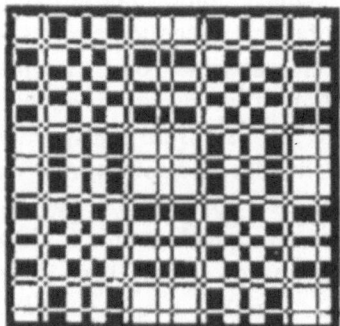

Fig. 12.1 Separable design for a Dammann grating (from [1]).

This grating has a period P_0 and a total width of R. If it is illuminated with a Gaussian beam of half-width w_0 at the $1/e^2$ intensity points then following McCormick [2] we may write the amplitude $I_0(x_0)$ immediately behind the grating as

$$I_0(x_0) = I_0\left[h(x_0) \otimes comb(\frac{x_0}{P_0})\right] x\left[gauss(\frac{x_0}{w_0})rect(\frac{x_0}{R})\right]$$ (12.1)

where h is the binary grating function, $gaus(x_0/w_0)$ is a Gaussian envelope with intensity profile $exp(-x_0^2/w_0^2)$, $rect(x_0/R)$ is the square pupil defined by the edge of the grating, and $comb(x_0/P_0)$ is the lattice of delta functions which define the periodic replication of the grating function. The convolution operation is represented by $\otimes$ (See Fig.12.2). Assuming that the paraxial approximation is valid then in the back focal plane of the lens we obtain the Fourier transform of this amplitude distribution. This is

$$I_1(x_1) = I_1 \left[\text{gauss}(\pi x_1 w_0 \lambda f) \otimes \text{sinc}\left(\frac{x_1 R}{\lambda f_1}\right) \right] \otimes \left[\text{comb}\left(\frac{x_1 P_0}{\lambda f}\right) H(x_1) \right] \qquad (12.2)$$

where sinc(x) represents the sinc function $\sin(\pi x)/\pi x$ and $H(x_1)$ is the Fourier transform of the hologram envelope. The diffraction orders are separated by a distance of $\lambda f/P_0$ and have a profile determined by the convolution of the Gaussian function and the sinc function in equation 12.2. If the width w_0 of the Gaussian beam is sufficiently small so that it is not significantly attenuated by the aperture of the hologram then the width of the diffracted orders will be dominated by the Gaussian term. In this case the half-width of each diffracted spot is $w_1 = \lambda f/\pi w_0$.

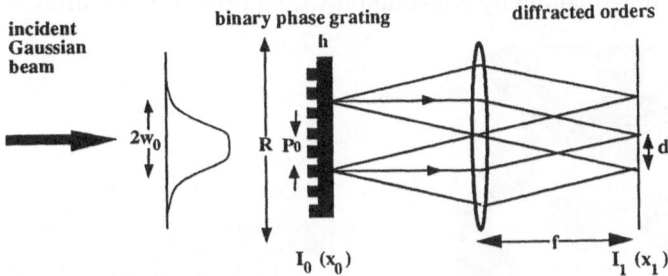

Fig. 12.2 The use of a binary phase grating as a Fourier plane array generator.

Fourier plane array generators have many advantages for use in parallel optoelectronic systems. One of the most important of these is that the overall intensity envelope of the diffracted orders is independent of the profile of the illuminating beam. This is in contrast to image plane array generators such as microlens arrays, for which the intensity of each spot depends upon the illuminating beam profile [3]. For Fourier plane array generators it can be seen from equation 12.2 that if a sufficient number of grating periods are illuminated then the spot intensity envelope depends only upon the grating function h_0. By a suitable choice of grating function a Fourier plane array generator can be used to define a uniform fanout or a weighted interconnection. Because of the nature of the Fourier transform operation these array generators display shift invariance. Each diffracted order has the same intensity profile which is equal to the Fourier transform of the illuminating beam profile. If the grating is illuminated by a Gaussian beam then each diffracted order is a tightly focused Gaussian beamlet. This is important in reducing crosstalk between parallel channels. Similarly Fourier plane array generators can be used to perform multiple imaging operations [1,4,5]. This is shown in Fig.12.3. The incident image is convolved with the Dirac-δ function defined by the grating diffraction orders. Dammann gratings were originally developed in order to perform this operation for photolithographic replication within integrated circuit manufacture [1]. However, this operation also has important applications within optoelectronic processing systems and will be discussed in detail in sections 13.4 and 13.5.

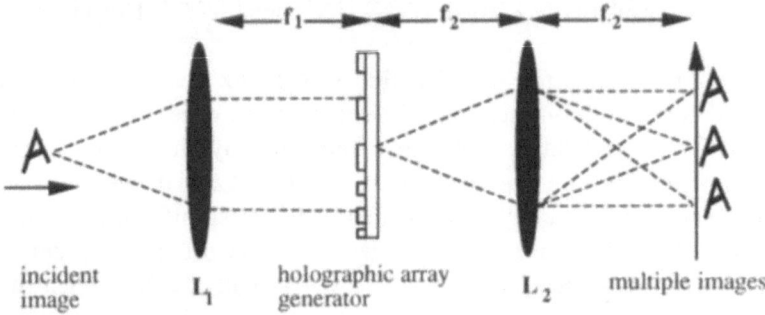

Fig.12.3 Multiple imaging by convolution with a Fourier plane array generator.

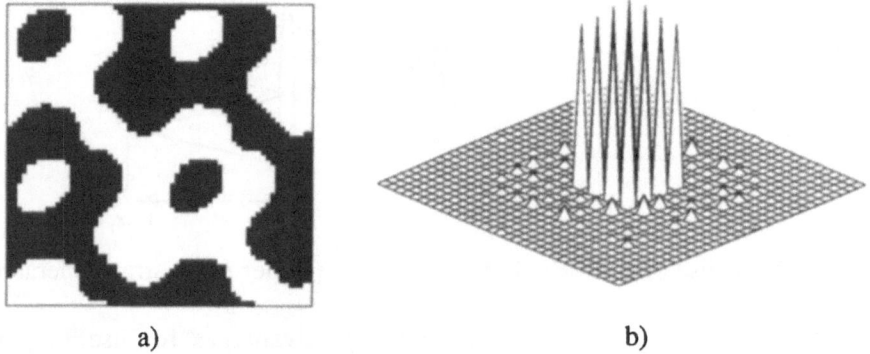

a)   b)

Fig. 12.4 A 64 x 64 pixel non-separable binary phase hologram a) which diffracts light into 4 x 4 odd orders b).

Binary phase gratings have a maximum efficiency of 81% if inversion symmetric diffraction patterns are used. While fanout patterns are usually inversion symmetric, more complex diffraction patterns are not and in these cases binary phase gratings have a maximum efficiency of 40.5% [6]. Typically Dammann gratings have efficiencies in one dimension of 70% to 80%. However, when these gratings are crossed to obtain two dimensional fanout patterns these efficiencies fall to around 60%. Greater efficiencies can be obtained by designing the binary function in two dimensions with the use of global optimization techniques [7-11]. In these cases efficiencies of up to 78% have been reported. These techniques also allow a non-separable fanout pattern to be obtained. An example of a non-separable design which diffracts into the central 4 x 4 odd orders is shown in Fig.12.4. This has a design efficiency of 70.0% and a fanout non-uniformity of 0.7%. Alternatively higher diffraction efficiencies can be obtained by using a greater number of phase levels [12]. Binary phase gratings can be fabricated by the use of electron beam lithography. In this way very small feature sizes can be obtained which will give large diffraction angles. Splitting ratios of up to 128 x 128 have been obtained [13] in this way, although some studies show that the tolerance for phase-depth error is very low for large splitting ratios [3]. Despite this errors of less than 1% in spot intensity have been reported in a 4 x 4 array generator [9]. Greater splitting ratios can also be obtained by cascading Fourier plane array generators [2].

12.3 Space variant Interconnections

12.3.1 1 to N^2 Mappings

One of the simplest applications for a binary phase grating is as a 1 to N^2 interconnect device. One demonstration of this has been given by Killat et al [14]. A binary phase grating was used to uniformly split light emitted by a 50 μm core graded index fibre to as many as 35 output fibres. (see Fig.12.5). Overall losses of 2.3 dB and a non-uniformity of less than 10% were reported. The advantage of using Fourier plane array generators for this operation is that the modal form of the input beam is conserved from the input port to the output port. Consequently the output beams are well matched to the output fibres and so give rise to the low insertion losses reported.

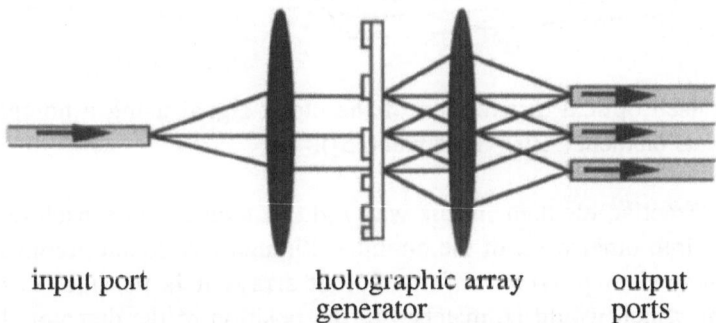

input port   holographic array   output
                              generator   ports

Fig.12.5 Holographic star coupler (After Killat et al. [14])

Another important application for 1 to N splitting devices is the routing of clock signals within micro-electronic circuits. The clock signal is required to synchronize the operation of the many individual components of the system. The period of the clock signal determines the operation rate of the system. This period cannot be reduced below the time required to propagate the signal from the clock to the most distant part of the system. At this point `clock-skew' occurs and the components of the system cannot all switch at the same time. There has been considerable research into the implementation of optically propagated clock signals [15-17]. In this application light is modulated with the clock signal and directed to the sub-systems of the device. Each sub-system has a photodetector which regenerates the clock signal as an electronic signal. The maximum size of each sub-system is determined by the maximum distance over which the signal can be propagated electronically without significant skew. A splitting component is required to direct the light from the source to each sub-system. Goodman [15] has suggested the use of diffractive elements to perform this operation and one possible scheme is shown in Fig.12.6.

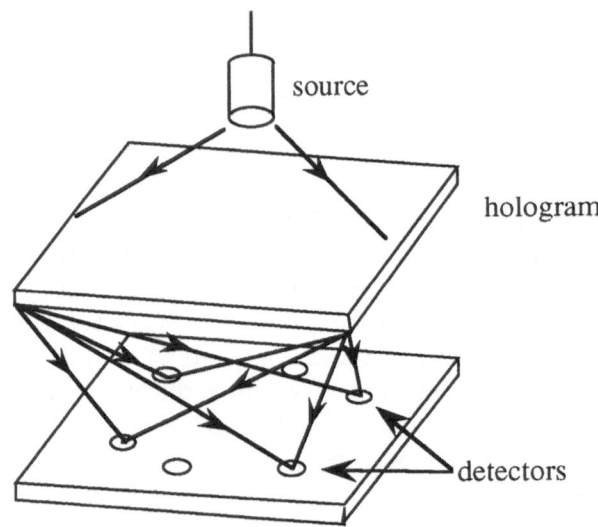

Fig. 12.6 Focused optical distribution of the clock signal using a holographic optical element (After Goodman [15]).

By using a diffractive element in this way light is focused onto each detector with little loss into other parts of the circuit. Although Dammann gratings were originally designed to provide uniform fanout arrays it is possible to obtain fanout patterns which would be matched to the position of the detectors by the use of non-separable design techniques. Fourier plane array generators have the advantage of shift invariance which will ease alignment problems. However if the clock distribution system is to be made as compact as possible it may be preferable to use Fresnel holograms. One example of this is given by Prongue [18]. Another alternative may be the use of the planar integrated array generator developed by Downs and Jahns [19]. In their design a glass substrate is used as a planar wave guide in which light is first reflected from a grating and then focused with the use of a Fresnel lens. This is shown in Fig.12.7.

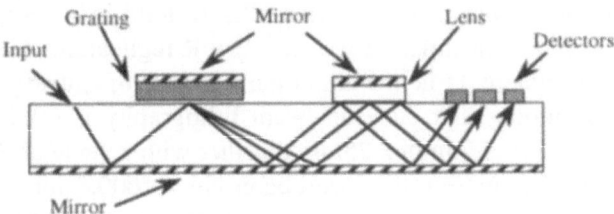

Fig. 12.7 Planar array generator (after Downs and Jahns [19]).

12.3.2 N to N Mappings

In the applications discussed so far in this chapter a single input channel is interconnected to many output channels in a 1 to N^2 mapping. Although these applications are important they do not make full use of the high degree of interconnectivity which may be obtained through the use of diffractive elements. There are many applications within the field of parallel information processing in which many input channels must be interconnected to many output channels. One example of this is the inner product operation used in many neural network models. This operation is schematically represented in Fig.12.8. An array of N input nodes x_1 .... x_N is connected to an array of N output nodes y_1 .... y_N with connection weights W_{ij} between the ith and jth output node. At each output the signals from each input are summed so that the operation may be written as

$$y_j = \sum_{i=1}^{N} W_{ij} x_i \qquad (12.3)$$

This operation is required in associative memory models such as the Hopfield model [20]. Many different optical implementations of this model have been suggested and investigated (e.g. [21,22]). One solution is to use a diffractive element to define the weighted interconnection pattern from the input to the output nodes. A multifacet hologram is required to perform this operation. Light from an input node falls upon a single facet of such a hologram and is then diffracted with the correct interconnection weight to the output nodes. This is shown in Fig.12.9.

Because of the shift invariant nature of the Fourier transform array generators all the sources are correctly imaged to all the detectors. This scheme has been investigated and implemented by several authors [23,24]. It should be possible to construct very large systems with complex interconnections by the use of this method. For ease of construction and alignment it may be preferable to fabricate all the holograms in a single substrate. If we wish to interconnect N x N input channels to N x N outputs then it will be necessary to fabricate N x N facets. If p is the smallest feature size which may be fabricated then binary phase gratings typically require an area of 4Np x   4Np in order to encode a N x N interconnection. In order for equation 12.2 to hold it is necessary to replicate each facet a number of times. If we have R x R replications then we can see that $N2 = L/4Rp$ where L is the maximum length of substrate which may be patterned. Assuming that electron beam lithography is used with a relatively small substrate size of 25mm x 25mm together with a value of R=4 and minimum feature size p of 1 micron, then we obtain $N2 = 1600$. Thus it should easily be possible to interconnect an array of 40 x 40 input nodes to 40 x 40 output nodes in this way. The size of each facet and hence of the input and output channels may be calculated as 640 x 640 microns. If a larger substrate size is used (L = 100mm) then as many as 6400 channels would be available. In such a system 40 x 10^6 interconnections would be defined although the resolution of the optical system used may become a limiting factor at this point (see Chapter 11).

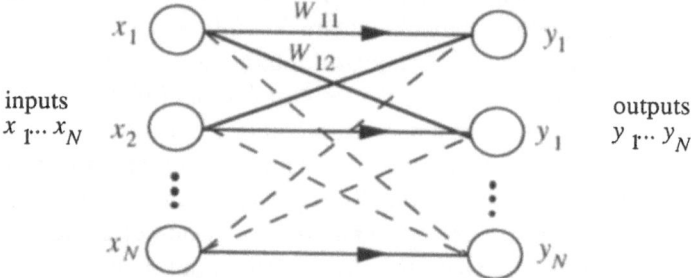

Fig. 12.8 Inner product operation - each input xi has a weighted interconnection to all the outputs yj.

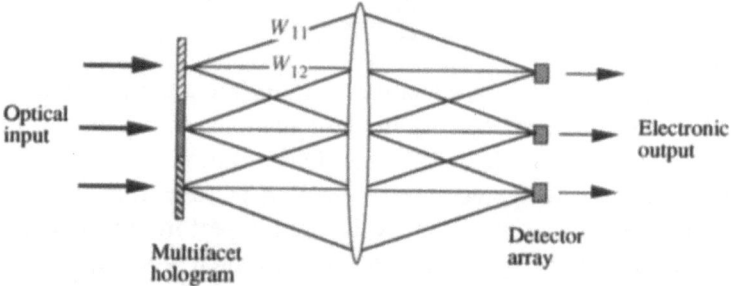

Fig. 12.9 Shift variant interconnection by use of a multifacet hologram.

12.4 Space invariant interconnections

The interconnections described in the previous section are characterized by possessing a unique interconnection pattern for each input channel. This allows very complex and dense interconnection patterns to be defined. One major disadvantage of this scheme however is that each input channel requires a single hologram. Consequently the area of the multifacet hologram grows as the square of the number of channels. Although arbitrary interconnections can be obtained in this way the scalability of the scheme may be limited to at most a few thousand channels. Many parallel processing applications however do not require such an arbitrary interconnection scheme. Instead it may be required to perform the same relatively simple operation on all the input channels with a greatly reduced amount of interconnections. As each interconnection pattern is identical for each input these interconnections are shift invariant. This allows a fixed-weight N^2 to N^2 mapping to be performed.

With the use of a single Fourier plane hologram an entire array of parallel channels can be interconnected in a shift-invariant manner. This is achieved by use of the multiple imaging operation described in section 13.2. In Fig.12.3 the separation of each diffraction order is sufficient to accommodate each multiple image without overlap.

Shift-invariant processing operations can be performed if the images are allowed to overlap. Fig. 12.10 Use of a Fourier plane computer generated hologram to perform a convolution operation. Hologram H encodes the convolution kernal h, f is the input image and g is the result of the convolution. If the input channels are mutually incoherent then the intensity at each output channel is equal to the convolution of the input channels with a kernal defined by the interconnection weights of the hologram. This is shown in Fig.12.10. Jenkins et al [25] have investigated the practicality of this scheme for use in digital optical processors. They have shown that by placing several logic gates at each output port more complex operations can be performed. Fig.12.11 shows a scheme for a binary Laplacian edge detector similar to one discussed in reference [25]. The input to the system consists of an inverted version of the image. This is then multiply imaged using a hologram with the impulse response shown in Fig.12.11(a) onto the output plane. Each output plane has 3 gates; x, y and z. Gate y is a NOR gate which has the value of the nearest-neighbour channels A B C D as its inputs. Gate x is a buffer which takes the value of the input channel E. The outputs from the two gates are used as the input to gate z. This returns a 1 if the point E is set to 1 and at least one of its neighbours is zero, otherwise it returns a zero. The width of each input pixel must be sufficiently small so that in the output plane each multiple image does not spread out onto neighbouring gates in each output channel.

Edge detection is an example of the type of preprocessing question for which shift-invariant optical interconnections are ideally suited. Such operations which require a limited degree interconnection have many applications within image processing. Within the mammalian visual system the retina performs many of these operations, including edge and motion detection as well as higher order operations such as the location of corners and curves. There is much on-going research in this area within electronics, in particular the well known work of

Mead [26]. The advantage of specifying a shift invariant solution to the preprocessing operation is that the same operation is carried out in parallel over the entire image simultaneously. In this way a complex noisy image can rapidly be transformed into a much simpler image which is specified in terms of lines and primitive shapes. By preserving important information within an image while reducing its complexity, the task of subsequent image processing operations such as pattern recognition is greatly simplified.

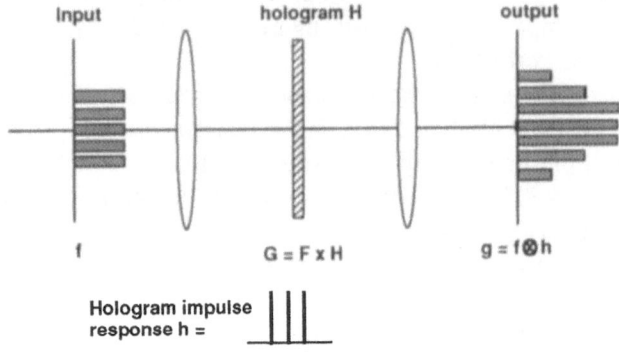

Fig. 12.10 Use of a Fourier plane computer generated hologram to perform a convolution operation. Hologram H encodes the convolution kernal h, f is the input image and g is the result of the convolution.

The Laplacian edge detector requires only nearest neighbour interconnections and so may be relatively simple to implement electronically. Reference [27] describes a shift invariant architecture for texture recognition which demonstrates a greater degree of interconnectivity. This design is based around the multilayer perceptron which is a well known neural network architecture. A simple multilayer perceptron is shown in Fig.12.12. The network in this example consists of 3 layers of neurons, an input layer a hidden layer and an output layer. The input layer and the hidden layer are densely interconnected, while the output layer is connected to the hidden layer by a simple fan-in relationship. Multilayer perceptrons can perform many complex pattern recognition tasks. The connection pattern shown in Fig.12.12 is not, however, shift-invariant. Fig.12.13 shows a similar multilayer network which can be specified as a shift-invariant interconnection pattern [27]. Each neuron in the input layer is connected to 9 neurons in the hidden layer. Each central group of 3 neurons in the hidden layer corresponds to the hidden layer in the perceptron for a single output neuron. The 1 to 9 fanout operation W_1 is shift-invariant and so could be performed by a fixed weight holographic fanout element. The interconnection operation W_2 is also shift invariant. This is a simple weighted fan-in operation and so could be perfomed electronically. A posible opto-electronic implementation of this system is shown in Fig.12.14. The first hologram H_1 creates a uniform Gaussian beamlet array. Each beamlet illuminates a single pixel in the electrically addressed spatial light modulator (SLM) in order to provide the input image. The second hologram H_2 performs the shift-invariant weighted interconnection operation W_1 of this image onto the detector plane. Most useful processing operations will require bipolar interconnection weights. This can be achived by the use of two photodetectors for each connection weight in W_1 [22,28]. If each pair of photodetectors is configured so that a differential output is obtained then a negative weight is equivalent to the routing of a beamlet to the negative response photodetector. Fig.12.15 shows such a circuit for differential operation. It is assumed that such a system is used in the system shown in Fig.12.14 so that the hologram H_2 must encode two fanout weights for each bipolar weight in W_1. The fan-in operation W_2 is then performed via electronics. Thresholding at each output node will then give the correct output from the multi-layer perceptron.

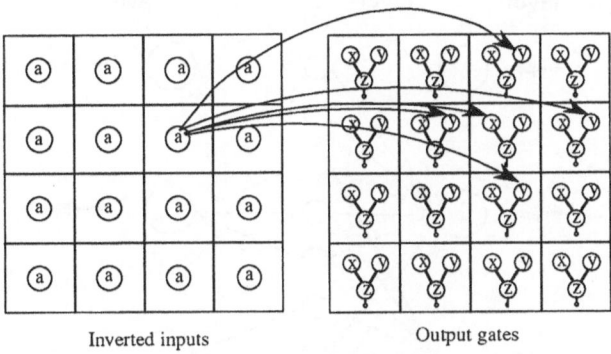

Fig. 12.11 a) Input and ouput planes of the edge detector. The arrows show the point spread function of the hologram. Each input channel connects to the 'y' gates of its nearest neighbours and its own 'x' gate. The 'z' gate gives the edge detected output.

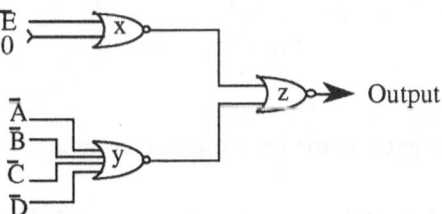

Fig. 12.11 b) Arrangement of logic gates x, y and z for each output channel.

Fig. 12.11 Binary Laplacian edge detector (From Jenkins et al [25]).

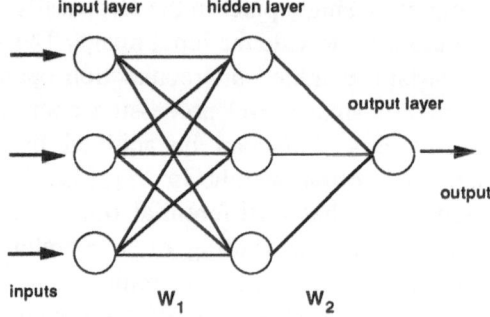

Fig. 12.12 A multilayer perceptron.

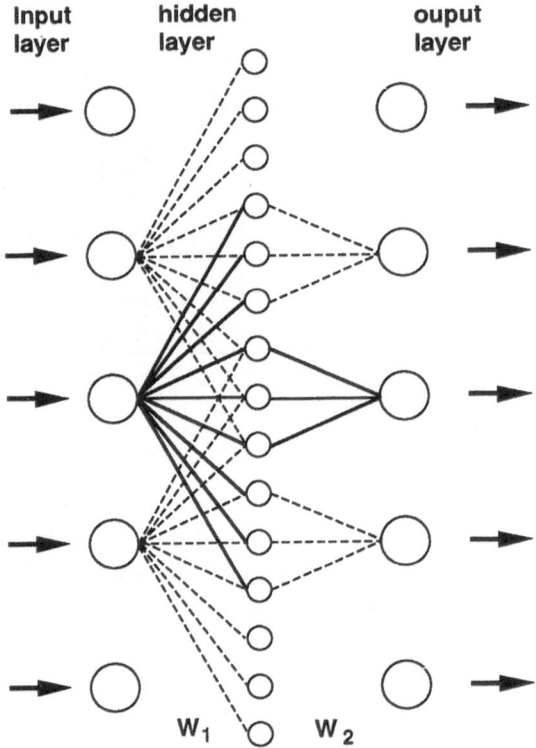

Fig. 12.13 Shift invariant scheme for a multilayer perceptron.

The ability to perform negative or inhibitory interconnections is very important in many image processing applications. These include operations such as the `on-centre off-surround' function used for edge detection within the retina [26,29]. The Laplacian edge detector discussed earlier in this section is a simple example of this scheme for binary images. In addition to greatly increasing the range of operations which may be performed by the shift-interconnection schemes which employ bipolar detectors there are further advantages to be gained. It has been shown by de Groot [30] that by specifying an interconnection as the sum of a positive and a negative value a much greater immunity to degradation in the contrast ratio of the input device is gained. In this way it is not the overall intensity which is important but the difference in intensity between neighbouring image points. Such a differential scheme is believed to occur in the retina [31] allowing optimum performance under a wide variety of illumination conditions.

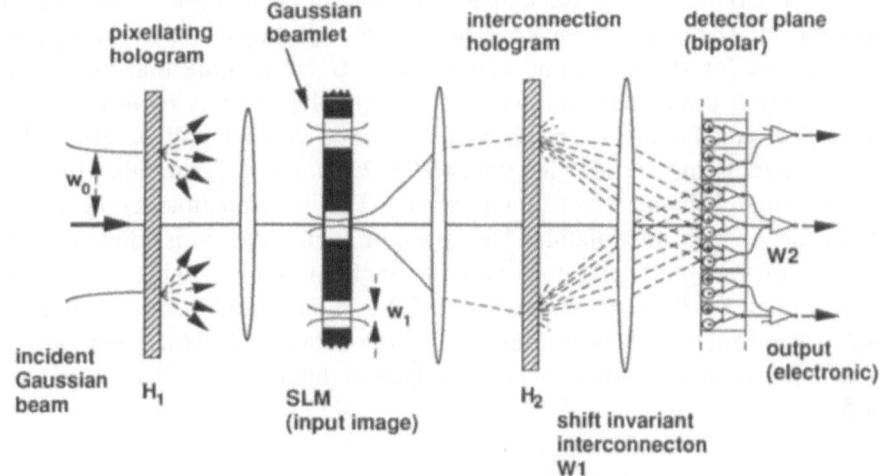

Fig. 12.14 Implementation of shift invariant multilayer perceptron with diffractive components.

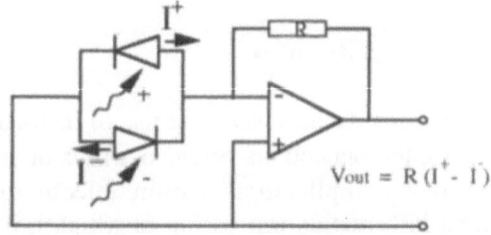

Fig. 12.15 Bipolar photodetector circuit which allows inhibitory interconnections.

In contrast with the shift variant interconnections discussed in section 13.3 the number of channels which can be processed simultaneously is not limited by the size of the hologram. Instead the limit is now determined by the resolution of the optical system and the size of the input and output arrays. A good optical system may allow as many as 10^8 resolvable points. However this is unlikely to be achieved in practice. For bipolar interconnections each output channel requires twice as many photodetectors as there are nodes in the hidden layer. Photodetector dimensions of around 10 microns should be achievable. The minimum separation of the photodetectors is determined by crosstalk between beams in the output plane and by the space required for the electronic components. From equation 12.2 we can calculate that for a hologram with a width of 25 mm the minimum spot size of Gaussian beams in the output plane will be $w_{min} \approx 2 \mu m$. The amount of power falling outside the 5 micron radius of the photodetector is $\eta = \exp(-2r^2/w^2) \approx 3 \times 10^{-6}$. Hence crosstalk between channels from the overlap of beams in the output plane should be negligible if aberrations are minimized. The width of each output channel will thus be determined by the space required for the electronic components. If we assume that an area of approximately 5 times that required for the photodetectors is required for the amplification and fanout wiring then each output channel will be 100 x 100 microns in size. Using 1:1 imaging with a 25 x 25 mm hologram, this will allow 250 x 250 parallel channels to be implemented. This is compatible with currently available spatial light modulators. The speed of such a device is limited by the sensitivity of the photodetectors and the incident power. This is strongly dependent upon the implementation method chosen. The high parallelism of this architecture means that even if a relatively slow data rate of 1MHz per channel was used the total throughput would be greater than 5×10^{10} channels per second.

12.5 Free-space optical switching systems

12.5.1 Optical crossbar architectures

In the previous two sections we have seen the use of diffractive optical elements to obtain fixed weight interconnections between arrays of sources and detectors. There are however many applications within telecommunications, parallel computing and optical information processing in which it is necessary to switch information from one channel to another. In this section we will demonstrate some of the ways in which thin diffractive optical elements can play a key role in optical space switching systems. By extending the concepts of multiple imaging as a method of routing parallel data arrays and the use of Gaussian beamlets to define optical channels we will show that very large scale high-speed parallel switching systems may be obtained. These will be reconfigurable N to N mappings.

One of the best known schemes for optical space-switches is the vector-matrix multiplier architecture, shown in Fig.12.16. The input channels are arranged in a column as shown. Light from these channels is spread out horizontally using a cylindrical lens onto an array of switches (typically a liquid crystal spatial light modulator). Light transmitted through the switch array is focused using a second cylindrical lens into a row of output channels. The transmittance values of the pixels in the switch array can be expressed as a matrix Wij. If the light intensities of the input and output channels are expressed as vectors x and y respectively, then the operation of the switch can be written as a vector matrix multiplication

$$y_j = \sum_{i=1}^{N} W_{ij} x_i \qquad (12.4)$$

In this way any combination of input channels may be connected to any combination of output channels. This architecture was originally investigated as a crossbar switch for the routing of telecommunications channels [32] and as an element within optical information processing systems [33]. Although the vector matrix multiplier architecture allows an arbitrary interconnection to be defined with only a single switching plane the architecture as shown in Fig. 12.16 has many drawbacks which limit the size and speed of switching systems which are based around it. Recent improvements in the performance and availability of spatial light modulators have prompted a re-evaluation of this architecture [27,34].

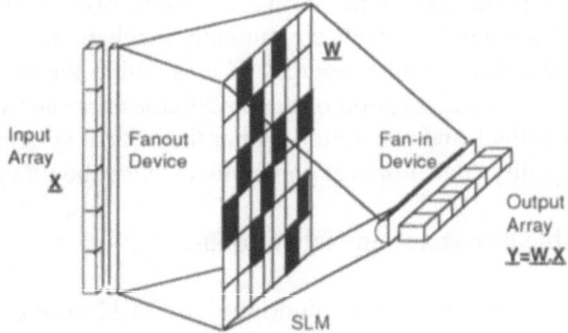

Fig. 12.16 Schematic diagram of optical vector-matrix multiplier.

In evaluating a switching system there are two main criteria to be considered. The first of these is the insertion loss of the system and the second is crosstalk. Insertion loss, receiver sensitivity and the required bit error rate (BER) determine the signal bandwidth of the switch [35]. The second criterion, crosstalk, usually limits the maximum size of the system. Crosstalk occurs when information in one channel spills into other channels. The maximum number of channels is reached when the noise within a channel arising from crosstalk from all the other channels becomes too great to allow the signal carried by the channel to be reliably discerned.

It is important to separate intrinsic sources of loss which are fundamental to the system from excess losses which may be controlled. The greatest source of intrinsic loss is fanout loss which arises from the broadcast nature of the scheme. Light from each input channel is divided uniformly amongst all the output channels. Consequently for an N channel switch only 1/N of the power from each input channel is received at any output channel. This contributes a factor of $3\log_2 N$ dB to the total insertion loss.

In addition to this loss, intrinsic loss may also occur when attempting to collect light from many input channels into a single output channel. This concept has been introduced in Chapter 12. The concept of intrinsic fan-in loss was first introduced within the context of optical switching systems by Goodman [36]. Intrinsic fan-in loss is a consequence of Louiville's theorem [37]. If N beams which subtend a solid angle ω_1 and which occupy an area of A_1 are collected into a port which can accept light froma solid angle of ω_2 and which has an area of A_2 then the maximum fraction of light which can be collected is $\omega_2 A_2 / N \omega_1 A_1$. Thus if the input and output channels possess identical numerical apertures and cross-sectional areas, only 1/N of the light will be collected and so this intrinsic fan-in loss will contribute a factor of $3 \log_2 N$ dB to the total insertion loss. This factor is particularly important in for example, telecommunications systems, in which light must be switched between two identical arrays of optical fibres. It should also be pointed out that fan-in loss as described here occurs only for the case in which the input channels are mutually incoherent.

There are also several other sources of loss within the crossbar design. These include the non-zero absorption of the modulator elements when in the 'on' state and stray reflections and absorption within the optical system. With care and the use of high quality coated optical components these should be made negligible.

12.5.2 Scalable free-space optical crossbars

If we return to the vector matrix multiplier in Fig.12.16 it can be seen that there are several disadvantages in this design. The most obvious of these is that light from each input channel is smeared out over the SLM by the cylindrical lens. Unless the active switching region of each SLM pixel occupies the entire area of the pixel, light will be lost into the 'dead space' of the SLM. If the dead space is absorbing then the insertion loss of the switch will be increased. Alternatively if the dead space allows transmission of some or all of the light then the crosstalk of the switch will be increased. As many high speed SLMs have small active switching areas this may be a serious problem. A second disadvantage arises from the one-dimensional arrangement of the input and output channels. The cylindrical lenses must spread the light out N-fold in one dimension but not at all in the other. If the input channels were to be split into two dimensions rather than one, the tolerances on the fanout elements would be greatly reduced. An N channel crossbar would then require the fanout of beam-spreading element to spread the light out $\sqrt{N} \times \sqrt{N}$ times rather than 1 x N.

The one-dimensional architecture also has important consequences if such switches are used within telecommunications applications. It was stated previously that if a crossbar was used to switch between two identical arrays of optical fibres a 1/N fan-in loss would be experienced. This could be avoided by using an input array which consisted of monomode fibres and an output array of multimode fibres. A standard multimode fibre may support as many as 5000 modes. Thus it is in principle possible to collect the light emitted from 5000 monomode fibres into a single multimode fibre without experiencing intrinsic fan-in loss. In the vector-matrix multiplier architecture however, fan-in occurs in one-dimension only. Consequently only a small fraction of the available modal volume of the multimode fibre can be filled and so in this case the maximum number of channels would be reduced to the square root of the number of modes, i.e. approximately 70.

After considering these various points it can be seen that many advantages would ensue from redesigning the vector matrix multiplier as a two-dimensional crossbar. In order to achieve this we require a method by which light from the input channels may be split into two dimensions rather than one. In addition we would like to be able to image the input channels directly onto the SLM pixels rather than spreading the light out into dead areas of the modulator. One possible solution is to use a Fourier plane array generator to multiply image a two-dimensional array of input channels onto a spatial light modulator. For a crossbar with N x N input and output channels a single hologram which diffracts light into N x N uniform fanout spots is required. In this way the input image of N x N channels is replicated as an N x N array of images on the SLM. In order to fully interconnect two two-dimensional arrays a four dimensional tensor is required as the weight matrix. If this is to be represented on a two-dimensional spatial light modulator it is necessary to format it as an N x N array of N x N subarrays. Each subarray defines the interconnection pattern from all the input channels to a single output channel. This spatial coding technique has been suggested by several authors in different contexts [38,39]. After the SLM has allowed light to be transmitted or blocked a microlens array is used to collect the light from each submatrix into the corresponding output channel.

This architecture has been investigated for use as a telecommunications crossbar switch [34] and as a matrix-matrix multiplier within optoelectronic neural networks [27,40]. A similar fanout scheme was implemented by Weible et al [41] with a fixed weight mask rather than a spatial light modulator for use as an optical associative memory. Fig. 12.17 shows the use of this scheme in a telecommunications crossbar switch in which two arrays of optical fibres are interconnected. In this case a reflective spatial light modulator is used. This is an active-backplane ferro-electric liquid crystal on silicon device [42]. Each pixel may be directly addressed in a similar way to the memory cells within a DRAM and so very fast switching is possible. Polarised light from the monomode fibre input array is multiply imaged through the polarising beam splitter onto the SLM. Each monomode fibre emits a beam which has an approximately Gaussian profile with a very small beam waist. Consequently when light from this array is reimaged onto the SLM each multiply imaged channel is a tightly focused Gaussian beamlet. This limits the crosstalk between adjacent channels and means that the active switching area of each SLM pixel can be very small. The SLM rotates the polarisation of those beamlets which are to be transmitted to the output channels and the light is reflected. At the polarising beam splitter the light in these channels is reflected towards the output array. This consists of an array of multimode fibres. In order to efficiently collect light into these fibres an array of microlenses is used to collimate each beamlet. An array of larger lenses is used to collect and focus all the light from each subarray into the corresponding output fibre. In this way the beamlet arrays are correctly matched to the multimode fibre. The action of the microlenses and the collecting lenses is the reverse of a microlens array generator [3].

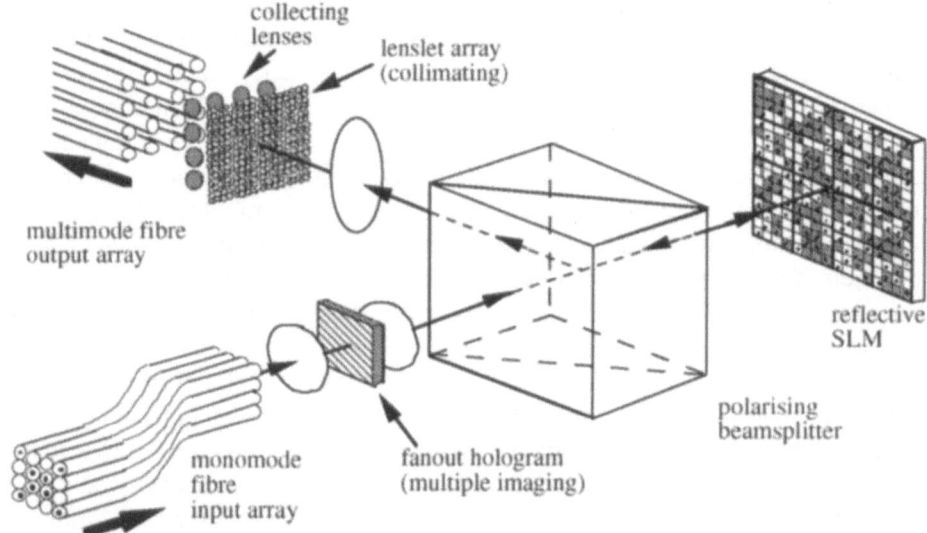

Fig. 12.17 Telecommunications crossbar switch incorporating holographic fanout.

The speed and scalability of this crossbar is determined by the insertion losses and interchannel crosstalk. In this system the main contributions to crosstalk will be the overlap of adjacent beamlets at the SLM and the non-zero transmission of the SLM in the off-state. Each beamlet at the SLM is a Gaussian beam with e^{-2} beam waist w. The fraction of power η from this beam falling outside a radius r is $\eta = \exp(-2r^2/w^2)$. This can be conservatively estimated to be equal to the total power falling within this radius from all the other beamlets in an array. Hence the maximum array pitch p expressed in units of w for a given value of crosstalk η is

$$p = 2\sqrt{\ln\eta/2} \tag{12.5}$$

In a digital system a crosstalk of 10^{-2} should easily be tolerable at this point. This would then allow an array pitch of p=3.0. For a typical monomode fibre this allows a minimum beamlet spacing of 8 microns for this level of crosstalk.

The SLM is designed to operate at normal incidence. Each Gaussian beamlet however is composed of plane waves with a range of angles. The FELC SLM discussed here operates by switching the optical axis within the liquid crystal. By aligning the off-state optical axis with the input polariser the off-state transmission of the SLM will not be affected. The on-state transmission will be affected but this will have only a minor effect upon crossbar performance. A more important factor is the transmission of light through the polarising beam splitter. Shamir et al [43] have shown that the transmission of light through two crossed polarisers is $\xi \approx \theta^4$ for small incidence angles θ. An acceptable value for ξ of 10^{-3} gives a maximum divergence angle of 0.17 radians.

The total size of each multiple image is Np x Np for a crossbar which interconnects N^2 channels and where p is the pitch of the beamlet array. For simplicity 1:1 imaging is considered here. Hence p is equal to the outside diameter of the monomode fibre - typically 125 microns. The maximum off-axis distance is along the diagonal of the SLM and is equal to $\sqrt{2Np/2}$. From this the maximum divergence angle of the diffracted waves after the hologram is found to be $\theta_m = \sqrt{2}Np/2nf$ where f is the focal length of the lens placed after the beam splitter. In order to preserve contrast ratio this must be less than the maximum allowed divergence angle for a given crosstalk ξ. Hence $f > \sqrt{2}Np/2n \, \theta_m$. For the values of f, θ_m and p given so far this gives f=0.32N mm. Hence for a 256 channel crossbar this gives a focal length of f=84 mm and SLM side dimension of 32 mm.

The data rate of the crossbar is determined by the total insertion loss. The system has an intrinsic fanout loss of $3\log_2 N$ dB and we assume that no intrinsic fan-in loss is experienced. The expected excess insertion loss is 10dB. If the input power per channel is 10dBm then for a 256 channel crossbar the total received power per channel will be -24 dBm. Dias et al [35] have calculated that if pin diode receivers are used then this reived power level will allow a data rate of more than 1 Gbit per second per channel. Thus very high data rates should be possible, although reconfiguration time depends upon the switching speed of the modulator. This requires that messages are sent in packets which are at least as long as the modulator reconfiguration time. In order to implement this crossbar a hologram which produces a 16 x 16 fanout is required. As fanout holograms of up to 128 x 128 have been reported it can be seen that the limiting factor to the scalability of this crossbar is the size of the SLM and the crosstalk arising from the finite contrast ratio of the SLM. For an SLM with a contrast ratio of X dB then the signal to crosstalk noise ratio for N channels will be X - $\log_{10}$ (N-1) dB assuming equal power in all channels. As current ferro-electric liquid crystal SLMs have contrast ratios of between 300 and 1000 [44] the maximum number of channels which may be interconnected in this way is of the order of 1000.

The strength of this scheme lies in the use of tightly confined Gaussian beamlets to define the channels. Because a Gaussian beam is a minimum uncertainty wavepacket diffraction is minimised. This results in very low crosstalk between adjacent channels and allows very tight packing of beamlets and thus a maximum number of channels. A second advantage in using Gaussian beamlets is that the active switching area of each modulator pixel can be very small. In the example discussed here the size of each SLM pixel may be less than 10 microns. The remaining area of the SLM is available for control and logic circuitry. Alternatively if the beamlets were used to directly address a photodetector array then this would allow photodetectors with very small areas to be employed. As the optical power required to charge a photodetector to a particular voltage is determined by the capacitance of the detector which is in turn proportional to its area, small photodetectors allow switching at higher speeds or lower light levels.

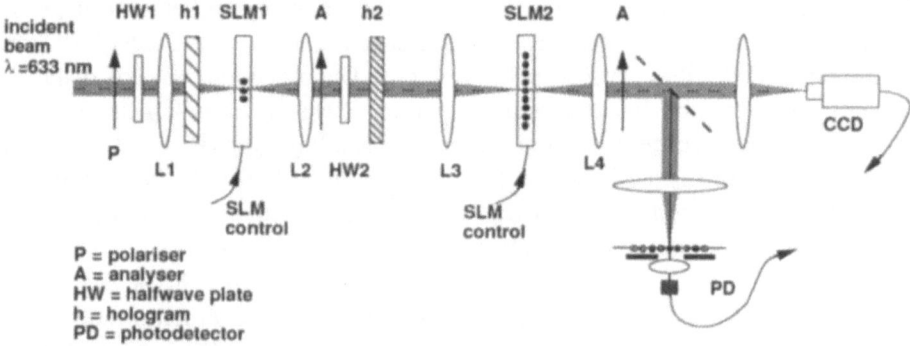

Fig. 12.18 Experimental scheme for the matrix-matrix interconnect.

In the telecommunications crossbar switch the original array of beamlets is obtained from an array monomode fibres. An alternative method is to use a Fourier plane array generator to split the light from a single Gaussian beam into many tightly focused spots. The intensity of each beamlet generated in this way may then be modulated by an SLM. A 25 channel optoelecronic matrix-matrix multiplier has been implemented in this way [45,46]. The experimental system is shown in Fig.12.18. The Fourier plane array generator h_1 (a binary computer generated hologram) creates a 5 x 5 beamlet array. Each beamlet iluminates a single pixel in SLM1 which provides the input to the system. The second Fourier plane array generator h_2 multiply images this pattern 5 x 5 times onto the weight matrix SLM (SLM2) which has 25 x 25 pixels. The image is then split into two paths. In one path an image of the 25 x 25 modulated beamlet array is obtained at the CCD camera. In the other the output for a single channel is obtained by aperturing off all the 5 x 5 sub-arrays except for one. The light within this sub-array is focused into a photodetector. The operation of this system is shown in Fig. 12.19 which shows a multiply imaged pattern presented at SLM1 after the multiple imaging operation. By switching pixels at SLM2 different channels can be selected. This is shown for two sub-arrays in Fig.12.20. It has been found [46] that when an array of beamlets is used to encode optical channels (rather than a plane wave) the signal to noise ratio is increased from 4dB to better than 9dB. The worst-case signal to noise ratio for the 25 channel system was found to be 6dB. This system is an important demonstration of the application of thin diffractive elements as optical array generators and as parallel signal routeing devices.

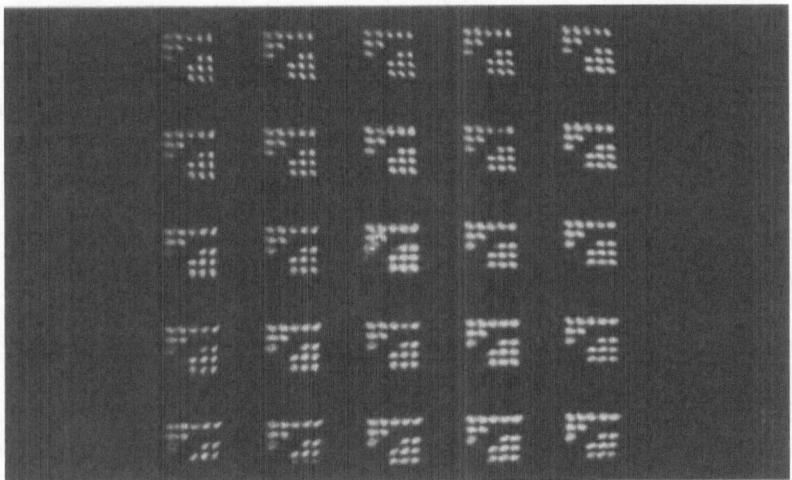

Fig. 12.19 The output of the experimental system observed at the CCD camera. All pixels in SLM2 are set to 'on' and a pattern presented at SLM1 is multiply imaged.

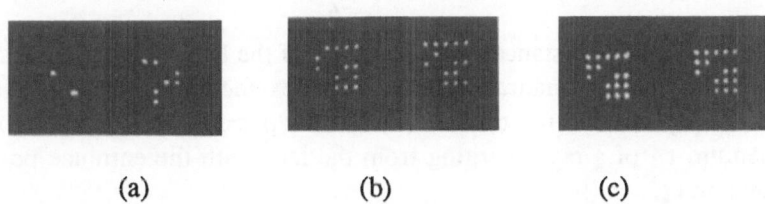

(a)   (b)   (c)

Fig. 12.20 Detail of two sub-arrays at the output of the system when pixels in SLM2 are switched.

12.6 Compact parallel optical interconnections

12.6.1 Minimum optical interconnection volume

A claim often made for optoelectronic interconnection technology is that very compact, high density interconnections 0are possible. An investigation of the lower limit for the volume of an optically interconnected array of points has been made by Ozaktas and Goodman [47]. By considering the total number of degrees of freedom available for optical communication within a closed volume they have shown that the minimum volume required for an optical channel of length L is given by

$$V = \left(\frac{\lambda^2 L}{2\pi} \right) \tag{12.6}$$

Thus in principle an optical channel can occupy a volume considerably smaller than that required for an equivalent insulated electric wire. In addition it is of course possible to route optical signals through free-space rather than over a two-dimensional surface. At present however, optoelectronic interconnection technology has not yet begun to approach this limit. There is thus considerable scope for the miniaturization of optoelectronic interconnection schemes. This is important if optical interconnection technology is to be compatible with VLSI systems.

12.6.2 Compact Fourier plane optical interconnections

One criticism which is often made of Fourier plane array generators is that the requirement which they have for a bulky Fourier transform lens limits their scope for miniaturisation. Recently however, small gradient refractive index (GRIN) lenses have become widely available. By using elements such as these it is possible to obtain very compact Fourier plane optical interconnections [48-50]. GRIN lenses differ from bulk glass lenses in that the refractive index profile $n(r)$ of the lens medium varies as

$$n(r) = n_0(1 - \frac{A}{2}r^2) \tag{12.7}$$

where r is the radial distance from the centre of the lens and A is a constant of the lens. Within such a quadratic refractive index medium rays follow sinusoidal paths [37. By use of a ray transfer matrix it is possible to relate the position and momentum r_2 of a ray emerging from the lens with the entrance position and momentum r_1.

$$\begin{pmatrix} r_2 \\ r_2' \end{pmatrix} = M \begin{pmatrix} r_1 \\ r_1' \end{pmatrix} \tag{12.8}$$

where r_1 and r_1' are the entrance position and momentum respectively and M is the ray transfer matrix. For a meridional ray r is the radial distance from the entrance position of the ray to the axis of the lens and r' is the sine of the angle of incidence. M is written as

$$M = \begin{pmatrix} a & b \\ c & d \end{pmatrix} \tag{12.9}$$

For a lens with the refractive index profile given by equation 12.7 the values of a,b c and d are given by [51]

$$a = \cos(l\sqrt{A}) \tag{12.10}$$

$$b = \frac{\sin(l\sqrt{A})}{n_0\sqrt{A}} \tag{12.11}$$

$$c = -\frac{A\sin(\sqrt{A})}{\sqrt{A}} \tag{12.12}$$

$$d = \cos(l\sqrt{A}) \tag{12.13}$$

where l is the length of the lens. If the length of the lens is related to the constant A by

$$l = \frac{1}{\sqrt{A}}(2n+1)\frac{\pi}{2} \tag{12.14}$$

where n is an integer then the exit position r_2 of any ray depends only upon the entrance angle r_1'. Thus the lens acts a Fourier transform lens with an effective focal length of $f = 1/n_0\sqrt{A}$. Such lenses are referred to as 0.25 pitch lenses.

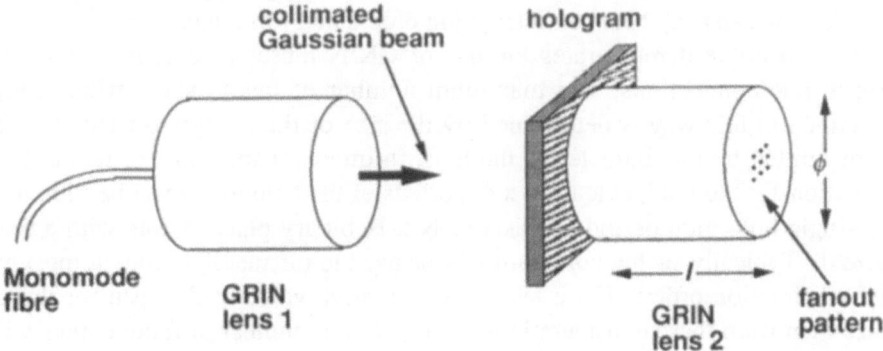

Fig. 12.21 Compact holographic interconnect using a GRIN lens.

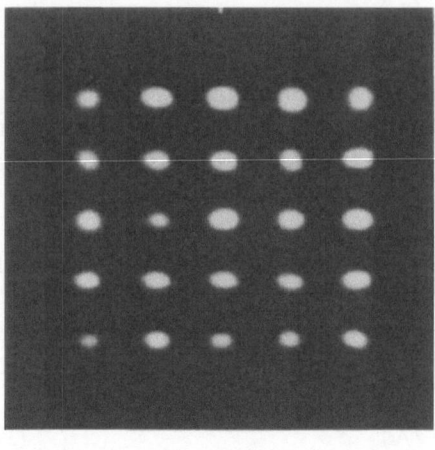
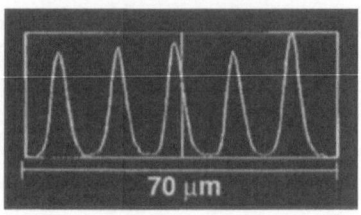

a)   b)

Fig. 12.22 Observed fanout array **a)** and intensity profile **b)** at the back face of
the GRIN lens (from [50]).

Fig.12.21 shows the use of a GRIN lens together with a Fourier plane array
generator to form a compact free-space optical interconnection [50]. Light
emitted from a monomode fibre was collimated using the first GRIN lens L_1. This
collimated beam was incident upon a binary phase grating H which was
designed to reconstruct a 5 x 5 array of fanout spots. A second GRIN lens was
placed immediately behind the hologram. Because this was a 0.25 pitch lens the
fanout pattern from the hologram was obtained at the exit face of the lens. The
experimentally observed diffraction pattern is shown in fig.12.22. This resulted in
a very compact optical interconnection device. The lens L_2 had a diameter ϕ of
1.8 mm and a length l of 3.7 mm. From the manufacturer's data the quadratic
constant of the lens A had a value of 0.183 mm^{-2} and the refractive index of the
lens n_0 was 1.658. The hologram used had a grating period of 64 microns which
resulted in a spacing between diffraction orders of 14 microns.

This example demonstrates the use of GRIN lenses to obtain compact free-
space interconnections. The maximum number of fanout spots which may be
obtained in this way is determined by the size of the hologram used. This is in
turn limited by the diameter of the lens. In order for the approximation used in
equation 12.2 to hold, at least 4 x 4 periods of the hologram must be illuminated.
A single hologram period consists of N x N binary phase pixels with a feature
size d. Typically such a hologram will be used to diffract light into at most N/4 x
N/4 diffraction orders. If we take a conservative value of d=2 μm for the pixel
size then from these considerations the maximum number of fanout spots will be
approximately 25 x 25 for a lens with diameter used here.

A more useful application of this system is to use the GRIN lens together with a
Fourier plane array generator to allow compact multiple imaging operations to be
performed. For example the large bulk glass lenses in the optical crossbar

discussed in section 12.5 may be replaced with GRIN lenses. In this application the CGH and lenses are required to multiply image an M x M array of input channels up into an M x M array of M x M output channels. The number of channels is determined by the resolution of the GRIN lens and the minimum size of the detector/modulator elements at the focal plane of the lens. If the lens gives diffraction limited performance it should focus an incident Gaussian beam into a beamlet with a beam waist of 1.2 microns. Aberrations and defects in the lens will result in a somewhat larger beam waist than this however, particularly at the edges of large multiply-imaged arrays. The detector/modulator arrays with active pixel sizes of 10 μon a pitch of x=20 μmay be envisaged. Each beamlet must be separated by this distance. The spacing of each diffraction order s must therefore be equal to Mx. For a hologram with N pixels with feature size d, s=n λf/Nd where f is the focal length of the GRIN lens and n is the spacing of diffraction orders used in the array. For example if the hologram diffracts into odd orders only then n=2. In order to reconstruct an M x M array such a binary phase hologram typically requires at least 4M x 4M pixels. Hence M ≤ N/4. From these considerations we obtain

$$M \leq \sqrt{\frac{n\lambda f}{4dx}} \qquad (12.15)$$

In order to increase the spacing between the multiple images a lens with a larger focal length should be used. Choosing f=6 mm, n=2, d=1 μm and x= 20 μm this gives M ≤ 10. Hence such a multiple imaging system could be used to globally connect 100 parallel channels using a lens with a diameter of 4 mm and a length of 5 mm. This scheme is shown in fig.12.23. The system is reduced to the smallest possible size by cementing the hologram directly to the lens faces. Because of the shift-invariant nature of Fourier holograms there should be reasonable tolerance to lateral positioning errors. Alignment tolerance will be a critical factor however in positioning the input array and the output detector or modulator arrays. Fig. 12.23 Matrix-matrix interconnection scheme with a GRIN lens.

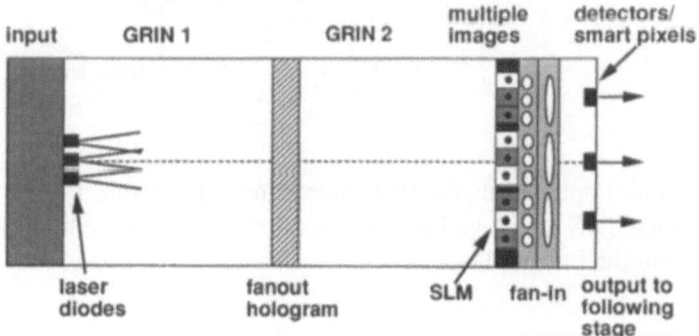

Fig. 12.23 Matrix-matrix interconnection scheme with a GRIN lens.

A similar scheme to the one discussed here has been investigated by researchers at Nippon Sheet Glass Co. [48,49]. In this case arrays of microlenses were used to perform the multiple imaging operation. A very compact interconnection architecture was again obtain by the use of GRIN lenses. One suggested use for this system was as an optical bus for the interconnection of stacks of VLSI devices. This is shown in fig.12.24. An ingenious method of overcoming alignment difficulties was also suggested. All the GRIN lenses within one parallel bus start as a single SELFOC[1] gradient index rod. This is placed in a substrate and then cut into sections of the correct length. Because the rod sections remain positioned within the substrate they remain in perfect alignment. The VLSI boards which contain arrays of sources and detectors then slot into the gaps. This is shown on fig.12.25. At the beginning of this section it was stated that the minimum volume for an optical channel is $V=\lambda^2 L/2\pi$. At the output face of the crossbar described here as many as 100 x 100 channels may be obtained. If the lens diameter is 4 mm and two lenses with a length of 5 mm each are required then the total volume of the system is $1.3 \ 10^{-7} \ m^3$. Hence the total volume required for each output channel is $1.3 \ 10^{-11} \ m^3$. The minimum volume which an optical channel of this length requires is $6.4 \ 10^{-16} \ m^3$. Hence even if the factor of 2π is ignored the volume of the system discussed here is still 4 orders of magnitude greater than the theoretical limit. Thus although it may seem ambitious to consider a 100 channel free-space crossbar within a volume of $0.12 \ cm^3$ such a design is unnecessarily spacious in terms of the fundamental limits to optical interconnections.

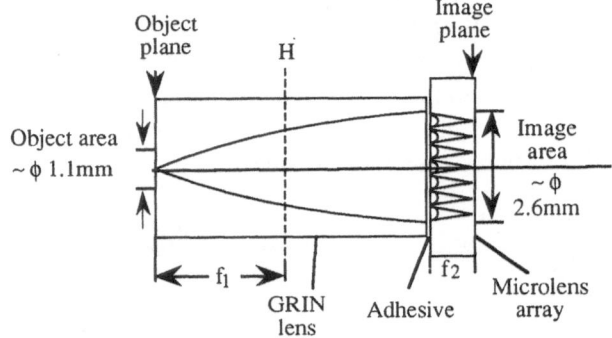

Fig. 12.24 Parallel optical bus for VLSI interconnection (from Hamanaka [48]). The GRIN lens together with the microlens array are multiply imaged onto the image plane.

---

1 SELFOC is a trademark of Nippon Sheet Glass Ltd.

12.6.3 Fresnel plane interconnections

An alternative approach to obtaining compact optical interconnections with diffractive elements is to employ Fresnel plane devices. In contrast to the Fourier plane devices discussed so far in this chapter these devices are designed to diffract light into the Fresnel plane and Fourier transform lenses are not required. Fresnel holograms may either be recorded optically or plotted as computer generated holograms.

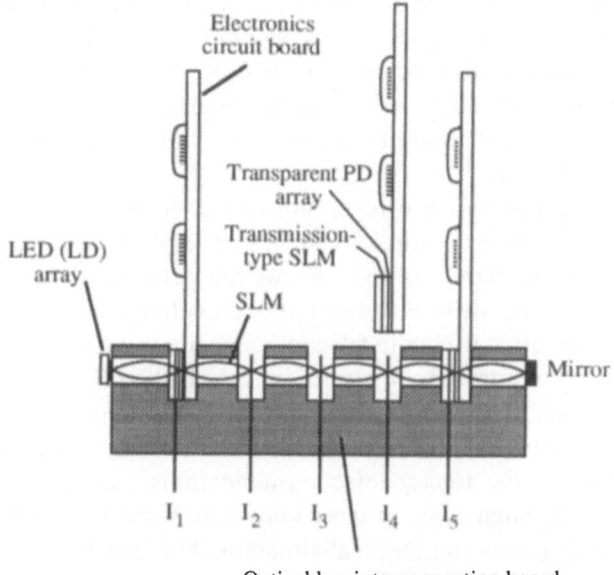

Fig. 12.25 System for optical bus (From Hamanaka [48]). Data is routed in parallel between the slot-in boards.

The methods by which optically recorded holograms may be obtained are covered thoroughly in chapters 4 and 5 of this book. Fig. 12.26 shows an example of an optical recording set-up for a simple interconnection hologram. In this case a hologram is required to collimate the light emitted by a monomode fibre. The hologram is recorded as the interference between the expanding spherical wave emitted by the fibre U' and a plane reference beam R. After the hologram has been developed and replaced in front of the fibre it will be reconstruct the plane wave R* when light from the fibre is incident upon it. It is a simple extension of this scheme to construct a holographic coupler for two monomode fibre by placing two such holograms back to back (see 12.27) The plane wave R_1^* reconstructed by the first hologram H_1 is incident upon the second hologram H_2 where it reconstructs the converging spherical wave U_2^*. This should then be perfectly matched to the second fibre and so little coupling loss should occur. This approach has been investigated by many researchers (for example [52-54]). Monomode fibres provide a very good point source for expanding spherical waves. If a single collimating hologram described above is illuminated with a plane wave it will act as a lens and focus the light down into a diffraction limited spot. If the holographic plate is laterally shifted after each exposure a multifacet holographic array generator will be obtained [55]. A single plane wave incident upon the hologram will be reconstructed into an array of spots. This has been extended by Prongué and Herzig [18] to obtain an array generator for optical clock distribution. The focal length of the array generator was 31 mm. A more compact array generator was fabricated by the use of a total internal reflection (TIR) recording technique. This is described in chapter 4. A 100 x 100 array of lenslets was fabricated in an area of 1 cm^2 with an array pitch of 100 µm. The focal length of the lens was 390 µm and the diffraction limited spot size was 10 µm.

Although interferometric recording allows high efficiency holograms to be obtained with relative simplicity it has several disadvantages. One problem commonly encountered is that the wavelength which is used to record the hologram may be different from the wavelength used to reconstruct the hologram. This often occurs in telecommunications applications where diffractive elements are required to operate in the 1200-1600 nm region. As most holographic materials are sensitive to the visible range region of the spectrum diffractive elements are often recorded at these wavelengths and then reconstructed at the longer telecommunications wavelengths. This results in aberrations. Although some of these may be reduced by modifying the recording geometry [52] this is not an ideal situation. The interferometric method is also limited by the complexity of the recording process. Although multifacet array generators can be obtained by a mechanised step and repeat recording operation it is difficult to obtain arrays of lenses with differing parameters.

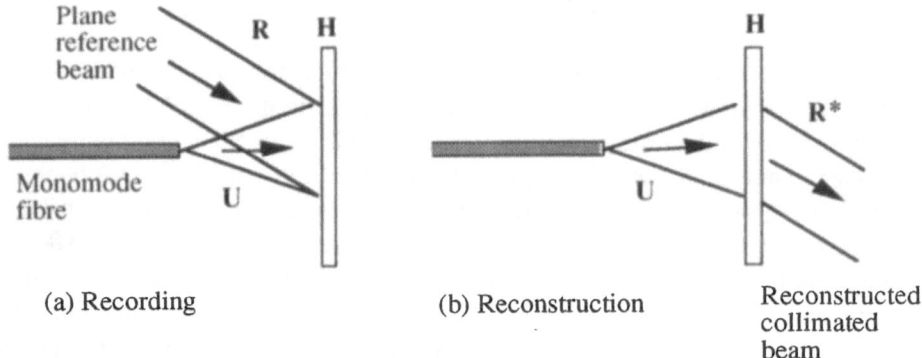

(a) Recording   (b) Reconstruction   Reconstructed collimated beam

Fig. 12.26 Recording and reconstruction geometry of a holographic lens

An alternative to interferometric recording is to calculate the required fringe patterns for the hologram and then fabricate the element using suitable a lithographic technique. One of the simplest examples of this is the fabrication of an array of Fresnel zone plates. Zone plates can be fabricated as a chrome amplitude mask on a glass substrate using electron-beam lithography techniques. However the diffraction efficiency of such elements is very low ($\approx$10%). Much greater diffraction efficiency can be realised by fabricating the zone plate as a phase structure. The theoretical efficiency of a binary phase zone plate is 41% if absorption and reflections are ignored. To improve efficiency beyond this it is necessary to increase the number of phase levels - i.e. to obtain a quantised approximation to a blazed element. This is described in more detail in Chapter 5. There is a considerable amount of on-going research into the fabrication of these elements for optical interconnect applications. One such application for these lenslet arrays is as beamlet waveguide arrays in architectures such as the optical crossbar switch discussed in section 13.5. This has been investigated by researchers at AT & T [56]. The use of microlens arrays as a beamlet waveguide is shown in fig.12.28. By using a separate lens for each channel there is little crosstalk between adjacent channels. Many thousands of lenses may be fabricated on the same substrate, thus giving the potential for a very large number of parallel channels. One advantage of the use of synthesised microlens arrays such as these is that the focal lengths and numerical apertures of the the lenses can be adjusted to match exactly the requirements of the sources and detectors which are to be coupled. A key issue in multiple beamlet architectures such as these is the tolerance of microlens arrays to errors in positioning. This has been investigated by McCormick [56] who found that for large NA lenses in the context of the S-SEED optical array processor the positional error tolerance was less than 10% and the angular tolerance was only a few arc minutes. This study however considered lenses placed in the confocal arrangement. While this minimises the modal volume of the optical channels and also the number of lenses required it is only marginally stable [57] and hence particularly vulnerable to misalignment. It should be possible to relax these severe alignment tolerances by reducing the distances between lenses.

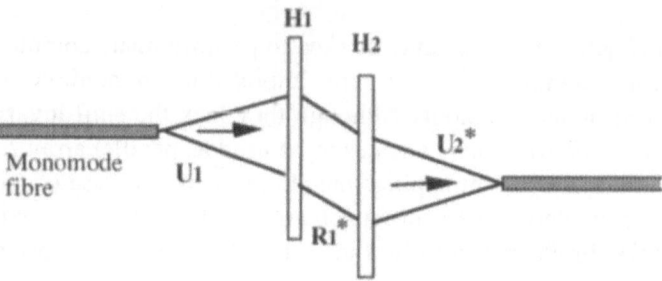

Fig. 12.27 Holographic fibre coupler

290

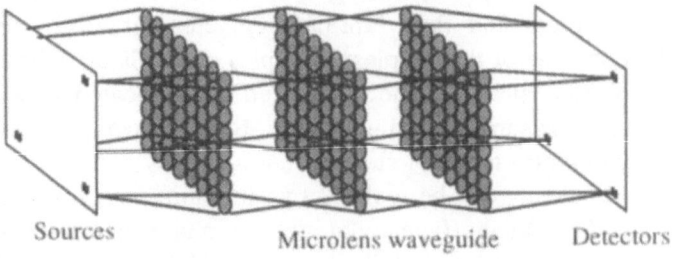

Sources   Microlens waveguide   Detectors

Fig. 12.28 Microlens array as a beamlet waveguide

It is relatively simple to modify the basic Fresnel lens design to obtain elements which perform more complex beam shaping operations. For example Fresnel lenses with different focal lengths may be fabricated on the same substrate. By changing the fringe pattern the position of the focal spots may be controlled [58]. Fig. 12.29 shows the fringe pattern for a hologram which reconstructs a circular focal line [58]. By a continuation of this technique it is possible to obtain line foci and focal curves which extend in three-dimensional space [59] One example of this is a 'light sword' element [60] which is shown in Fig. 12.30. This has a helical focus which extends in space directly in front of the hologram.

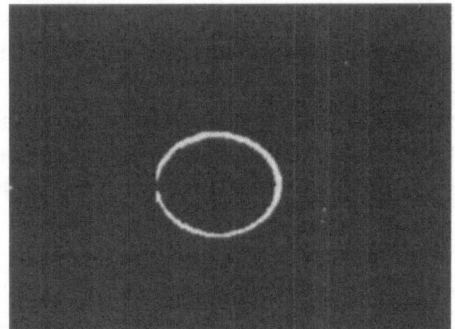
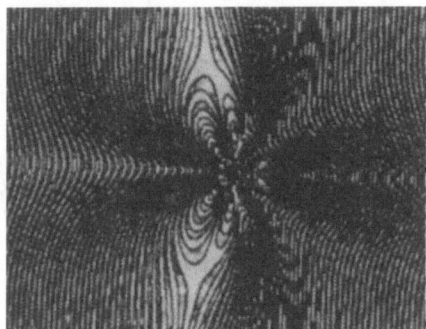

Fig. 12.29 a) Circular focal line fringe pattern (from [58]). **b)** Diffraction pattern.

Fresnel diffractive elements may be used to perform many complex beam shaping and routing operations which are impossible to achieve by the use of conventional optical methods. Although they lack the shift invariance condition which enables Fourier plane devices to be used as parallel array routing elements they have many important applications. In particular the use of arrays of Fresnel microlenses to define very compact arrays of optical channels will become increasingly important within the context of parallel optical information processing.

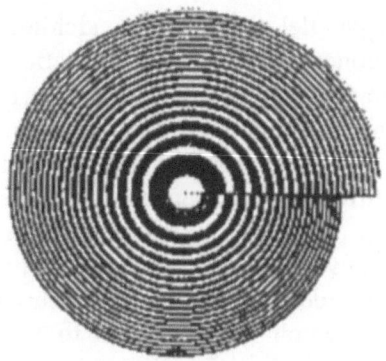

Fig. 12.30 Light sword element (from [59]).

12.7 Conclusions - future directions for parallel optical interconnections

Within this chapter we have presented some of the applications within parallel optical information processing for which diffractive optical elements are well-suited. Many different and important operations may be carried out by diffractive elements. These include optical array generation, weighted interconnections, focusing, beam shaping, beamlet array waveguiding, multiple imaging and the definition of shift-invariant interconnections.

Much early work in the field of optoelectronic interconnections was concerned with the replacement of electrical wires with optical 'wires'. Diffractive elements were used to provide point to point interconnections within conventional architectures. More recently however, new concepts of parallel optoelectronic processing systems have emerged. Diffractive optical elements allow operations to be performed which are not possible within conventional electronic systems. For example the use of Fourier plane array generators as multiple imaging devices allows the routing and manipulation of two-dimensional arrays of data in parallel. In the same way shift invariant convolution filters allow complex non-local interconnections to be defined simultaneously across an entire two-dimensional array. Weighted fanout holograms allow arbitrary interconnection patterns to be defined between hundreds or thousands of data channels. All of these are parallel operations which allow entirely new architectures and processing concepts to be formulated which differ radically from the conventional von Neumann serial computer design.

Many of the new parallel optoelectronic architectures which are emerging are based around the concept of the interconnection of arrays of optoelectronic processing and logic elements through free-space. There are many different forms which such elements may take and there is a considerable on-going research effort in this area. However many of the basic functions of these devices are common to many different implementation technologies. As the complexity of a parallel processing system increases each element has an increasing independence from any central control unit. At its most extreme this results in the neural processing model in which each neuron is an autonomous decision maker. One device which should prove to be particularly useful in the optoelectronic implementation of such architectures is the `smart pixel' spatial light modulator [61,62]. These devices consist of pixels which can receive and transmit information both optically and electronically. They will possess a significantly higher degree of functionality than current SLM devices and will allow complex optoelectronic information processing operations to be performed. The exact form of these devices will depend upon the application in which they are to be used and there are many alternative schemes. For example the modulator could consist of an S-SEED or the device may directly drive an LED or a laser diode. Bipolar optical interconnections could be obtained by using two photodetectors at each pixel, connected as shown in fig.12.15. The electronic inputs may come from neighbouring devices or they may be signals from a master control unit located elsewhere. The electronic logic unit may in fact be a serial processor with local memory and arithmetic and logical unit. In this case the electronic unit may be thought of as an 'electronic island' [63] which interacts with neighbouring islands optically. The size if the island depends upon the nature of the processing task which is implemented and also on the break-even point between optical interconnections and electrical interconnections. This last point is still an on-going subject for debate (see for example [64]). However, regardless of the exact form of these devices, the basic concepts of parallel optical interconnections will remain unchanged.

Many of the individual elements which are required for future optoelectronic processing and communication systems are already avalailable. Smart pixel SLMs and similar devices are being designed and developed together with arrays of laser diodes and high speed light modulators. Although there are still many issues to be addressed in this area, for example concerning the alignment of large parallel optoelectronic systems, they are now a practical reality. Diffractive elements have a vital part to play in these new architectures and as such will become increasingly important in the future.

References

1. H Dammann and K Gortler, `High-efficiency in-line multiple imaging by means of multiple phase holograms', Opt. Commun. 3 , 312-315, 1971.
2. F B McCormick, `Generation of large spot arrays from a single laser beam by multiple imaging by means of binary phase gratings', Opt. Eng. 28 (4), pp 299-304, 1989.
3. N Streibl, `Beam shaping with optical array generators', J. Mod. Opt. , 36 , pp 1559-1573 ,1989.
4. G Groh, `Multiple imaging by means of holograms of correlated objects', Appl. Opt. 7 , pp 967-969, 1969.
5. L P Boivin `Multiple imaging using various types of simple phase grating', Appl. Opt. 11 (8) p. 1782, 1972.
6. F Wyrowski and O Bryngdahl, `Digital holography as part of diffractive optics' , Reports on Progress in Physics , 54 (12) pp 1481-1572, 1991.
7. MA Seldowitz, JP Allebach, DW Sweeny, `A synthesis of digital holograms by direct binary search', Appl Opt 26 (14) p. 2789, 1986.
8. M.R.Feldman and C.C.Guest, `Iterative Encoding of high-efficiency holograms for generation of spot arrays', Opt. Lett., 14 (10), pp 479-481, 1989.
9. M.P. Dames, R.J. Dowling, P. McKee, D. Wood, `Design and fabrication of efficient optical elements to generate intensity weighted spot arrays', Applied Optics 30 (19), pp 2685-2691, 1991.
10. A. G. Kirk and T. J. Hall, `Design of computer generated holograms by simulated annealing: coding density and reconstruction error', sumitted to Opt. Commun. , 1992.
11. A G Kirk and T J Hall, `Design of computer generated holograms by simulated annealing: observation of metastable states', sumitted to J. Mod. Opt. , 1992.
12. W-H Lee, `Computer-generated holograms: techniques and applications', Progress in Optics XVI, Ed. E Wolf, North Holland 1978.
13. M R Taghizadeh, J Turunen, B Robertson, `Passive Optical Array Generators', OSA 1991 Technical Digest Series, Vol 6, p 148, 1991.
14. U Killat, G Rabe, W Rave, `Binary phase gratings for star couplers with high splitting ratios', Fibre and Integrated Optics 4 (2) pp 159-167, 1982.
15. J W Goodman, F I Leonberger, S-Y Kung, R A Athale `Optical interconnections for VLSI Systems', Proc. IEEE 72 (7) p. 850, 1984.
16. L Bergmann, A Johnston, R Nixon, `Appications and design considerations for optical interconnects in VLSI' Proc. SPIE 625 , p 117, 1986.
17. D H Hartman, `Digital high speed interconnects: a study of the optical alternatives', Opt Eng. 25 (10) pp 1086-1102, 1986.

18. D Prongué and H P Herzig, `HOE for clock distribution in integrated circuits: experimental results', Proc SPIE 1281 , 1990.
19. M M Downs and J Jahns, `Integrated optical array generator', Opt. Lett 15 (14) pp 769-770, 1990.
20. J J Hopfield, `Neural networks and physical systems with emergent collective and computational abilities', Proc. Natural Acad. Sci. USA 79 pp 2554-2558, 1982.
21. D Psaltis, N Farhat , `Optical information processing based on associative memory model of neural networks with thresholding and feedback', Opt. Lett. 10 (2) p. 98, 1985.
22. N H Farhat, `Optical Implementation of the Hopfield Model', Appl. Opt. 24 (10) p. 1469, 1985.
23. H J White, W A Wright, `Holographic implementation of a hopfield model with discrete weightings', Appl. Opt. 27 (2) p 331, 1988.
24. N C Roberts, `Fixed holographic interconnects for neural networks', Congress on Optical Science and Engineering, the Hague, 12-15 March 1990.
25. B K Jenkins, P Chaval, R Forchheimer, A A Sawchuck, T C Strand, `Architectural implications of a digital optical processor', Appl. Opt. 23 (19) pp 3465-3474, 1984.
26. C Mead, Analog VLSI and neural systems , Addison-Wesley, MA, 1989.
27. A G Kirk, G D Kendall, M-Y Chan, T J Hall, `Holographic Interconnections within cellular optical processing systems', submitted to Optical Computing and Processing , 1992.
28. M G Robinson K M Johnson, D Jared, S Wichart, G Moddel, `Custom designed electro-optic components for optically implemented, multi-layer neural networks', in Optical Computing , OSA Tech. Digest Series 6, pp 84-87, 1991.
29. K T Spoehr and S Lehmkuhle, Visual Information Processing , W.H.Freeman, San Francisco, 1982.
30. P J de Groot, R J Noll, `Reconfigurable bipolar analog optical crossbar switch', Appl. Opt. 28 , pp 1582-1587, 1989.
31. R Nevatia, Machine perception , Prentice Hall, NJ, 1982.
32. J W Goodman, A R Dias, L M Woody, `Fully parallel high-speed incoherent optical method for performing discrete Fourier transforms', 2 , (1), p 1-3, 1978.
33. D Psaltis, `Two-dimensional optical processing using one dimensional input devices', Proc. IEEE 72 (7), pp 962-974, 1984
34. A G Kirk, T J Hall, W A Crossland, `A compact and scalable free-space optical crossbar', in Holographic Systems, Components and Applications IEE Publication No. 342, pp 137-141, September 1991.
35. A R Dias, R F Kalman, J W Goodman, A A Sawchuck, `Fibre optic crossbar switch with broadcast capability', Opt. Eng. 27 (11) pp955-960 1988.

36. J W Goodman, `Fan-in and fan-out with optical interconnections', Opt Acta , 32 (12), pp 1489-1496, 1985.

37. D Marcuse, Light Transmission Optics , Van Nostrand Rheinhold, New York, 1972.

38. M Ishikawa, N Mukhozaka, H Toyoda, Y Suzuki, `Optical associatron: a simple model for optical associative memory',   Appl. Opt. 28 , pp 291-301, 1989.

39. F T S Yu, T Lu, X Yang, `Optical neural network with pocket-sized liquid crystal televisions', Opt. Lett. 15 , pp 863-865, 1990.

40. A G Kirk, G D Kendall, H Imam, T J Hall, `An optical neural network with reconfigurable holographic interconection', Proc. SPIE Vol. 1402 , 1991.

41. K J Weible, G Pedrini, W Xue, R Thalman, `Optical implementation of a neural network associative memory using diffraction gratings', Japanese J. Appl. Phys. 29 (7), pp L1301-L1303, 1990.

42. D J McKnight, D G Vass, R M Sillito, `Development of a spatial light modulator - a randomly addressed liquid crystal over nMOS array', Appl. Opt. 23 pp 4757-4762, 1989.

43. J Shamir, H J Caulfield, R B Johnson, `Massive holographic interconnection networks and their limitations', Appl. Optics , 28 (2), pp 311-324, 1989.

44. W A Crossland, M J Birch, A B Davey, D G Vass, `Ferroelectric liquid crystal/silicon VLSI backplane technology for smart spatial light modulators', in Two Dimensional Optoelectronic Device Arrays, IEE Digest 1991/158, October 1991.

45. A G Kirk, H Imam, K Bird and T J Hall, The design and fabrication of computer generated holographic fanout elements for a matrix-matrix interconection scheme', Proc. SPIE Vol. 1574 , 1991.

46. A G Kirk, S Jamieson, H Imam, T J Hall, Experimental implementation of an optoelectronic matrix-matrix multiplier which incorporates holographic multiple imaging', submitted to Optical Computing and Processing , 1992.

47. H Ozaktas, J W Goodman, Lower bound for the communication volume required for an optically interconnected array of points', J. Opt. Soc. Am. A 11, pp 2100-2106, 1990.

48. K Hamanaka and T Kishimoto, `Multiple imaging and multiple Fourier transformation using planar arrays', Japanese J. Appl. Physics 29 (7) pp L1277-L1280, 1990.

49. K Hamanaka, H Nemoto, M Oikawa, E Okuda, T   Kishimoto, `Multiple imaging and multiple Fourier transformation using planar microlens arrays', Appl. Opt. 29 (28) pp 4064-4070, 1990.

50. A G Kirk, H Imam, T J Hall, An efficient holographic interconnect in 0.01 cm^3, in Holographic Systems, Components and Applications , IEE Publication 342, pp 161-165, September 1991.

51. W Shaomin, L Ronchi, `Principles and design of optical arrays', Progress in Optics XXV , pp 280-348, Ed. Wolf, North Holland, Amsterdam, 1988.

52. J S Leggatt, `Small aperture lenses for optical interconnect application' in Holographic Systems, Components and Applications , IEE Conference Proceeding No. 76, Cambridge, 1987.

53. J S Leggatt, G R Chamberlin, D E Sheat, D J MacCartney, `Holographically generated optical components for routing and wavelength division multiplexing applications', Proc. SPIE 1281 pp 227-243, 1990.

54. O D D Soares, `Holographic coupler for fibre optics', Opt. Eng. 20 (5) pp 740-745, 1981.

55. J Andreassen, R Gibson, M G Scott, `Generation of optical interconnect elements', in Holographic Systems, Components and Applications IEE Publication No. 76, pp 153-158, September 1987.

56. F B McCormick, F A P Tooley, J M Sasian, H S Hinton, `Microbeam interconnection using microlens arrays for free-space photonic systems', in Photonic Switching 1991 , Technical Digest Series 6 OSA, p 244.

57. H Kogelnik, T Li, `Laser beams and resonators', Appl. Optics 5 (10) pp 1550-1567, 1966.

58. B Kress, `Fresnel computer generated holograms', French DEA Diploma Thesis, Department of Physics, King's College London and Laboratoire des Systemes Photoniques, Universite Louis Pasteur, Strasbourg, 1991.

59. S Bara, C Frere, Z Jaroszewicz, A Kolodziejczyk, D Leseberg, `Modulated on-axis circular zone plates for the generation of three-dimensional focal curves', J. Mod. Opt., 37 , (8), pp 1287-1295, 1990.

60. A Kolodziejczyk, S Bara, Z Jaroszewicz, M Sypek, `The light sword elements - a new diffractive structure with extended depth of focus', J. Mod. Opt. 37 (8), pp 1283-1286,1990.

61. N Collings, W A Crossland, P J Aycliffe, D V Vass, I Underwood, `Evolutionary development of advanced spatial light modulators', Appl. Optics 23 pp 4740-4747, 1989.

62. S H Lee, S A Esener, M A Title, T J Drabik, `Two-dimensional silicon/PLZT spatial light modulators: design considerations and technology', Opt. Eng 25 (2) pp 250-260, 1986.

63. J E Midwinter, `Photonics in switching: the next 25 years of optical communications ?', IEE Proceedings Paper 8366J, October 1990.

64. M R Feldman, S C Esener, C G Guest, S H Lee, `Comparison between optical and electrical interconnects based on power and speed considerations', Appl. Opt. 27 (9), pp 1742-1751, 1989.

13 Interconnects with optically thick elements

G. Pauliat and G. Roosen
Institut d'Optique, CNRS, Orsay

13.1 Introduction

From chapter 11, we know that the number of degrees of freedom of a space invariant interconnect linking two 2D arrays of N elements is equal to N. For such interconnections, thin holographic elements are thus well suited. Indeed, their required surface scales proportionally to the number of the input and output channels, i.e. scales as the input and output surfaces: the system geometry does not depend on capacity N.

For a space variant system each channel needs a separate portion of the thin hologram surface which, for instance, may lead to the use of multifacet holograms (chapter 12). Thus, the required number of degrees of freedom of a space variant system is N^2. The surface of the thin hologram must grow as the square of the number of input channels. For large N, this surface may very quickly become prohibitive. Similarly, a given optical system (with a given space-bandwith product of N possible channels) may interconnect all N channels in a space invariant way whereas its capacity is reduced to $N^{1/2}$ useful channels for space variant interconnects. The correlator shown in figure 13.1 illustrates this capacity reduction [1].

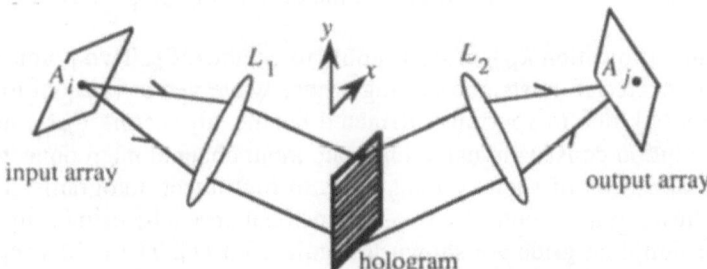

Fig.13.1. Correlator used to implement a space variant interconnect between two arrays of input A_i and output A_j channels [1], only two channels are represented.

In this system, the hologram realizes the interconnect between two 2D arrays. Each spherical wave issued from each point source is transformed into a plane wave by means of first lens L_1. It is then diffracted on gratings superimposed in the thin hologram. The diffracted beams are then focused by second lens L_2 on the output channels. However, with thin holograms, a beam is diffracted not only by the grating required to interconnect this channel but is also diffracted into spurious beams by all the other gratings used for the other channels. Thus, one must arrange the input and output channels so that these spurious beams are not directed toward output channels. These special arrangements, called *sampling grids*, are determined by observing the diffraction process. For a thin grating of wave vector k_g, the diffraction of the input plane wave with wave vector k_R to the output plane wave with wave vector k_S corresponds to the conservation of the tangential components in the grating plane. Moduli k_R and k_S of wave vectors k_R and k_S are determined by the propagation equation of light inside the medium. For the isotropic media we consider in this chapter, we have $k_R=k_S$. We get:

$$k_{R,//} - k_{S,//} = p\,k_g \qquad (13.1)$$

where integer p defines the order of diffraction and notation $//$ represents the tangential components.

This conservation law is illustrated in figure 13.2 in the thin hologram plane and for the first positive diffracted order.

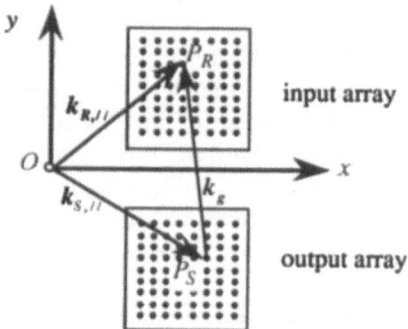

Fig.13.2. Wave vector diagram drawn in the thin hologram plane.

In this plane, projection $k_{R,//}$ of k_R is equal to vector OP_R. Two points P_R and P_S are interconnected if exists one grating whose wave vector is equal to P_RP_S. To avoid cross talk due to spurious diffracted beams, all vectors P_RP_S must differ. The first solution consists in using only one input channel interconnected to a 2D array. Juxtaposition of such systems leads to multifacet holograms. The second solution allows to interconnect 2 two-dimensional arrays by using sampling grids. Two such sampling grids are shown in figure 13.3 [1]. The points represent the channels useful for space invariant interconnects while the thick dots are the remaining channels for space variant interconnects. One may remark that if the incident and emergent angles are small enough then the patterns of points P

shown in figures 13.2 and 13.3 are equivalent to the input and output arrays of channels *A* in figure 13.1.

These two examples illustrate that only the square root of possible input and output channels can be used. This capacity limitation arises from the diffraction process for thin gratings which makes no difference between input beams sharing wave vectors with the same tangential projection (Eq. 13.1). In the following, we will demonstrate that for thick gratings another condition, the *Bragg condition*, must be added to Eq. 13.1 so that the interconnect density can be increased.

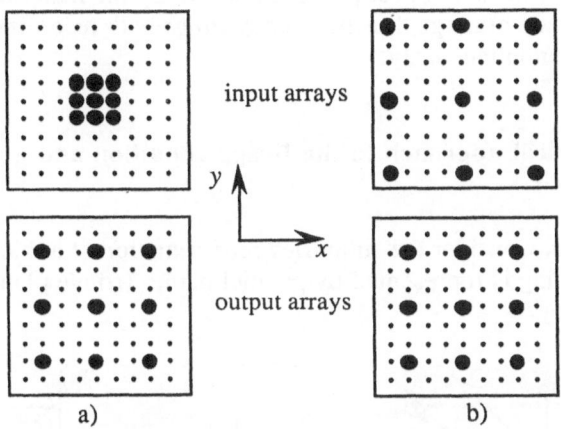

Fig.13.3. Example of sampling grids for thin holograms, the thick black dots represent the sampled channels [1].

Hereafter, in order to introduce this Bragg condition, we first describe the diffraction process in thick gratings. If the reader is already familiar with thick gratings, he can go directly to paragraph 13.3 where we examine how to enlarge the sampling grid density by using this condition. This increase in channel capacity is gained at the expense of some drawbacks which are detailed in paragraph 13.4. Some limitations inherent to existing volume holographic materials are underlined in paragraph 13.5. We then give some examples of the use of thick holographic interconnects in paragraph 13.6. In paragraph 13.7, before concluding, we underline another advantage of thick dynamic gratings which is their ability to establish dynamic self aligning interconnects. They can thus compensate for any misalignment.

13.2 Diffraction properties of thick gratings

Diffraction properties of gratings vary significantly according to the thickness of the material in which those gratings are recorded. When the material thickness becomes larger than the grating period, we are dealing with the so-called "thick gratings" (see chapter 4). Note that this is relevant even for rather thin materials (few tens of µm) when recording gratings with spacings of a few wavelengths. Then, the incidence angle of the readout beam cannot be arbitrarily chosen as it was for thin gratings. As it is shown later, this readout angle depends on the grating spacing and on the readout wavelength. Moreover, only one principal diffracted beam exists.

In the following, we present the conditions corresponding to the thick grating regime and derive some important characteristics. We however limit our presentation to plane wave diffraction on *static gratings*. The term static gratings is used when the diffraction process does not modify the material characteristics. Conversely to static gratings, for *dynamic gratings* both recording and diffraction processes occur simultaneously.

13.2.1 Geometrical approach to the Bragg condition and grating thickness criterion

Bragg condition. Consider the simplified representation of a thick grating (figure 13.4.a). The grating is represented by parallel planes (fringe planes) separated by the fringe spacing Λ.

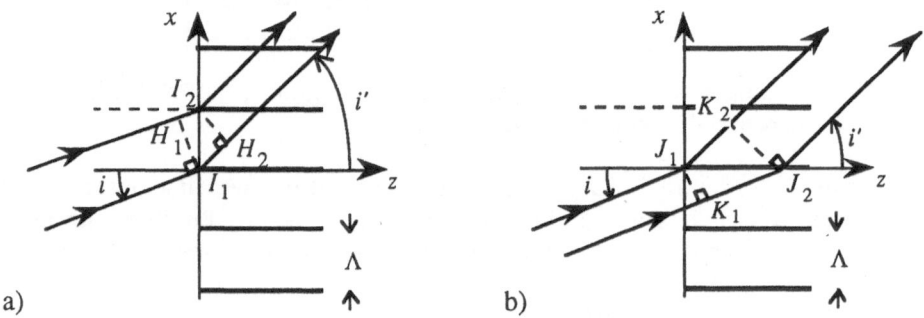

a)   b)

Fig.13.4. Diffraction of two light rays on a static grating.

Two incoming light rays are diffracted at points I_1 and I_2 that are at the same depth in the material. For sake of symmetry, the two diffracted beams are lying in the plane defined by both the grating and the incident beam wave vectors. In isotropic media ($k_R = k_S$), they are in phase and interfere constructively when the path difference δ verifies:

$$\delta = H_1 I_2 - I_1 H_2 = \Lambda(\sin i - \sin i') = p\lambda / n \qquad (13.2)$$

where notation $H_1 I_2$ stands for the algebraic length, λ is the readout wavelength in vacuum, n is the average refractive index of the material and integer p defines the order of diffraction. Note that k_R, k_S and k_g are contained in the same plane. Thus this equation is the one found for diffraction by a thin grating and is equivalent to Eq. 13.1. It represents the conservation of the wave vector tangential components.

For a thick grating, a second relation has to be satisfied: two light rays diffracted by the same fringe plane at different grating depths have to be in phase (figure 13.4.b). The path difference must now obey:

$$\delta = K_1 J_2 - J_1 K_2 = J_1 J_2 \left(\cos i - \cos i' \right) = q\lambda / n \qquad (13.3)$$

This equality can be fulfilled for all depths if integer q equals 0 only. Then, $i'=i$ is related to the transmitted beam and $i'=-i$ to the diffracted beam. This diffraction corresponds to a simple reflection of the incident beam on the grating fringes. This equation is thus equivalent to the conservation of the normal wave vector components:

$$k_{R,\perp} - k_{S,\perp} = 0 \qquad (13.4)$$

One concludes that in thick gratings the first order of diffraction exists only. Combining Eq. 13.1 with Eq. 13.4 we obtain the conservation law for the wave vectors:

$$k_R - k_S = k_g \quad \text{with} \quad \| k_R \| = k_R = \| k_S \| = k_S \qquad (13.5)$$

An equivalent equation is found by introducing $i'=-i$ and $p=1$ in Eq. 13.2 which leads to:

$$\sin i = \lambda / 2n\Lambda \qquad (13.6)$$

Equations 13.5 and 13.6 are two different ways to express the so called *Bragg condition*. To get the maximum diffraction efficiency, the reading beam has to be incident at a given angle that is defined by the reading beam wavelength, the grating fringe spacing and orientation. In the following, we will make use of formulation 13.5 because it allows very simple graphical representations.

Compared to the thin grating conservation law, Eq. 13.1, the thick grating conservation law, Eq. 13.5, is more restrictive. Less possible input and output channels of the interconnect system of figure 13.1 may share the same interconnection gratings. We can thus increase the useful number of independent channels in thick media. Larger densities for sampling grids can be achieved.

Before calculating this gain in density, we will first define more precisely the minimum grating thickness which represents the boundary between thin and thick gratings.

Grating thickness criterion. To determine this thickness criterion, we must calculate the diffraction efficiency versus grating thickness. We take the simplest case of a uniform transmission grating, see figure 13.4, of low diffraction efficiency. If s represents the amplitude diffracted by a layer of unit length then the total diffracted amplitude for a grating of thickness l is expressed by:

$$S = \int_0^l s \exp\left(j \frac{2\pi n \left(z \cos i + (l - z) \cos i' \right)}{\lambda} \right) dz \qquad (13.7)$$

Because the diffraction efficiency is low, the reading beam amplitude is not attenuated and thus, the diffracted amplitude s is constant. We immediately get:

$$S = s\,l\exp\!\left(j\,\frac{\pi n l(\cos i + \cos i')}{\lambda} \right) \text{sinc}\!\left(\frac{\pi n l(\cos i - \cos i')}{\lambda} \right) \tag{13.8}$$

The relative intensities of the beams diffracted in various directions can now be calculated. The diffracted angles i' for various orders are given by Eq. 13.2. The corresponding diffracted amplitudes are then obtained by inserting these values in Eq. 13.8. For a beam incident at Bragg angle, and if all angles are not too large (<45°), a limited development gives the ratio of the diffracted energy in the two first diffracted orders corresponding to $p=+1$ and $p=-1$. One gets:

$$\frac{\|S_{-1}\|^2}{\|S_{+1}\|^2} = \text{sinc}^2 Q \quad \text{with} \quad Q = \frac{\pi \lambda l}{n \Lambda^2} \tag{13.9}$$

This equation shows that for $Q>10$ the $p=+1$ order is much larger than the $p=-1$ order, and all other orders as well: only one diffracted beam exists. *This Q factor is the grating thickness criterion* (see also section 4.2.1).

13.2.2 Phase and absorption gratings

A refractive index grating (or equivalently a phase grating) or an absorption grating is represented as a periodic spatial modulation of the material refractive index or absorption. If the x axis is along the grating wave vector then we can write:

$$n = n_0 + n_1 \cos(k_g x + \varphi) \quad \text{or / and} \quad \alpha = \alpha_0 + \alpha_1 \cos(k_g x + \varphi) \tag{13.10}$$

where n_0 and α_0 are the average refractive index and absorption of the medium and n_1/n_0 and α_1/α_0 are the index and absorption modulations.

These gratings may have very complex structures. The modulations may be non uniform and may vary along the x, y and z directions. Sometimes, the fringes are tilted if the phase φ depends on the z coordinate. Some index gratings are also birefringent and n_1 becomes a tensor which depends on both the reading and diffracted beam polarizations. This is for example the case for gratings recorded in photorefractive crystals (chapter 9). Furthermore, the boundary conditions may vary, so that we can distinguish reflection from transmission gratings. There exists no general formula describing all these kinds of gratings. Only simple general cases lead to analytical expressions.

Two approaches have been developed to compute the diffraction efficiency: the coupled wave theory [2] and the modal approach [3]. The modal approach is a direct calculation of the medium modes that represent optical eigen waves propagating without any deformation in the material. The diffraction efficiency is obtained by computing the coupling of the external input and output optical electric fields with theses modes. Although being rigorously exact, this approach is complex and not very intuitive. The coupled wave theory offers superior physical insight and is usually preferred. This is a perturbational approach where one assumes that the waves propagating inside the medium are in first approximation equal to the optical eigen waves (usually plane waves) of the medium in the absence of any grating. The grating is treated as a perturbation which couples the eigen waves. These two approaches can lead to exact results and are in that case rigorously equivalent [4]. However, the approximations which are usually performed to obtain analytical expressions may produce different results [5].

In the next paragraph, as an example, we will use the coupled wave theory to compute the diffraction efficiency and angular Bragg selectivity of a uniform index grating where the fringes are normal to the material entrance face. For more complex cases, the reader could refer to literature. For instance, in Ref. [2], the diffraction efficiency of uniform gratings is presented for an index or absorption modulation, whose fringes may have any orientation with respect to the grating entrance face. The case of non uniform gratings written with arbitrary-profile plane waves is developed in Ref. [6]. Index gratings attenuated along the direction perpendicular to the grating wave vector are analyzed in Ref. [7]. This last case corresponds to some holograms recorded in dynamic photorefractive media.

13.2.3 Example of uniform index transmission gratings

Coupled wave analysis. We restrict this presentation to uniform index gratings whose fringes are perpendicular to the entrance surface of the material (unslanted gratings). We also assume that the material is not birefringent and that the light polarization is perpendicular to the plane of incidence. In first approximation, we consider that only two beams propagate in the medium: the reading beam and the main diffracted beam. Compared to multiwave coupled wave analysis, this approximation is called the two coupled wave calculation. The reading beam, with electric field R, is linearly polarized along y and is incident with an angle i on the phase grating (figure 13.5).

By diffraction, beam S is generated at the grating entrance and energy is continuously exchanged between beams R and S as they propagate in the volume of the grating. The wave propagation equation is necessarily fulfilled, i.e.:

$$\nabla^2 E + k^2 E = 0 \tag{13.11}$$

where $E(x,z,t) = \frac{1}{2}\{\varepsilon(x,z)\exp(j\omega t) + \text{cc.}\}$ is the total optical field and $k = 2\pi(n_0 + n_1 \sin(k_g x))/\lambda$. One sets the propagation constant $\beta = 2\pi n_0/\lambda$. $\kappa = \pi n_1/\lambda$ is called the coupling constant.

The total electric field in the material is:

$$\varepsilon(x,z) = \left(R(z)\exp(jk_R.r) + S(z)\exp(jk_S.r) \right)\hat{e} \tag{13.12}$$

where k_R and k_S are the wave vectors of the reading and diffracted beams respectively, $\hat{e}$ is the unit polarization vector. Note that $\|k_R\| = k_R = \beta$. The propagation wave vector k_S of diffracted beam S is defined by:

$$k_S = k_R + k_g \tag{13.13}$$

When the Bragg condition is satisfied $k_R = k_S$.

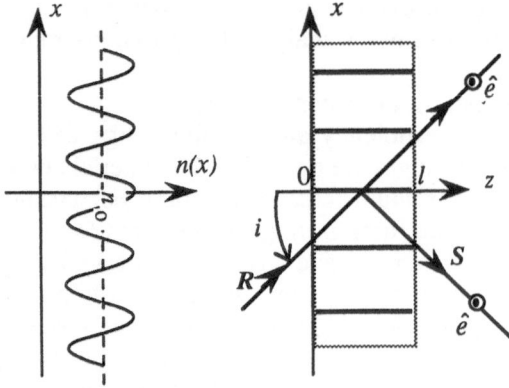

Fig.13.5. Notations used for the coupled wave analysis.

Inserting (13.12) into (13.11), using the slowly varying envelop approximation and comparing terms with the same phase exponents lead to the well known two coupled wave equations [2]:

$$
\begin{cases}
\cos i \dfrac{dR}{dz} = j \kappa S \\[2mm]
\cos i \dfrac{dS}{dz} - j\left(\dfrac{k_R^2 - k_S^2}{2\beta} \right) S = j \kappa R
\end{cases}
\tag{13.14}
$$

Parameter $\Theta = \left(\dfrac{k_R^2 - k_S^2}{2\beta} \right)$ represents the phase mismatch when the Bragg condition is not fulfilled. One finds $\Theta = 0$ when the Bragg condition is satisfied.

The amplitude of the diffracted and transmitted beams at the grating output are found by integration of the coupled wave equations with the boundary condition $S(0) = 0$. Note that these two beams are $\pi/2$ out of phase. This has important consequences when the thick grating is recorded in dynamic materials such as photorefractive crystals.

The diffraction efficiency, defined as $\eta = S(l)S^*(l)/R(0)R^*(0)$, is given by:

$$
\eta = \left[\frac{\sin\left(v^2 + \xi^2\right)^{1/2}}{\left(v^2 + \xi^2\right)^{1/2}} \right]^2 v^2, \quad \text{with} \quad v = \frac{\pi n_1 l}{\lambda \cos i} \quad \text{and} \quad \xi = \frac{l}{2\cos i}\Theta
\tag{13.15}
$$

The diffraction efficiency dependence on parameter ξ is represented figure 13.6. This figure demonstrates that the Bragg selectivity slightly depends on grating strength v.

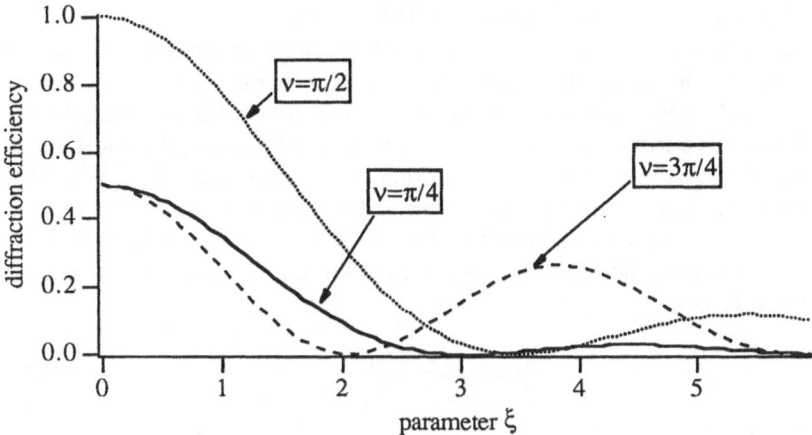

Fig.13.6. Diffraction efficiency versus parameter ξ for various values of v.

For a reading incidence at Bragg angle, parameter ξ equals 0 and the diffraction efficiency $\eta = \sin^2\left[\dfrac{\pi n_1 l}{\lambda \cos i}\right]$ can reach 100%. For example in dichromated gelatin, $n_1 \approx 2.10^{-2}$, $\lambda \cos i = 0.3\mu m$ and $l = 15\mu m$ gives $v = \pi/2$ and $\eta = 100\%$. For photorefractive crystals n_1 ranges from 10^{-3} to 10^{-5} and the required grating thickness is at least 10 times larger for reaching 100% diffraction efficiency.

Note that if a material has an absorption coefficient α, the efficiency in Eq. 13.15 has to be multiplied by $\exp(-\alpha l/\cos i)$. Thus, Eq. 13.15 indicates that the increase of η with l is balanced by a reduction due to absorption. Deriving η versus l one finds that, for low efficiencies, the maximum value of η is obtained for $\alpha l \approx 2$. This optimum value is the one preferred for optical holographic memories.

Obviously, the maximum diffraction efficiency can be also reached out of Bragg incidence. For example if $v \approx \pi$ then $\eta \approx 0$ at Bragg incidence. However, by a correct choice of the phase mismatch, we can set $\sqrt{v^2 + \xi^2} \approx 3\pi/2$ leading to $\eta \approx 50\%$.

Angular selectivity. For small deviations δi or $\delta\lambda$ from the Bragg condition, parameter Θ can be written:

$$\Theta - k_g \,\delta i \cos i - \frac{k_g^2 \delta\lambda}{4\pi n_0} \quad \text{that gives} \quad \xi = \frac{k_g l}{2}\delta i - \frac{k_g^2 l}{8\pi n_0 \cos i}\delta\lambda \qquad (13.16)$$

For $v \ll 1$, η reaches 0 when ξ equals π. For $\delta\lambda = 0$, one derives the angular selectivity of a thick grating:

$$\xi = \frac{k_g l}{2}\delta i \approx \pi \quad \Rightarrow \quad \delta i \approx \Lambda / l \qquad (13.17)$$

With $\Lambda = 2\mu m$, $l = 1mm$, one finds $\delta i \approx 2.10^{-3}$ radians.

The increased interconnect capacity of thick gratings originates from this very strong angular selectivity. However, this selectivity may also be a source of problems. A very high mechanical stability is required and, for instance, the set up must be insensitive to any temperature variation. Furthermore, during the system assembling, the positioning of a pre-recorded hologram can be very difficult and all the components must be designed with a high precision. For the correlator based interconnect system sketched in figure 13.1, we can easily calculate the precision required for the focal length f of the two lenses. If the angles are small, we immediately get:

$$\frac{\delta f}{f} \approx -\frac{\delta i}{i} \approx \frac{2 n_0 \Lambda^2}{\lambda l} \tag{13.18}$$

For $l=5mm$, $n_0=2$, $\lambda=0.5\mu m$ and $\Lambda=1\mu m$ we find $\delta f/f \approx 1.6~10^{-3}$. Such a precision is not always achievable. In the following, we will see that the recording of thick gratings at the very same place and with the very same beams that are used to implement the interconnect system, is a very simple way to alleviate these positioning problems. Furthermore, in section 13.7, we will detail some interconnect geometries which make use of the dynamic properties of some real time media so that the interconnect system adapts itself to any mechanical change.

Wavelength selectivity. Setting $\delta i = 0$ in Eq. 13.16, one gets the wavelength selectivity of a thick grating:

$$\delta\lambda = \frac{2 n_0 \Lambda^2 \cos i}{l} \tag{13.19}$$

With $\Lambda = 2\mu m$, $l = 1mm$, $n_0 = 1.5$, one obtains $\delta\lambda \approx 10nm$.

This $\delta\lambda$ corresponds to a frequency bandwidth δv equal to 10^{13}Hz for $\lambda=0.5\mu m$. This result demonstrates that, except for very special arrangements [8], only *laser sources must be used* with thick holograms. Indeed, for a broadband source, the diffraction efficiency decays because only a very narrow spectral bandwith can "see" the grating.

Reading out complex holograms. From Eq. 13.16 one sees that a wavelength change $\delta\lambda$ may be compensated by an angle variation δi. This compensation is possible for a single grating only. Complex holograms are indeed equivalent to a superposition of several single gratings. Therefore, Eq. 13.16 must be fulfilled for all these single grating components. One solution exists only: $\delta i \approx 0$ and $\delta\lambda \approx 0$.

13.2.4 Recording and reading out thick gratings

Until today, optical recording is the only way to produce thick gratings. In the previous paragraph, we have seen that the wavelength and angular selectivity condition of thick gratings is very severe. However, one should remark that two beams R and S incident on a photosensitive medium, with wave vectors k_R and k_S, record a grating whose wave vector k_g is exactly defined by Eq. 13.5. Without changing the recording set up, if beam R illuminates the material then it is diffracted into the reconstructed beam S. The reconstruction geometry exactly corresponds to the recording geometry. Therefore, the recording of holographic interconnects "in place" is a very simple way to overcome the repositioning problem of thick gratings. For that reason, most of the experiments are performed with dynamic materials. In these real time media, no grating development is required: the recording and reading out may be done without moving the sensitive medium and thus without any alignment problem. For those recording and readout configurations, the recording beams S are called the *training waves*.

Furthermore, for materials in which a development procedure is required, the physical parameters of the material are modified during the development procedure. In that case, the readout geometry may not correspond to the recording geometry. For instance, with dichromated gelatine, the development process may change the grating fringe slant and may thus modify the Bragg incident angle.

13.3 Superimposed gratings recorded in thick media: limitation of the interconnect capacity by the Bragg degeneracy

We have seen that the thick grating diffraction process is more restrictive than the thin grating diffraction process. As a result, the space variant interconnect capacity can be enlarged. Before describing how to use this increased capacity, we will first introduce the useful concept of *Bragg degeneracy*. Afterwhile we will show how to implement space variant interconnects between first a 1D input and a 2D output array and second, between two 2D arrays.

13.3.1 Bragg degeneracy

The concept of Bragg degeneracy, also called *conical diffraction*, is a direct consequence of the Bragg condition expressed by Eq. 13.5. This equation demonstrates that different wave vector pairs (k_R, k_S) can be interconnected by the same grating k_g. This Bragg degeneracy is illustrated in figure 13.7. The grating has been for instance written by the waves with wave vectors $(k_{R,1}, k_{S,1})$. The wave vector $k_{R,1}$ generates a cone whose axis is k_g. Then all other wave vectors $k_{R,i}$ lying from the cone apex O to the cone base are also at the Bragg angle for k_g. The corresponding optical waves are diffracted with maximum diffraction efficiency in beams with wave vectors $k_{S,i}$. In order for a space variant interconnect system to present no cross talk, all different couples of input and output channels must not be connected through the same common gratings. Their wave vectors must thus not define the same cone. Equivalently, we see from figure 13.7 that *the tips of wave vectors must not form a rectangle* [1], this rectangle is represented here by the thick grey line.

We will now see two different ways to take into account this Bragg degeneracy.

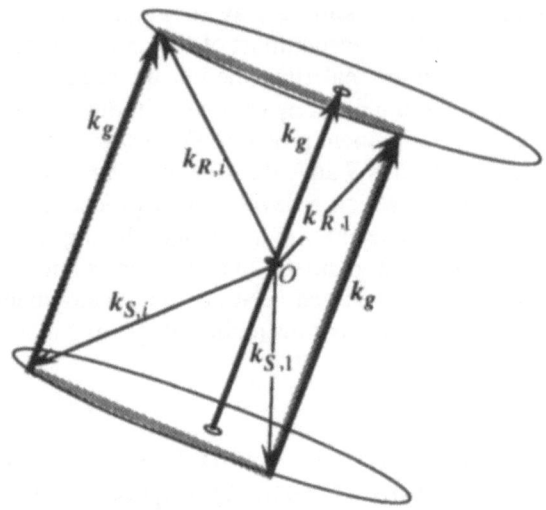

Fig.13.7. Illustration of conical diffraction.

13.3.2 Interconnects between a 1D input and a 2D output array

The simplest way to avoid conical diffraction consists in aligning the tips of the input wave vectors along a circle as shown in figure 13.8. The output wave vector tips can form the most compact possible arrangement, i.e. which is diffraction limited. The maximum number of useful channels for the output array is thus equal to space bandwidth product N of the optical system. In order to avoid cross talk, the angular spacing between the input wave vectors must be larger or equal to both the angular Bragg selectivity ($\delta i_B = \Lambda / l$ given by Eq. 13.17) and the minimum resolvable angle δi_D between wave vectors due to diffraction. If the medium is thick enough we get $\delta i_B < \delta i_D$, and the number of input channels reaches its maximum $N^{1/2}$. If $\delta i_B > \delta i_D$ then the number of input channels is limited by the Bragg selectivity and is equal to $N_B^{1/2} \approx N^{1/2} \delta i_D / \delta i_B$.

For the optical interconnect system shown in figure 13.1, the corresponding input and output channel arrangements are depicted in figure 13.8.b. In case where the incident and emergent angles are small enough, the input pattern is a line while the output array is a dense rectangular square matrix.

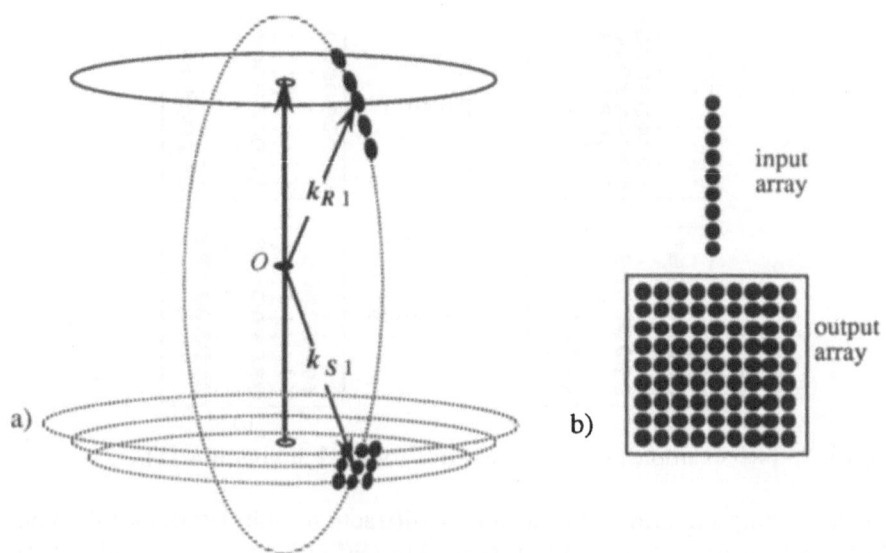

Fig.13.8. a) first possible arrangement of input and output wave vectors; **b)** corresponding input and output planes for the correlator based interconnect system.

This kind of arrangements is commonly employed to implement optical memories (paragraph 13.6.2). Usually, each input channel is the address of one stored image. Each output channel may represent one image pixel.

13.3.3 Interconnects between two 2D arrays: sampling grids

If the arrangement described above may be useful for optical memories, quite often it is more convenient to present both input and output arrays in a bidimensional format. In that case, sampling grids must be used. We can design such sampling grids according to the principle given in paragraph 13.3.1: *the tips of wave vectors must not form a rectangle*. Once again, for the correlator system if the incident and emergent angles are small enough, we can easily represent these sampling grids in the input and output channel arrays. Two examples of such sampling grids are given in figure 13.9 [1]. These sampling grids are less energy efficient than the previous arrangement shown in figure 13.8. Indeed, similarly to the sampling grids used for thin holograms, input beams may be diffracted by gratings corresponding to other channels. These diffracted beams are of course directed outside the useful output channels so that no cross talk appears.

The comparison of these sampling grids with the sampling grids for thin holograms previously shown in figure 13.3 demonstrates the increase in capacity.

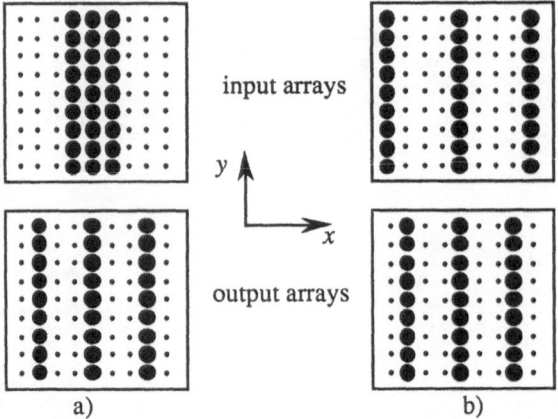

input arrays

y

x

output arrays

a)   b)

Fig.13. 9. Two examples of sampling grids for thick holograms [1].

For these sampling grids, once again the diffraction limit determines the number of channels of the output array. It is equal to $N^{1/4}$ along the x axis and $N^{1/2}$ along the y axis. For the input array the diffraction limit also gives the angular spacing between the channels along the x direction. It is equal to $N^{1/4}$. Similarly to the previous arrangement shown in figure 13.8, the angular spacing between the input channels along the y axis must be larger or equal to both the angular Bragg selectivity δi_B and the minimum resolvable angle δi_D between wave vectors due to diffraction. It is thus equal to the minimum value between $N_B^{1/2} \approx N^{1/2} \delta i_D / \delta i_B$ and $N^{1/2}$.

One must remark that if the two lenses are identical, then the maximum interconnect capacity C does not depend on the chosen arrangement. By definition, this capacity is equal to the product of the number N_1 of input channels times the number N_2 of output channels. For instance, if N is the space bandwidth product of each lens (i.e. if the space invariant interconnect capacity is NxN *input*x*output* channels) then, we compute the following maximum capacities.
If $N_B < N$ then:
-1D input toward a 2D output, $N_1 \approx N_B^{1/2}$, $N_2 \approx N$ so that $C \approx N_B^{1/2} N$,
-2D input toward a 2D output , $N_1 \approx N_B^{1/2} N^{1/4}$, $N_2 \approx N^{3/4}$ so that $C \approx N_B^{1/2} N$.
If $N_B > N$ then:
-1D input toward a 2D output, $N_1 \approx N^{1/2}$, $N_2 \approx N$ so that $C \approx N^{3/2}$,
-2D input toward a 2D output , $N_1 \approx N^{3/4}$, $N_2 \approx N^{3/4}$ so that $C \approx N^{3/2}$.

This value for C is thus larger than the capacity $C=N$ obtained for thin holograms.

13.3.4 Gratings recorded at one wavelength and read out at another wavelength

Reconfigurable dynamic optical interconnects of high bandwidth signals at a given wavelength can be also implemented by recording multiplexed gratings at another wavelength [9]. Transmission gratings are written or erased using planar wavefronts of wavelength λ_w to which the crystal is sensitive. To provide non destructive readout of the interconnection gratings the signal beams have a longer wavelength outside the sensitivity domain of the recording material.

To diffract n input signals toward m output channels $n.m$ gratings are required so that, in general, $2n.m$ writing beams are needed to record these gratings. However, because of the conical Bragg diffraction, if the writing beams are plane waves whose wave vectors are on the surface of a single cone, the resulting set of gratings may verify the Bragg condition for plane waves of longer wavelength propagating on a different cone. As shown in figure 13.10, the axes and bases of both cones must meet exactly. In this configuration, each signal beam, whose wave vector is k_{s1}, is related to one writing beam k_{w1}. With this geometry, the required number of writing beams is equal to $n+m$ only. Indeed, the interference pattern of the two writing beams, k_{w1} and k_{w2}, induces a grating with wave vector K_{12} which diffracts the signal beam k_{s1} in a beam k_{s2}. By adding a third writing beam k_{w3}, the signal beam k_{s1} is diffracted in a second direction given by k_{s3}.

In order to write more than 1 set of gratings inside the material, a sequential time division multiplexing is used. For instance, to write two independent sets of gratings for diffracting the two input beams k_{s1} and $k_{s1'}$ toward different directions, then the two sets are written by alternately switching on each group of writing beams at a rate much faster than the buildup or erasure rates of the recording material.

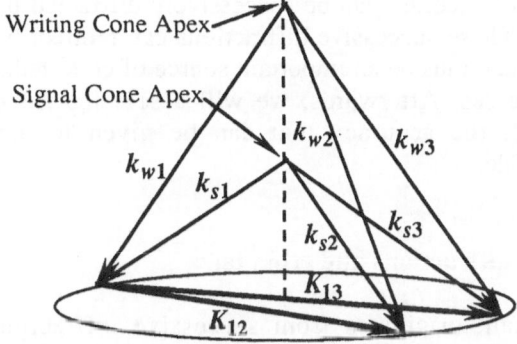

Fig.13.10. Wave vector diagram for the the conical diffraction geometry. k_{wi} and k_{si} are respectively the wave vectors of the writing and signal beams [9]

Demonstration of this recording geometry was performed with the experimental set up depicted in figure 13.11. The writing beams were issued from an Argon laser tuned at 514nm and the signal beams were from an HeNe laser at 633nm. The dynamic holographic medium was a photorefractive $Bi_{12}GeO_{20}$. Diffraction of 2 signal beams in 3 different possible directions was demonstrated. The authors expect that the capacity of such a system may be extended to about 128 interconnects in a 1cm^3 $Bi_{12}GeO_{20}$ crystal and up to 5000 in SBN samples.

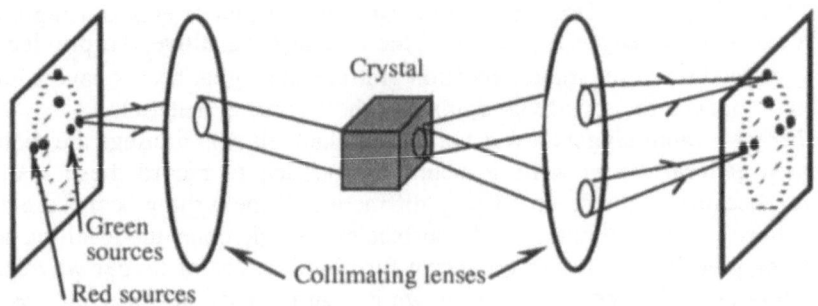

Fig.13.11. Two-wavelength conical interconnect system based on multiplexed
volume gratings stored in a photorefractive crystal [9].

13.4 Limitations inherent to volume holographic interconnects

From the previous section, we know that the input and output points must be
arranged in sampling grids in order to avoid cross talk due to conical diffraction.
This condition is not the only one required to assure noiseless interconnections.
Indeed, conversely to thin holographic elements, an optical wave travelling in a
volume holographic medium can be successively diffracted by several layers of
different gratings. These successive diffractions can redirect light in some of the
output points and can thus be an important source of cross talk. We will hereafter
detail these processes. Afterwards, we will assess the fidelity of the optical
interconnects, i.e. the accuracy that can be given to the weight of each
interconnection node.

13.4.1 Successive diffractions and cross talk

n^{th} order cross talk originates from successive diffractions on n different
superimposed gratings. This source of cross talk is described in details in Ref.
[10]. We only sketch here the mechanism of this cross talk. We take the simplest
example of the holograms required to randomly interconnect an array of two input
points **1** and **2** (corresponding to waves with wave vectors k_1 and k_2) to an array
of two output points **a** and **b** (corresponding to k_a and k_b).

Second order cross talk. Second order cross talk appears when the most efficient
recording technique is employed: the simultaneous exposure of each input wave
with all the desired training waves [10]. Successive exposure of each input wave
with one training wave at a time indeed leads to lower diffraction efficiencies. For
instance if we want to set the interconnection between input **1** with output **a** and

b, and input 2 with output a only, then all the following holograms will be recorded:
-the gratings with wave vectors $(k_a\text{-}k_1)$, $(k_b\text{-}k_1)$ and $(k_a\text{-}k_2)$ called interlayer [10] gratings;
-the intralayer grating with wave vector $(k_b\text{-}k_a)$.
These wave vectors are represented on the momentum conservation diagram in figure 13.12.

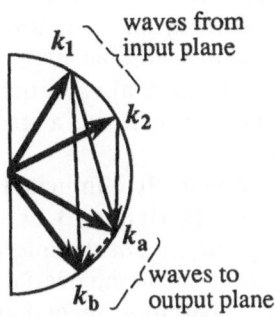

Fig.13.12. Momentum conservation diagram; the bold arrows represent the input and output wave vectors, the thin arrows are for the three desired interconnect grating wave vectors and the arrow in dashed line is the "parasitic" intralayer grating responsible for the second order cross talk.

From this diagram, we see that light issued from point source 2 is indeed connected to output a, as required, but is also connected to point b by successive diffraction on the two gratings whose wave vectors are $(k_a\text{-}k_2)$ and $(k_b\text{-}k_a)$. This source of noise scales as the product of the two diffraction efficiencies η_1 and η_2 of the two successive gratings involved while the signal scales as η_1. Therefore, the signal to noise ratio is inversely proportional to η_2. For large input and output arrays, this signal to noise ratio is also inversely proportional to the fanout: the number nb of intralayer gratings. If nb and η_2 become too large, the signal to noise ratio may become lower than unity.

Third order cross talk. Conversely to second order cross talk, third order cross talk does not depend on the recording procedure. Taking the previous interconnection example (figure 13.12) we see that the input point 2 is also connected to point b by three successive diffractions on gratings with wave vectors $(k_a\text{-}k_2)$, $(k_1\text{-}k_a)$ and $(k_b\text{-}k_1)$. The signal to noise ratio is now inversely proportional to the square of the diffraction efficiency η_1. Because many such third order diffraction processes may happen simultaneously, this source of noise imposes a limit on either the density of interconnections or on the diffraction efficiency. This limit is determined in more details for completely interconnected networks in Ref. [10].

314

Cross talk elimination. One should note that for a fanout of unity, i.e. if each input is linked to one output only, second order and higher order cross talks do not exist. For larger fanout, different methods may be used to completely or partly eliminate these cross talks.

We have seen, for instance, that second order cross talk only exists when using the most efficient recording procedure. Another way to decrease this second order cross talk is to take advantage of the physical recording properties of some holographic media. For example, in some photorefractive crystals, because of the dependence of the diffraction efficiency on the grating fringe spacing, efficiency η_2 of intralayer gratings can be made negligible compared to efficiency η_1.

Second and higher order cross talks always decrease when reducing the diffraction efficiency. For some applications, a trade off may thus be found between efficiency and cross talk.

A more definitive method to completely eliminate these cross talks consists in changing the recording geometry. In figure 13.13.a, we show the previously considered recording geometry for a densely interconnected network which presents cross talk. This cross talk is due to the overlaping of the different training waves in the recording medium. Therefore, second and higher order cross talks are avoided by using the recording geometry sketched in figure 13.13.b, where the training waves do not overlap.

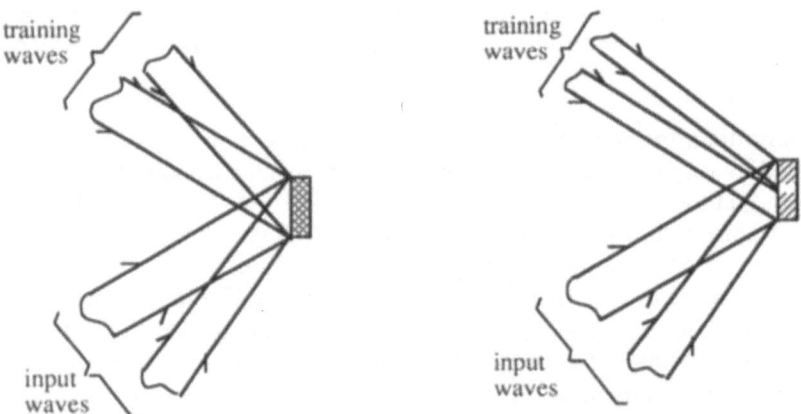

Fig.13.13. a) densely interconnected recording geometry presenting cross talk, b) recording geometry without the cross talk due to successive diffractions.

The improvement in the signal to noise ratio with this second geometry is of course obtained at the expense of a reduction in the interconnection density. Indeed, for gaussian training beams, the smallest required recording volume per beam is obtained with the beam waist located in the center of the holographic medium. In that case, if l is the medium thickness, λ the optical wavelength in vacuum and n the refractive index, the minimum beam diameter at the surface of the medium is $\sqrt{2\lambda l/n\pi}$ so that the minimum occupied surface per training wave is at least $\lambda l/2n$. In that case, at the maximum, we can put $N_{max} \approx 2Sn/\lambda l$ training waves in a crystal whose entrance surface is S. In the first geometry, this number is limited by diffraction only and is thus much larger. It is equal to $N_{vol} \approx 2\pi Sn^2/\lambda^2$.

13.4.2 Fidelity of the interconnections

The fidelity of the interconnect is defined as being the accuracy to which the desired weight of each interconnection node can be achieved. Two main phenomenons may reduce the fidelity.

The first originates from the physical recording mechanism in thick holographic media. When the sum of the index modulations of individual gratings approaches the maximum achievable index change, each recording of a new node modifies the previously recorded gratings. This effect is reduced when working far away from the saturation of the medium.

The second comes from conical Bragg diffraction and from second and higher order cross talks. The recording of a new node, corresponding to a new input channel, may not only increase the amount of noise but may also decrease (or even increase) the weight of previously recorded nodes by redirecting part of other input signals in various directions.

An example of this phenomenon is given in Ref. [11] in the simplest case of a 2 beam combining hologram recorded in a $LiNbO_3$ sample. The objective is to adjust the two superposed gratings to efficiently diffract the incident beams (of wave vectors k_1 and k_2) along a common direction where they are going to coherently add in a beam of wave vector k_a. When grating (k_1-k_a) is present only, the computed diffraction efficiency for beam 1 is $\eta_{1\rightarrow a} \approx 44\%$. Similarly, if grating (k_2-k_a) is present only, the diffraction efficiency for beam 2 is $\eta_{2\rightarrow a} \approx 53\%$. If both gratings are simultaneously present, after being written under the same conditions as previously, then the two diffraction efficiencies become $\eta_{1\rightarrow a} \approx 35\%$ and $\eta_{2\rightarrow a} \approx 33\%$. The diffraction efficiencies are not only reduced but the relative weights of the two interconnection nodes are now reversed: $\eta_{1\rightarrow a} > \eta_{2\rightarrow a}$. Of course, the diffraction efficiencies can always be a priori computed if the interconnection pattern is known. However, for large networks, the computation time may become too large.

A more detailed study of fidelity has been published recently [12]. This study is based on a two dimensional analysis of diffraction processes and therefore does not take into account the effect of conical diffraction. Nevertheless the following important results are reported:
i) the relative ratios of the amplitudes of the output of the interconnect depend on the modulation strength, the fidelity thus decreases when increasing the interconnection strength;
ii) in most cases, supervised learning adaptive networks may learn out their inherent imperfections and may thus partly compensate from their lack of fidelity.

13.4.3 Temporary conclusion

In previous sections, we pointed out that, because of the Bragg selectivity, volume holograms overcome thin holograms in the interconnect density. We have now just seen that this increase of density is gained at the expense of an increase of cross talk and of a loss of fidelity.

We must underline that the above discussed sources of cross talk may not be the only ones inherent to volume interconnects. For instance, in Ref. [13], the authors present an other possible source of noise due to off Bragg diffraction.

Unfortunately, very few experimental data are available.

Furthermore, the reader must keep in mind that other sources of noise may appear in the implementation of interconnects with non linear materials. For instance the noise may be induced by the beam fanning phenomenon (described in Chapter 9) when using high gain photorefractive materials. Non linearities of the recording response curve of the material may also produce some undesired gratings and must therefore be taken into account.

13.5 Interconnect capacities of existing materials

13.5.1 Introduction

In the above sections, we discussed the fundamental limits of interconnect systems using thick holograms. Now, we will assess the limitations imposed by existing materials. We will consider as an example the realization of an optical holographic memory built around a photorefractive crystal [14]. We will first estimate the bit capacity of such a memory and then we will demonstrate that a trade off must be found between this capacity and the output data rate. In order to estimate the storage capacity of the holographic memory, we take the first possible arrangement shown in figure 13.8: the input array is a linear array of N_1 angularly multiplexed reference beams and the output array is a compact bidimensional image of N_2 pixels. Each image is recorded by interfering the image beam with one of the possible reference beams and is reconstructed with this very same reference beam.

Despite the large variety of photorefractive materials, few of them only are adequate for building memories. Indeed, we need an index modulation as large as possible and a correct dark storage time. Only 3 kinds of crystals are thus easily available: $BaTiO_3$, $LiNbO_3$ and SBN. Photorefractive properties such as the maximum achievable index modulation δn_{max} and the dark storage time may strongly vary from sample to sample. This is particularly the case for $LiNbO_3$ and SBN. Less variations are observed in $BaTiO_3$, except for the dark storage time that varies from few seconds in some samples to few days in other crystals. However, reduced $BaTiO_3$:Co crystals present a better sensitivity than undoped and unreduced samples. Nevertheless, our numerical calculations will be conducted with typical parameters of BaTiO$_3$ crystals and for a vacuum wavelength of 514nm.

Figure 13.14 shows a plot of the maximum refractive index modulation that can be induced in a BaTiO$_3$ crystal. The two interfering beams have extraordinary polarizations (parallel to the crystallographic **c** axis) and travel in a plane perpendicular to the **b** axis. θ_1 and θ_2 represent the angles between the wave vectors of the beams and the **a** axis. The index modulation is symmetrical about $\theta_1 = \theta_2$ for which it vanishes. One sees that the index modulation strongly depends on the difference $\theta_1 - \theta_2$ that corresponds to the grating spacing. It also varies, but more smoothly, with $\theta_1 + \theta_2$, that is with the orientation of the grating planes.

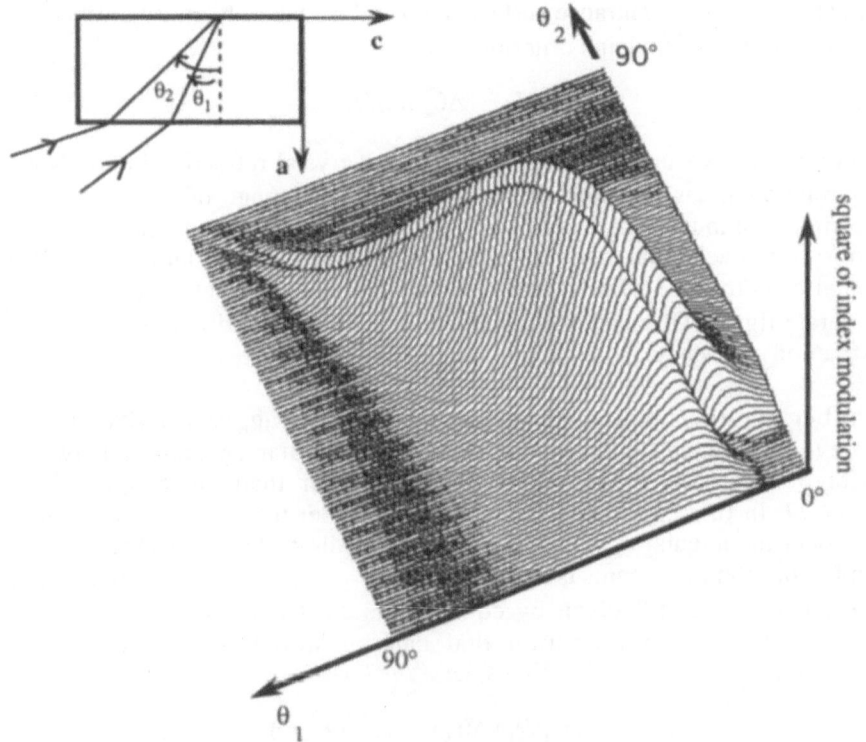

Fig.13.14. Evolution of the square of the photoinduced index modulation in a BaTiO$_3$ sample as a function of the propagation directions of the two interfering beams.

These dependencies are valid for all photorefractive materials. Figure 13.14 demonstrates that only a small part of the (θ_1, θ_2) plane is permitted if one wants a sizeable refractive index modulation. For example, with BaTiO$_3$ and $\delta n_{max} = 1.5$ 10^{-4}, if one admits a drop in the diffraction efficiency by a factor of 2 ($\sqrt{2}$ in index variation), the allowed angular aperture for the image beam is about $\Delta\theta_2 \approx$

3°. This does not depend on the multiplexing technique used. For the same reason, multiple storage by varying the angular direction of the reference beam is effective for an excursion of a few degrees only. For example, storage of 500 holograms was achieved by rotating the reference beam by 5° ($\Delta\theta_1 \approx 2°$ inside the crystal) [15].

13.5.2 Storage capacity of a BaTiO₃ photorefractive crystal

Number of pixels in the image beam. The number N_2 of pixels in a stored image is limited by diffraction and usable angular apertures $\Delta\theta_2 \times \Delta\Phi_2$ (paragraph 13.3.2). We assume for simplification a square image with $\sqrt{N_2} \times \sqrt{N_2}$ pixels, a crystal with a square entrance surface $d \times d$ and we take $\Delta\theta_2 = \Delta\Phi_2$. When limited by diffraction, the Rayleigh criterion gives :

$$N_2 \approx \left(\Delta\theta_2 \, d \, n/\lambda\right)^2 \qquad (13.20)$$

where λ is the vacuum wavelength and n the crystal refractive index. Note that this relation gives the number of pixels in each image of a stack of images superimposed in the same volume or, for images of m pixels, the number p (with $p = N_2/m$) of stacks that can be placed side by side in the volume. For a BaTiO₃ crystal of entrance surface 1 cm², we find $N_2 \approx 6.10^6$ pixels that is a relatively moderate figure. Of course, this number can be increased but at the expense of diffraction efficiency.

Number of superimposed images owed to the Bragg selectivity. In order to retrieve the images without cross talk, the angular separation between the different reference beam angles must be larger than the Bragg selectivity $\delta\theta = \Lambda/l$. In first approximation, we may consider that this selectivity does not vary over the angular aperture $\Delta\theta_1$. Given the allowed angular aperture $\Delta\theta_1$, the number of reference beams is at maximum $N_1 = \Delta\theta_1/\delta\theta$. The number N_2 of pixels in each image is still given by equation 13.20. Consequently, the maximum number of binary information that can be stored in a volume V of a photorefractive crystal is $C = N_1.N_2$, i.e.:

$$C \approx \Delta\theta_1 \, (\Delta\theta_2)^2 \, V \, n^2/(\lambda^2 \, \Lambda) \qquad (13.21)$$

For example, for a BaTiO₃ crystal of volume 1 cm³, at the maximum of index modulation, we find $C \approx 10^9$ bits. This number can obviously be increased by using larger values of $\Delta\theta_1$ and $\Delta\theta_2$, i.e. by working out of the maximum efficiency. It has also been proposed to place side by side multiple crystals in order to increase the total capacity, reaching some Gigabytes or even Terabytes.
This rapid calculation shows that the actual capacity is, by far, much smaller than the theoretical one given by Vn^3/λ^3 (10^{14} bits for 1 cm³) [16].

13.5.3 Data rate-Capacity trade off

In fact, we do not only look for large capacity memories but we also desire high data rates. The power P of the readout beam is limited. We thus must integrate the diffracted signal during a given time τ for detecting a sufficient enough photon number per pixel of the reconstructed image. If η^H is the diffraction efficiency per hologram, N_2 is the number of pixels in the image, n_b is the photon number required for deciding the pixel value, the data rate is thus expressed as:

$$\frac{N_2}{\tau} = \frac{P \, \eta^H \, n_J}{n_b} \tag{13.22}$$

n_J being the photon number in 1 Joule ($n_J = \lambda/hc$, h Planck's constant and c light velocity in vacuum).

The maximum index modulation is shared between all N_1 superimposed images. The diffraction efficiency thus scales as $\sin^2(A/N_1)$ or as A^2/N_1^2 for low efficiencies. If l is the crystal thickness then:

$$\frac{N_2 N_1^2}{\tau} = \frac{P \, A^2 \, n_J}{n_b} \quad \text{with} \quad A = \frac{\pi \, \delta n_{max} \, l}{\lambda} \tag{13.23}$$

The detection error rate depends on signal n_b and on detector noise [17]. We here choose a reasonable value $n_b = 10^3$ photons. Thus for a $BaTiO_3$ crystal of 1 cm^3 volume, a reading power of 10 mW at 515 nm and 256 recorded images, we get a maximum information rate of $N_2/\tau \approx 4.10^{10}$ s^{-1}. Consequently, for images of 256 x 256 pixels, i.e. a total capacity of 2.10^7 bits, the minimum image reading time becomes 2 µs.

Increasing the total capacity by increasing both $\Delta\theta_1$ and $\Delta\theta_2$ is still possible. However this will decrease the diffraction efficiency and thus increase the readout time.

13.6 Correlators and memories

13.6.1 Correlators

Correlation in thick holographic media. Thin holograms for optical correlators have been presented previously in chapter 12. Similar systems may also be built around thick holographic media using either a two wave mixing or a four wave mixing configuration.

Optical correlation using four wave mixing in a dynamic (Kerr) holographic medium was proposed in Ref. [18]. In these systems, one hologram is recorded and is simultaneously read out by an auxiliary beam. These four wave mixing configurations can therefore perform one correlation at the same time only. As shown in figures 13.15.a and b, two slightly different geometries may be used.

The first, figure 13.15.a, is an extension of the Vander Lugt correlator. The Fourier transform of the image is recorded by interfering the image beam with a plane wave. It is simultaneously read out by the Fourier transform of the filter. In the second configuration, called the *Joint Fourier Transform Correlator*, the Fourier transform of the image interferes with the Fourier transform of the filter. The resulting hologram is read out by a plane wave. The main advantage of this configuration compared to the previous one is that the two Fourier transforms are automatically superimposed because of the use of a single lens.

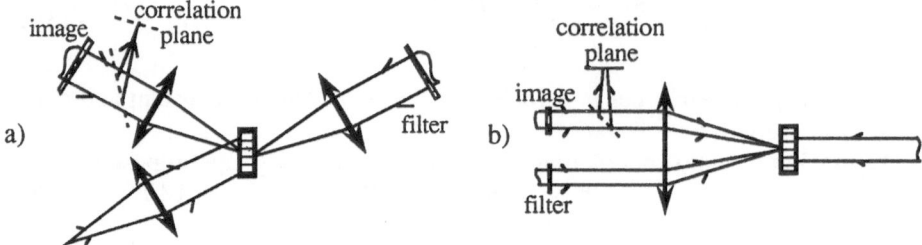

Fig.13.15. Four wave mixing based correlators, a) is equivalent to a Vander Lugt correlator and b) is the joint Fourier transform correlator;

Such systems were implemented using thick holographic media [19, 20]. However we must point out that these systems were constructed in thick media mainly because of the lack of efficient thin dynamic media. The thickness of the holographic media leads to the two following main drawbacks in correlators:
i) the shift invariance of Fourier holograms is considerably reduced because of Bragg selectivity,
ii) because of the medium thickness, holograms are recorded out of the Fourier plane so that the correlation function is distorted [21].
Optical Vander Lugt correlators may also be implemented with a two-wave mixing configuration [21-23]. In a first step, the Fourier transform of the filter is written in the material by interference with a reference plane wave. It is then read out in a second step by replacing the filter with the image. If the image corresponds to the filter then the image beam is diffracted and reconstructs the reference beam used during the recording. These two wave mixing configurations suffer from the same drawbacks as the four wave mixing configurations. However, several Fourier transforms of different filters may be superimposed in the same volume by using different reference beams and by taking advantage of the Bragg selectivity of thick media. The reconstruction of one of these reference beams indicates that the corresponding image is recognized. The multiplexing of the different Fourier transforms can be implemented with either the angular Bragg selectivity or, as shown below, with the wavelength selectivity.

Wavelength multiplexed shift invariant correlators. Reflection holograms present a much higher wavelength selectivity than transmission holograms while the angular selectivity remains similar. Such reflection holograms were proposed to build wavelength multiplexed optical correlators [22]. This system was demonstrated with the set up represented in figure 13.16.

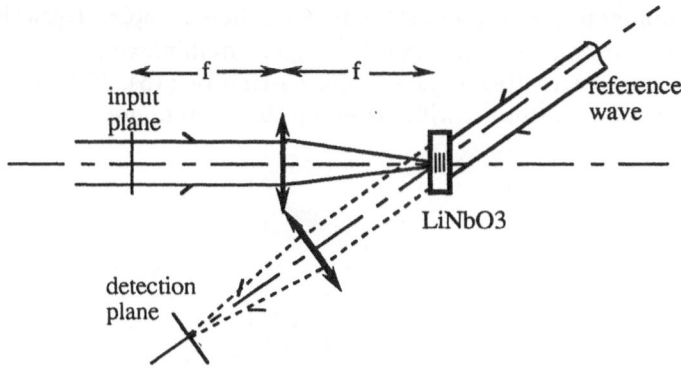

Fig.13.16. Geometry for reflection wavelength multiplexed correlator. The reference wave is used during recording only [22].

The two optical waves, the image and reference beams, are from the same tunable Argon laser. Input images (or filters during the training process) are Fourier transformed in the holographic media. The authors attributed a given wavelength to each of the filters to be recorded: the Fourier transform of the first input filter, representing the letter D, is recorded by interfering the image beam and reference beam with the laser tuned at 457.9nm, the Fourier transform of letter F at 465.8nm, K at 488nm and J at 496.5nm. Once all the filters are recorded, an unknown letter is presented to the correlator. The wavelength is then scanned. A correlation pic detected at a given wavelength means that the letter associated to this wavelength is recognized. The authors experimentally showed that this correlator is shift invariant with no apparent cross correlation.

13.6.2 Optical memories

Conversely to most optical correlators, high density optical memories benefit from the thickness of the holographic medium [16]. The Bragg selectivity allows for a large number of independently retrievable images to be stored in the same volume [14, 15, 24-26].

Several possibilities to multiplex holographic gratings have been discussed: wavelength encoding [27] and pure angular multiplexing [15, 24-26]. Angular reference beam multiplexing has become a widely spread method used in different schemes for associative memories [28-31] and high-capacity data storage [15, 24-26]. In most of these applications the reference beam is changed in its angle of incidence to get the additional parameter of addressing necessary for independent image retrieval. Acousto-optic deflectors were proposed but the necessity of using two acousto-optic devices [32] to compensate for the Doppler frequency shift increases the complexity of the optical set-up. Electro-optic translation devices were also proposed [32]. Mechanical changes of the reference beam angle are widely used [24-26]. A slightly different mechanical addressing was used [25] to store and fix (with the thermal fixing technique described in chapter 9) 500 superposed holograms in a photorefractive LiNbO$_3$ sample. These holograms were recorded with plane waves and the angular multiplexing was achieved by rotating the photorefractive crystal. More recently [15], 500 holograms of complex images were recorded with the set up shown in figure 13.17.

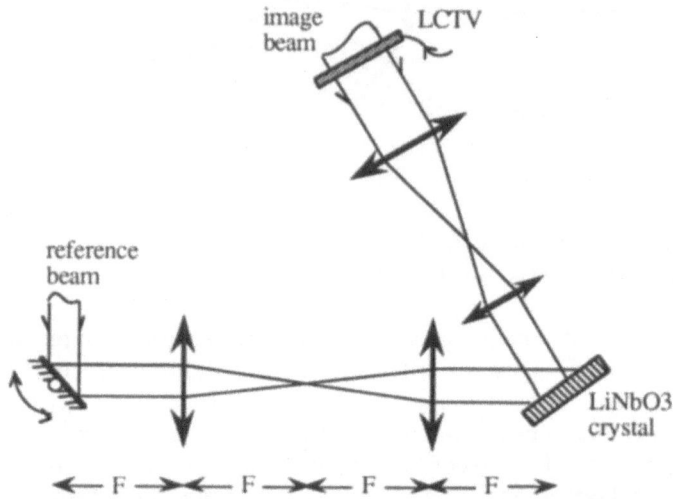

Fig.13.17. Angularly multiplexed memory. The incidence angle of the reference beam is adjusted by means of a rotating mirror placed at the focal point of a refractive telescope. The images are impressed on the image beam with the liquid crystal television set (LCTV) [15].

Multiple holograms were stored by changing the incidence angle of the reference plane wave. This incidence angle was changed by rotating a mirror located at the object focal point of the first lens of a refractive telescope while the LiNbO$_3$ photorefractive crystal was set at the image focal point of the second lens. Images were impressed on the object beam by means of a liquid crystal television set. Any hologram could be retrieved by rotating the mirror to the same orientation at which that specific hologram was recorded.

Alternatively, the use of an intensity spatial light modulator to define the various incident directions was demonstrated [33].

Nevertheless, these two last methods still suffer from some disadvantages: mechanically changing the incident angle of the reference beam requires a high reliability in the positioning and is therefore inherently slow whereas amplitude modulation is energy consuming.

Out of these reasons, phase encoding has been proposed [34-36]. Although a pure phase addressing multiplexing is promising, there has been only a few attempts to use pure phase encoding as means for data multiplexing in photosensitive media. For instance, sinusoidally phase modulated waves were employed [34] to record images in thick holographic media.

A new multiplexing technique, the deterministic phase encoding method [35, 36], was proposed recently to store M images with N_1 reference beams. To store each of these images, N_1 holograms of this image are simultaneously recorded by interfering the image beam with all N_1 reference beams. To record M images, $N_1.M$ holograms of these images are thus recorded. The phase of each of these beams relative to the phase of the image beam can be adjusted to be either 0 or π. This set of adjustable phases represents the address of the stored image. To take full use of the Bragg selectivity in thick recording media, the angular spacing between adjacent reference beams is chosen to be large enough to satisfy the Bragg condition. Therefore, retrieval can be achieved by reconstruction of the N_1 image waves of a given stored image with the N_1 reference beams carrying the set of address phases used during the recording of that image: because of Bragg selectivity, each reference wave can only reconstruct the holograms it has recorded. All these N_1 image waves, corresponding to the chosen phase address, interfere constructively. Theoretically, no cross talk and therefore noiseless reconstruction appears because the phase addresses of the images are chosen so that the reconstruction of the undesired images interfere destructively to produce zero intensity. This phase coding technique takes advantage of the selectivity of the Bragg condition and minimizes the space bandwidth product of the spatial phase modulator i.e.: the number N_1 of pixels of the binary modulator must be only equal to the number M of images to be recorded. Moreover, phase only addressing is naturally light efficient and provides simple and quick generated reference beam patterns, allowing thus a very easy image retrieval procedure with high storage capacity.

13.7 Self aligning interconnects

13.7.1 Introduction

As pointed out in chapter 11, an important experimental limitation of optical interconnects lies in the achievable density of output channels. Because of the micrometric size of the output detectors or fibers, the optical components must be positioned with a very high accuracy. A high stability (against mechanical vibrations and thermal expansion) of the mechanical holders is also required. Moreover, because of the dependence of the diffracted beam direction on the beam wavelength, all the laser sources must be wavelength stabilized.

A possible solution to overcome these problems may be found in the use of dynamic holographic media such as the photorefractive crystals presented in chapter 9. If the interconnection gratings are written in such real time media by beams issued from the input and output channels, any change in the optical set up is immediately compensated by a corresponding change in the holographic interconnection patterns. All the proposed self aligning systems we will present hereafter are derived from the original proposition described in the next paragraph.

13.7.2 Self aligning interconnect with a phase conjugate mirror

The corresponding optical set up is depicted in figure 13.18. This set up is designed to insure connection between the modulated source and the detector.

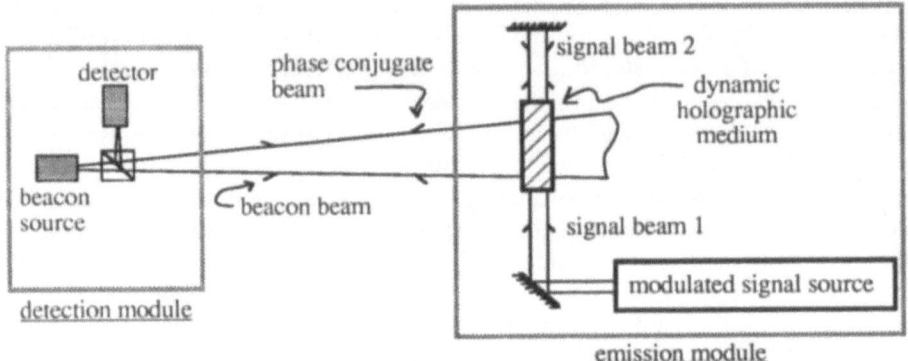

Fig.13.18. Principle of self aligning interconnection with a phase conjugate mirror.

Three coherent optical waves overlap in the dynamic holographic medium:
-two counter propagating beams issued from the same time modulated source (reflection of beam **1** on a mirror may be used for beam **2**);
-a beacon beam issued from a source located in place or close to the detector.
The beacon beam and signal beam **1** (respectively signal beam **2**), with wave vectors k_b and k_1, being coherent, interfere and create a grating (with wave vector k_1-k_b) inside the dynamic medium. Because signal beams **1** and **2** are exactly counterpropagating, k_2=-k_1, beam **2** (respectively beam **1**) reads out the induced grating and is diffracted in a beam with wave vector k_2+k_1-k_b=-k_b. The diffracted modulated beam is thus automatically directed toward the detector. Beacon light seems to attract light from the modulated source. The alignment task is thus simplified. The beacon light must only overlap the modulated beam in the holographic media. Furthermore, any change in the position of the detector (i.e the beacon source) is automatically corrected.

One must note that the interconnection strength is not sensitive to the temporal modulation of the signal beam (megahertz or gigahertz) because of the relatively slow response time of the holographic medium (typically in the millisecond or second range).

This proposed system looks very interesting for optical interconnections. However it can interconnect different sources to one detector only and all the beams need to be coherent.

In the following, we will present some improvements of this basic idea. First, we will depict its extension to a system able to interconnect two bidimensional arrays with coherent beacon and signal sources. We will then present another system where the beacon and signal sources may be mutually incoherent. This last system can be implemented with thick materials only, conversely to the two other ones which work with thin and thick holograms as well.

13.7.3 Self aligning interconnection with two dynamic media

This self aligning interconnection scheme has been proposed and demonstrated recently [37]. It is shown in figure 13.19.

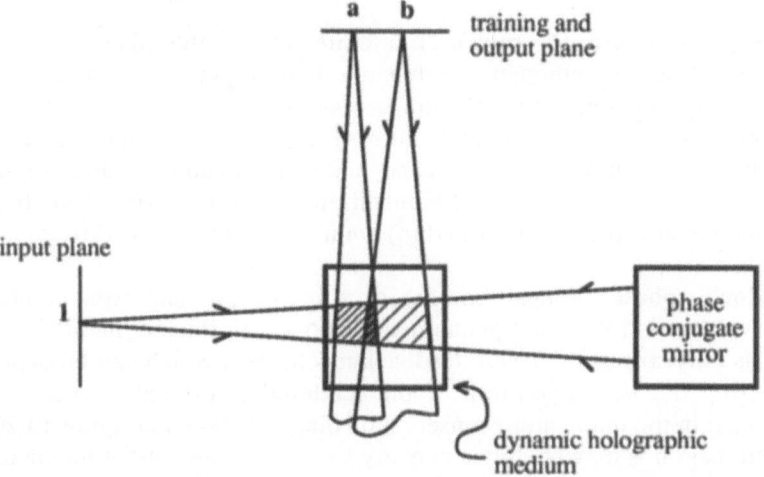

Fig.13.19. Self aligning interconnection set up during the recording process of the interconnects between input beam **1** and output channels **a** and **b** [37].

It uses two dynamic media, the first where the interconnection patterns are recorded (the interconnection medium), and the second used during a training cycle only as a phase conjugate mirror. The system works as follows. To interconnect input signal **1** to output channels **a** and **b** for example, the signal wave k_1 and the two training waves k_a and k_b are sent into the interconnection medium. All these waves must be coherent. After going through the interconnection medium, the signal wave is phase conjugated by the phase conjugate mirror. The generated phase conjugate wave, $-k_1$, goes back inside the interconnection medium, interferes with the two training waves and then creates the two required interconnection gratings with wave vectors $(-k_1-k_a)$ and $(-k_1-k_b)$. Once these two gratings are recorded, the input signal is diffracted in the two waves phase conjugated of the two training waves. The interconnection link between input **1** and output **a** and **b** is thus implemented. Other interconnection gratings, between input **2** and output **b** and **c** for instance, can be sequentially recorded in the same medium by switching on the corresponding input and training waves (**2, b** and **c**) and switching off all the other beams. Once the medium is trained, the interconnection system may be used, and all the input beams can be switched on simultaneously. To compensate either for the inherent erasure during readout of the dynamic holographic media, or for any mechanical change in the optical configuration, the entire training cycle may be repeated from time to time.

The main problem concerning this system is that it requires mutually coherent input and training waves. This necessity is avoided with the interconnection scheme presented in the following paragraph.

13.7.4 Self aligning interconnection with the double phase conjugate mirror

Double phase conjugate mirror principle. The double phase conjugate [38] mirror is a recently demonstrated dynamic holographic configuration in which two mutually coherent or incoherent beams induce phase conjugate replica of themselves. Such a phase conjugate mirror is represented in figure 13.20.a. Beam **1** and its scattered light on the one hand, and beam **b** and its scattered light on the other hand, write two identical transmission gratings. Therefore beam **1** is diffracted toward (phase conjugated to) point source **b** and **b** is diffracted to point source **1**.

The double phase conjugate mechanism is completely understood and has been studied in details [39]. This phenomenon appears in thick dynamic holographic materials only. It results from a feedback mechanism which can be explained by considering the time evolution of the scattered light. This scattered light is represented in the momentum conservation diagram shown in figure 13.20b. Such different diagrams may be drawn at every location in the bulk of the material. For sake of clarity, only two such diagrams are shown here: the first for beams propagating toward $z>0$ and located at $z \approx 0$, the second for beams propagating in the opposite direction ($z<0$) and located around $z \approx l$, l being the thickness of the material.

Beam **1** enters the medium by its left side, it is then scattered by medium imperfections such as inhomogeneities or surface scratches. This scattered light interferes with beam **1** and thus induces many gratings inside the medium. A similar process occurs for beam **b** at the opposite side. Among all the weak beams scattered from beam **1** at z=0, one exactly counterpropagates beam **b**. The wave vector of this beam, noted **b*** is thus $k_{b*}=-k_b$. During their propagations, beam **1** and this scattered beam **b*** write a weak grating which is exactly at the Bragg angle for beam **b**. Beam **b** is therefore diffracted on this grating from $z=l$. Its diffracted beam **1*** has a wave vector $k_{1*}=-k_1$ and thus counterpropagates beam **1**. Beam **b** and this diffracted beam **1***, reinforce the previously recorded grating in particular at $z=0$. This feedback process increases the scattered beams **b*** and **1*** until a steady state is reached. The other weak written gratings are completely erased during this process because no other common grating can be found between those gratings written by light propagating in both directions.

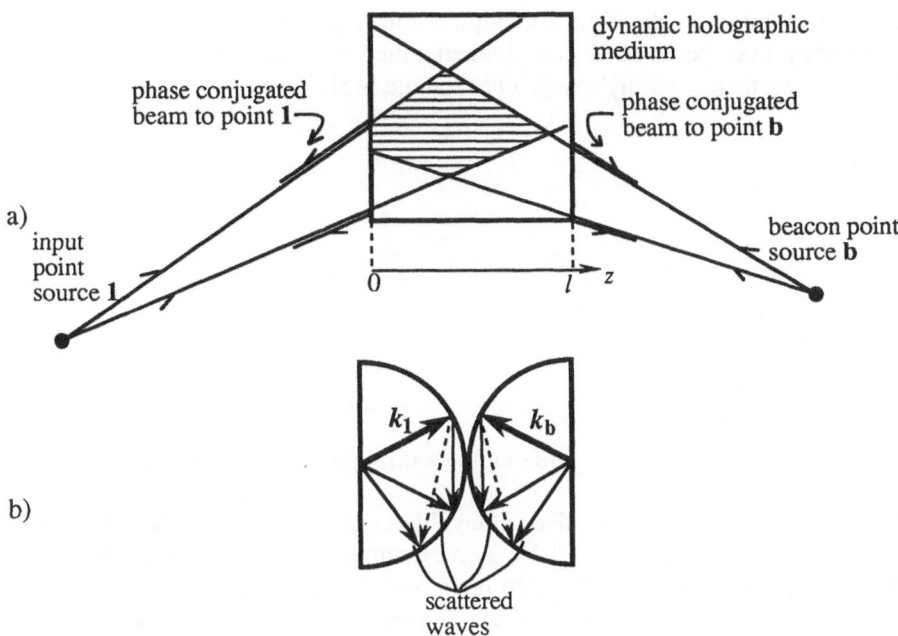

Fig.13.20. a) Double phase conjugate mirror, beams issued from points **1** and **b** generate a common transmission grating on which beam 1 is diffracted into the phase conjugate of beam **b** and beam **b** is diffracted into the phase conjugate of beam **1**. **b)** Momentum conservation diagrams of the double phase conjugate mirror. The only one common induced grating (vertical arrows) is reinforced by a feedback mechanism.

If input beams **1** and **b** are plane waves then, because of the Bragg degeneracy, the beams are phase conjugated in the plane of figure 13.20 only. True phase conjugation requires the insertion of images, aberrators or cylindrical lenses on the path of the two input beams [40].

Double phase conjugate mirrors were mainly demonstrated with the photorefractive effect, even if they can be implemented with other optical non linear processes such as Brillouin scattering.

Interconnections with the double phase conjugate mirror. Different schemes have been proposed to implement interconnections with photorefractive double phase conjugate mirrors [40-46]. For instance, a single point to a single point two way interconnect can be realized with the set up presented in figure 13.21. With similar set ups, light injection in optical fibers was demonstrated without any alignment problems. Interconnects through both multimode [41] or monomode [46] fibers were demonstrated. Similar designs may be used to broadcast one single signal into different channels by adding other fibers, detectors and sources on the right side of figure 13.21. We want to underline that even a 1 to 1 interconnect system such as the one depicted in figure 13.21 may be interesting because they may be used to complement other interconnect systems and can make them insensitive to any mechanical change [45].

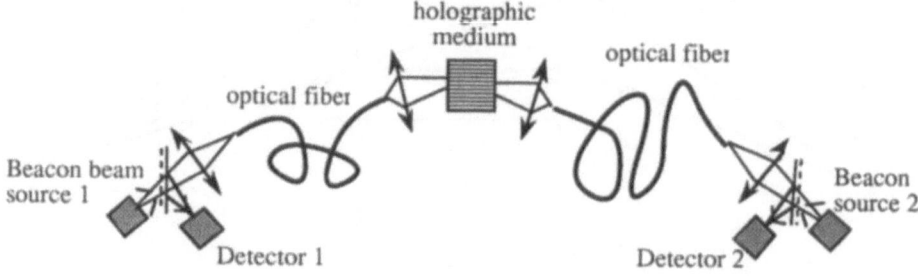

Fig.13.21. A 1 to 1 self aligning interconnect through optical fibers.

Arbitrary fan in and fan out is also achievable as demonstrated in Ref. [44]. The author gives the example of a 3 by 3 interconnection scheme with the inputs labelled **1, 2, 3** and outputs **a, b, c**. To interconnect **1** to **a** and **b**, and **2** and **3** to **c**, he proposed to cycle the lasers in the sequence **(1)-(a,b), (2,3)-(c)** until the interconnections are written. Then **1, 2** and **3** are used as transmitters and **a, b** and **c** as receivers. This scheme was demonstrated for a 2 by 2 interconnection system only.

13.8 Conclusion

We have theoretically shown that volume holographic interconnects should have larger capacities than thin interconnects but at the expense of more accurate positionings. However their ultimate capacity is limited by noise that has been shown to be inherent to volume holograms. Capacities currently reached with available volume holographic materials are by far lower than this maximum. This is mainly due to the low value of the photoinduced index modulation achievable. In fact similar *index modulation-interaction length* products are obtained in both thin and thick media so that demonstrated interconnect capacities are very close in both materials. For instance, 500 superposed holograms of complex images were recorded in volume photorefractive materials [15] and 1000 holograms of a single point source in photographic plates [47].

Progresses in new materials such as optical polymers [48, 49] make probable the development of high resolution holographic media with high index modulations. Full advantages of thick interconnects should thus be reached.

References

1. D. Psaltis, X. G. Gu, D. Brady: Fractal sampling grids for holographic interconnections. SPIE proceeding Vol. 963 Optical Computing 88 (1988), pp. 468-474.
2. H. Kogelnik: Coupled wave theory for thick hologram gratings. Bell Syst. tech. Journ. **48**, 1969, pp. 2909-2947.
3. C. B. Burckhardt: Diffraction of a Plane Wave at a Sinusoidally Stratified Dielectric Grating. J.O.S.A. **56**, 1966, pp. 1502-1509.
4. R. Magnusson , T. K. Gaylord: Equivalence of multiwave coupled-wave theory and modal theory of periodic-media diffraction. J.O.S.A. **68**, 1978, pp. 1777-1779.
5. M.G. Moharam, T. K. Gaylord: Rigorous coupled-wave analysis of planar-grating diffraction. J.O.S.A. **71**, 1981, pp. 811-818.
6. M. G. Moharam, T. K. Gaylord, R. Magnusson: Diffraction characteristics of three-dimensional crossed-beam volume gratings. J.O.S.A. **70**, 1980, pp. 437-442.
7. N. Uchida: Calculation of diffraction efficiency in hologram gratings attenuated along the direction perpendicular to the grating vector. J.O.S.A. **63**, 1973, pp. 280-287.
8. S. G. Rabbani, J. L. Shultz, G. J. Salamo, E. J. Sharp, W. W. Clark III, M. J. Miller, G. L. Wood, R. R. Neurgaonkar: Color imaging in photorefractive crystals. Appl. Phys. **B53**, 1991, pp. 323-329.
9. R. McRuer, J. Wilde, L. Hesselink, J. Goodman: Two-wavelength photorefractive dynamic optical interconnect. Opt. Lett. **14**, 1989, pp. 1174-1176.
10. H. Lee, X. G. Gu, D. Psaltis: Volume holographic interconnections with maximal capacity and minimal crosstalk. J. Appl. Phys. **65**, 1989, pp. 2191-2194.
11. E. N. Glytsis, T. K. Gaylord: Rigorous 3-D coupled wave diffraction analysis of multiple superposed gratings in anisotropic media. Appl. Opt. **28**, 1989, pp.2401-2421.
12. C. W. Slinger: Weighted volume interconnects for adaptive networks. Optical Computing and Processing **1**, 1991, pp. 219-232.
13. J. W. Lewis, L. Solymar: Spurious waves in thick phase gratings. Opt. Comm. **47**, 1983, pp. 23-26.
14. G. Pauliat, G. Roosen: New advances in photorefractive holographic memories. Int. Journ. Opt. Comp., issue on Photorefractive Materials 1993.
15. F. H. Mok, M. C. Tackitt, H. M. Stoll: Storage of 500 high-resolution holograms in a LiNbO$_3$ crystal. Opt. Lett. **16**, 1991, pp. 605-607.
16. P. J. van Heerden: Theory of optical information storage in solids. Appl. Opt. **2**, 1963, pp. 393-400.
17. K. Bløtekjaer: Limitations on holographic storage capacity of photochromic and photorefractive media. Appl. Opt. **18**, 1979, pp. 57-67.

18. D. M. Pepper, J. A. Yeung, D. Fekete, Amnon Yariv: Spatial convolution and correlation of optical fields via degenerate four-wave mixing. Opt. lett. **3**, 1978, pp. 7-9.
19. J. O. White, A. Yariv: Real-time image processing via four-wave mixing in a photorefractive medium. Appl. Phys. Lett. **37**, 1980, pp. 5-7.
20. J. E. Ford, Y. Fainman , S. H. Lee: Array interconnection by phase-coded optical correlation. Opt. Lett. **15**, 1990, pp. 1088-1090.
21. M. R. Weiss, A. Siahmakoun: Autocorrelation via two-wave mixing in barium titanate. Opt. Eng. **30**, 1991, pp. 403-407.
22. F. T.S. Yu, S. Wu, A. W. Mayers, Sumati Rajan: Wavelength multiplexed reflection matched spatial filters using $LiNbO_3$. Opt. Comm. **81**, 1991, pp. 343-347.
23. S. G. Faria, A. A. Tagliaferri, P. A. M. dos Santos: Photorefractive optical holographic correlation using a $Bi_{12}TiO_{20}$ crystal at $\lambda=0.633\mu m$. Opt. Comm. **86**, 1991, pp. 29-33.
24. L. d'Auria, J. P. Huignard, C. Slezak, E. Spitz: Experimental Holographic Read-Write Memory Using 3-D Storage. Appl. Opt. **13**, 1974, pp. 808-818.
25. D. L. Staebler, W. J. Burke, W. Phillips, J. J. Amodei: Multiple storage and erasure of fixed holograms in Fe-doped $LiNbO_3$. Appl. Phys. Lett. **26**, 1975, pp. 182-184.
26. H. Kurz: Photorefractive recording dynamics and multiple storage of volume holograms in photorefractive $LiNbO_3$. Opt. Acta **24**, 1977, pp. 463-473.
27. G. A. Rakuljic, V. Leyva, A. Yariv: Optical data storage by using orthogonal wavelength-multiplexed volume holograms. Opt. Lett. **17**, 1992, pp. 1471-1473.
28. D. Z. Anderson: Coherent optical eigenstate memory. Opt. Lett. **11**, 1986, pp. 56-58.
29. A. Yariv, S. K. Kwong: Associative memories based on message-bearing optical modes in phase-conjugate resonators. Opt. Lett. **11**, 1986, pp. 186-188.
30. Y. Owechko: Nonlinear Holographic Associative Memories. IEEE QE-**25**, 1989, pp. 619-634.
31. H. Xu, Y. Yuan, Y. Yu, K. Xu, Y. Xu: Performances of real time associative memory using a photorefractive crystal and liquid crystal electrooptic switches: Appl. Opt. **29**, 1990, p.p. 3375-3379.
32. L. D'Auria, J. P. Huignard, E. Spitz: Holographic Read-Write Memory and Capacity Enhancement by 3-D Storage. IEEE Magnetics **9**, 1973, pp. 83-94.
33. E. S. Maniloff, K. M. Johnson: Dynamic holographic interconnects using static holograms. Opt. Eng. **29**, 1990, pp. 225-229.
34. D. Z. Anderson, D. M. Lininger: Dynamic optical interconnects: volume holograms as optical two-port operators. Appl. Opt. **26**, 1987, pp. 5031-5038.
35. Y. Taketomi, J. E. Ford, H. Sasaki, J. Ma, Y. Fainman, S. H. Lee: Incremental recording for photorefractive hologram multiplexing. Opt. Lett. **16**, 1991, pp. 1774-1776.
36. C. Denz, G. Pauliat, G. Roosen, T. Tschudi: Volume hologram multiplexing using a deterministic phase encoding method. Opt. Comm. **85**, 1991, pp. 171-176.

37. K. Wagner, D. Psaltis: Multilayer optical learning networks. Appl. Opt. **26**, 1987, pp. 5061-5075.
38. M. Cronin-Golomb, B. Fisher, J. O. White, A. Yariv: Theory and Applications of Four-Wave Mixing in Photorefractive Media. J. Quantum Electon. **QE-20**, 1984, pp. 12-30.
39. S. Weiss, S. Sternklar, B. Fisher: Double phase-conjugate mirror: analysis, demonstration, and applications. Opt. Lett. **12**, 1987, pp. 114-116.
40. M. P. Petrov, S. L. Sochava, S. I. Stepanov: Double phase-conjugate mirror using a photorefractive $Bi_{12}TiO_{20}$ crystal. Opt. Lett. **14**, 1989, pp. 284-286.
41. H. J. Caufield, J. Shamir, Q. He: Flexible two-way optical interconnections in layered computers. Appl. Opt. **26**, 1987, pp. 2291-2292.
42. S. Weiss, O. Werner, B. Fisher: Analysis of coupled photorefractive wave mixing junctions. Opt. Lett. **14**, 1989, pp. 186-188.
43. J. Shamir, H. J. Caulfield, B. M. Hendrickson: Wavefront conjugation and amplification for optical communication through distorting media. Appl. Opt. **27**, 1988, pp. 2912-2914.
44. M. Cronin-Golomb: Dynamically programmable self-aligning optical interconnect with fan-out and fan-in using self-pumped phase conjugation. Appl. Phys. Lett. **54**, 1989, pp. 2189-2191.
45. M. P. Schamschula, H. J. Caulfield, C. M. Verber: Adaptative optical interconnection. Opt. Lett. **16**, 1991, pp. 1421-1423.
46. N. Wolffer, V. Royer, P. Gravey: Self-aligned connection between single mode fibers with mutually pumped phase conjugate mirrors. Technical Digest of the ICO International Topical Meeting on Optical Computing, Minsk 1992, paper 30D7.
47. J. T. LaMacchia, D. L. White: Coded Multiple Exposure Holograms. Appl. Opt.7, 1968, pp. 91-94.
48. Y. Shi, W. Steier, L. Yu, M. Chen, L. R. Dalton: Large stable photoinduced refractive index change in a nonlinear optical polymer with dispersive red side groups. Appl. Phys. Lett. **58**, 1991, pp. 1131-1133.
49. Z. Sekkat, M. Dumont: Polarization Effects in Photoisomerization of Azo Dyes in Polymeric Films. Appl. Phys. **B 53**, 1991, pp. 121-123.

14 Theory of interconnection networks

D. Fey and W. Stork
Physikalisches Institut Universität Erlangen-Nürnberg, Angewandte Optik

Interconnection networks have been widely discussed in the past and are reviewed in this book. Of particular interest are switching networks, which are promising applications for optical interconnect technology. Although the purpose of every interconnection network is the same, to connect one or more sender/receiver pairs, different solutions have been found. To introduce the subject we discuss applications of interconnection networks (Section 14.1). In the second section a classification scheme of these different interconnection techniques is given (Section 14.2). Finally we present concepts and experimental results of optical interconnection networks (Section 14.3).

14.1 Applications of interconnection networks

Interconnection networks have the task of connecting a number of communication participants, which are spatially separated from each other but would like to exchange information. The goal is to provide a connection from each communication participant to any other one. In the case of large numbers of participants the connection will almost certainly not be direct. This is particularly true for the most popular and important contemporary network, the telephone system. A full connection between all telephone participants is technically and economically impossible. Therefore in complex networks with heavy traffic and many data sources and destinations, like the telephone system messages, have not only to be transferred but also routed via intermediate stages.

With rising requirements for the communication performance in computer systems, especially in multiprocessor or in multicomputer systems, it is becoming increasingly necessary to organize communication in parallel data streams in such systems. Examples are the communication among processors in a multiprocessor system and the access of processors and controllers to common shared periphery within a large computer system. In the past the common approach for speeding up computers was increasing the clock frequency with a new improved technology. The higher clock frequencies of today's supercomputers require big technical and economic efforts. Also for basic physical reasons a further speed-up of electronic computers by shortened clock cycles is limited to one order of magnitude. In order to achieve a speed-up of several orders of magnitude, for applications such as weather forecasting or climate modelling, alternative architectures for parallel processing systems, with a large number of low speed processing units, seem to be a more promising possibility. However with distributed processing techniques complex and powerful interconnection networks are necessary to fulfil the demanding communication tasks in the parallel computer architecture.

Depending up on the actual computational problem which has to be solved by the multiprocessor, different interconnection schemes are preferable. The solution of partial differential equations needs frequent data exchange among nearest neighbours. In contrast to this in neural nets a comparably irregular communication structure is required. Processing elements may need to exchange data and messages not only to nearest neighbours. A general-purpose multiprocessor system needs an interconnection network which is dynamically adaptable to different problems. Because there is no interconnection network which is definitely the best one for all applications the design of an interconnection network is always a compromise among a set of interconnection schemes with different advantages and disadvantages in terms of speed, costs, effectivity and reliability. In the following chapter the different types of interconnection networks are classified.

14.1.1 Classification of interconnection networks

In order to cooperate on a common problem, processing elements have to communicate with each other. For example to exchange partial results from one processor to one or more other processors. or to synchronize the calculation among the processors. The most important question in relation to the efficiency of an interconnection network and therefore the performance of the whole multiprocessor or telephone transmission system is after Almasi/Gottlieb [1]: "How quickly it can deliver how much of what's needed to the right place reliably and at low cost and good value?" In order to measure the performance of an interconnection network in computing systems the following characteristics are determined.

- **Latency.** This is the time needed for the transmission of a single message from sender to receiver.

- **Bandwidth.** How much information can be transmitted by the network ?
- **Connectivity.** The degree of a node, i.e. how many directly connected neighbours can be connected to a processing element and how often can a neighbour be reached.
- **Hardware cost.** How much are the costs of the interconnection network in relation to the costs of the whole system in which the interconnection network is embedded.
- **Reliability.** How many different ways are available to reach a node, how high is the probability of a failure of a switching element or of a failure during the data transmission. The latter can be expressed by the bit error rate.
- **Functionality.** This includes functions provided by the switching elements beyond the switching functions, such as combining of messages and solving conflicts.

Besides these efficiency parameters there are five fundamental characteristics of an interconnection network which allow a classification and which a designer has to consider.

1) **Topology.** This topic is concerned with a static or dynamic network setup. A static network is unchangeable after the machine is built. Static networks are used for example in designs ranging from small systolic arrays, which are used for the design of special purpose VLSI architectures for example, to large supercomputers like the ILLIAC IV and the GoodyearMPP. For multipurpose applications it is desirable to make use of dynamically reconfigurable interconnections among the processors during operation. This reconfiguring is performed by switching elements within the network.

2) **Control.** The control system of an interconnection network can be realized with a centralized or distributed strategy. In the latter the switching element itself routes an incoming message to a certain outgoing port. Such networks are called self-routing or self-organized. Central controlled systems calculate the actual switch settings by a central unit before the data transfer.

3) **Switching method.** The transmission of huge amounts of data from one node to another node over a wide distance can be realized by two different switching methods. Either the transmission is performed by circuit switching or by package switching. In circuit switching networks a physical path between sender and receiver node is built up and after that the data is transferred in one closed bit stream. Package switched networks are characterized by a "store and forward" technique, meaning that the data are switched and transmitted node by node. In wide area nets the whole data stream is divided in smaller packages which are all transmitted separately from the sender to the receiver. Each different package can reach the destination via different paths which depends on the actual traffic

in the network. In multiprocessor systems often a special kind of "store and forward" technique, so-called message switching, is used. The transfer of the data is performed in one block of fixed length from node to node. Package switching networks are more economical if short messages are sent because in this case a more efficient use of the available bandwith is possible. In circuit switching networks the beginnig of the data transfer has to be delayed until the whole line is built up. The disadvantage of package switching is that more complex administration tasks have to be performed to assemble and disassemble the data packages.

4) Blocking or non-blocking. A distinction can be drawn between blocking and non-blocking interconnection networks. A blocking network has the feature that a sender and a free receiver pair cannot in any case be connected because the necessary transmission lines are occupied by another connection. Non-blocking networks are universal, which means that every disjunctive permutation between senders and receivers is possible at all times.

5) Single-stage or multi-stage. A further criterion to classify interconnection networks is the number of switching stages. A single stage network is characterized by a single step to get from source to destination. On the other hand multi-stage networks have multiple switching stages which have to be passed in order to reach the destination port.

In the following, static and dynamic topologies will be discussed in more detail, referring to the parameters and characteristics listed above.

14.1.2 Static connection topologies

Static connection topologies are appropriate for problems whose communication requirements are comparably easy to divide into regular patterns and are mostly restricted to neighbouring processing elements. A nearest neighbour connection is well-suited for numerical solutions of partial differential equations. Data exchange is only necessary for the data arranged at the edge of a two-dimensional data array which is mapped to a processing element. For example the application domain for such a connection topology is the analysis of events in space, ranging from robot vision to weather modelling to VLSI design [1]. The disadvantage of this interconnection scheme is that data transfer over larger distances has to pass more intermediate nodes. This means that processing elements will be blocked for the transfer time. With the exception of an all-to-all connection (full connection) such topologies tend to be more suited for special-purpose application classes in contrast to the switching networks which will be examined later.

Next we will give an overview of some typical static topologies as they have been used in parallel computers. We will start with an extreme case (in terms of global interconnection and price), this is the fully connected nodes, and we will finish at the end of this scale with meshes and rings.

Fully connected nodes. This is the most powerful and of course the most expensive interconnection network. Each node is connected to all the others (Fig.14.1). Totally $N(N-1)/2$ connections are necessary, if N is the total number of nodes in the network. A realization is only practicable for moderate values of N. The routing algorithm is trivial because a message which is to be sent from node A to node B is transmitted via the path connecting the two nodes.

Stars. In a star network (Fig.14.2) one node works as a central node which is connected to all the other nodes. Therefore in a network with N nodes in total the central node has a degree of N-1, the other nodes a degree of 1. Such a topology is often used in commercial networks where all the information is collected in one central mainframe computer. For a large value of N it is better to construct a hierarchy of stars because a solution with one central node is technically difficult and expensive.

The routing in star networks is quite simple. A message to be sent from A to B is transmitted along the connecting edge if either A or B is the central node, otherwise it has to pass two edges via the central node. The transfer power in this case is limited by the performance of the central node.

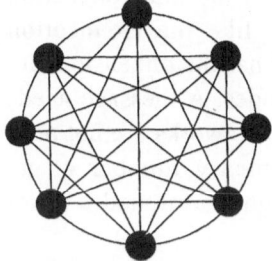

Fig.14.1 Full connection

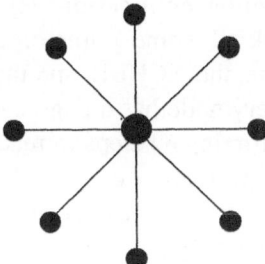

Fig. 14.2 Connection with a star

N-fold binary trees. The n-fold binary tree (Fig.14.3) is characterized by a hierarchical structure of different layers in which every layer is a binary tree. In a binary tree every node has two children and one parent node, with the exception of the root node which has of course no parent. The interconnection network needs N^2 switches. A message has to pass $\log_2 N$ switches and connection lines to reach the destination node.

The routing algorithm in a tree without intermediate connections of nodes in the same hierarchy level is realized as follows: In order to work effectively the algorithm needs a definite identification of the nodes in the tree. Therefore every node in the tree gets a binary number which is defined recursively. The root for example is characterized by a 1. The node number for the left and the right children is calculated by appending a 0 or a 1 to the parent number for the left and right successor, respectively. If we would like to send a message from A to B the lowest ancestor X of A and B has to be found. After this the message can travel up from A to X and afterwards it has to descend from X to B. The lowest ancestor of A and B is that node X with the longest common prefix of A and B. The path from X to B is calculated by removing the node number of X from the node number of B, the remaining suffix is read successively from left to right. By reading a zero the left child is selected, otherwise the right one.

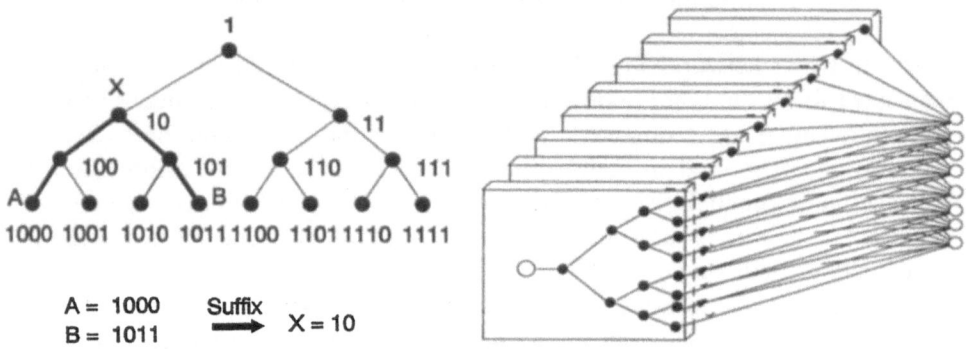

Fig.14.3 N-fold binary trees.

Hypercube connection. Hypercubes are often realized in the interconnection network of some commercial multiprocessor systems, like the Connection Machine, the NCUBE, and the FPS. A hypercube network has the dimensionality n if every node has a degree of n and consists of $N=2^n$ nodes. A message needs at most $n=\log_2 N$ steps to reach its destination. Hypercube networks are not non-blocking. Each node in the cube (Fig.14.4) gets a binary number as an identification address in such a way that addresses of neighbouring nodes differ by a power of 2.

Because the hypercube network can be regarded as a n-dimensional mesh in which each dimension contains only two nodes the routing algorithm can easily be implemented. The Exclusive-Or of the addresses of two nodes A and B has only a 1 those bit positions where the two numbers are different. Only along the dimensions corresponding to the bit positions which contain a 1 does a message have to be passed to get from A to B. A message is sent during the ith cycle to the neighbour in the ith dimension if the Exclusive-Or contains a 1 in its ith position otherwise it will remain at the node where it currently is. Therefore a message needs at most $n=\log_2 N$ steps if the result of the Exclusive-Or contains in every bit position a 1.

Meshes and rings. A very cost effective solution is to arrange all the nodes in one dimension along a line (Fig.14.4). The nodes are connected to their left and right neighbours with the exception of the boundary nodes which have of course only one outgoing connection line. If the two boundary nodes are also connected together the network becomes a ring. Higher dimensional meshes or rings are constructed in exactly the same way. Every node has 2n connections to its nearest neighbours where n is the dimension of the mesh or the ring. Two-dimensional meshes, which have been used for example for the connection topology in the ILLIAC IV, the DAP and the MPP, are very popular.

The routing is performed by sending data to the neighbour along the dimension in which the destination node is located. For example in a two-dimensional mesh a message from a node with the coordinates k,l is sent to the node m,n by transporting the message along m-k steps horizontally and n-l steps vertically. Such an architecture is not recommendable for applications in which data transfer often occurs between nodes which are not nearest neighbour, i.e. which are far apart, as the latency and the probability of a blocking of the nodes increases.

To conclude the discussion of static topologies we refer to a table published in [1] which contains a comparison of network architectures, with the exception of the star connection, shown according to their performance criteria, latency, bandwidth, connectivity and the wiring costs. The table itself is organized such that the topology with lowest latency is listed at the top, and that with lowest costs at the bottom.

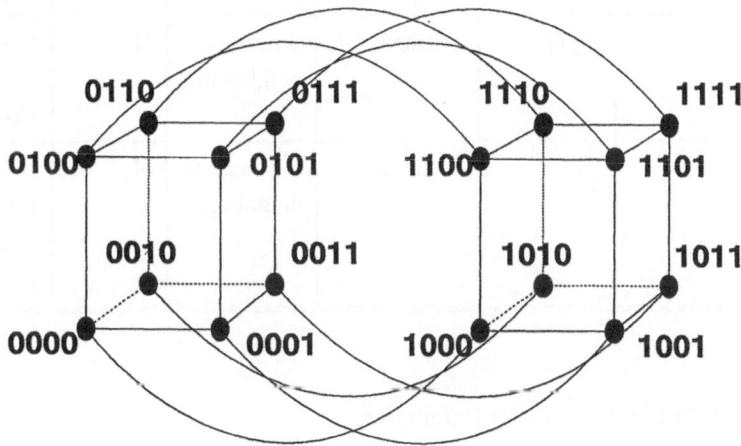

Fig. 14.4 Hypercube connection

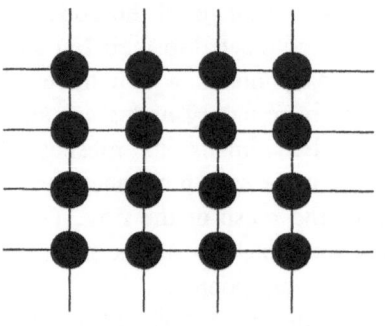
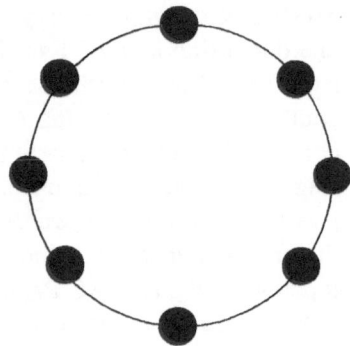

Fig. 14.5 Meshes and rings

Tab. 14.1: Comparison of different interconnection networks

Network	Minimum Latency	Maximum Bandwidth per PE	Connecti- vity	Wire Costs	Application
fully interconnected	constant	constant	any to any	N^2	
hypercube	logN	constant	3 nearest neighbours	N logN	Connection Machine
tree	logN	constant	3 nearest neighbours	N	Non Von Neumann, Cellular
mesh	$\sqrt{N}$	constant	4-8 nearest neighbours	N	ILLIAC IV, DAP, Systolic Arrays

14.1.3 Dynamic connection topologies

The topologies in the last section have a fixed interconnection scheme which can be very efficient for special-purpose applications. In contrast more general applications demand a possibility for a flexible interconnection among the processing elements. The need for a flexible mapping of different classes of algorithms on a parallel computer architecture motivated the design of dynamic interconnection networks. These networks can be divided into three classes: bus networks, crossbar networks and multistage networks. In the following we present a bus network as a low-cost solution with low communication performance and a crossbar network which is at the opposite extreme because of its high communication performance and its high costs. Multistage networks fall in between these two extremes and will be discussed in more detail because of their importance for optical realizations.

Crossbar networks. A crossbar network (Fig.14.7) is topologically equivalent to the fully connected nodes presented previously, in the case when the output is fed back to the input. Therefore it is also non-blocking. The number of paths is reduced to 2N and each path has N junctions with connection to N nodes. The crossbar needs N^2 switches. One disadvantage of the crossbar, as in the case of the fully connected nodes, is the high cost. Although the cost of the wiring is of the order of N the cost for the switches rises with N^2. As will be seen this problem can be solved at the cost of additional latency in multistage networks which operate mostly with 2x2 crossbars as elementary switching boxes. But before doing this let us examine a cheaper solution which is preferable if the number of communication nodes is small.

Bus networks. Bus networks (Fig.14.7) are the simplest and most popular networks among the blocking networks. A bus consists of one large communication line to which all communication participants are connected. If the bus is free a node can occupy the line and send its data. Only one node can access the line for data transfer at one moment to avoid conflicts. After gaining access the sender node gives the receiver address on the line. All the other nodes listen to the line to see if that address is their own. If this is the case the receiver node will take away the subsequent data from the line. Because in a bus network no parallel data transfer is possible the average bandwidth for each node is inversely proportional to the total number of participants.

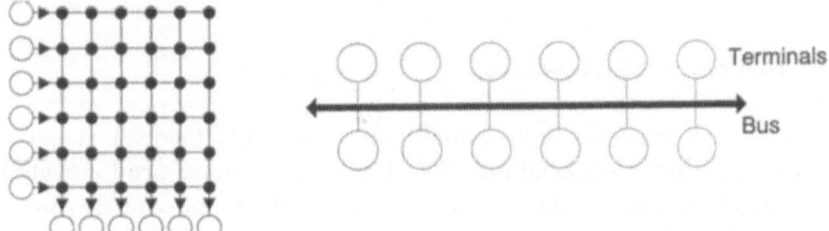

Fig. 14.6 Crossbar connection   **Fig. 14.7** Bus networks

The bandwidth of a bus is simply the product of the system clock frequency and the bus width. Because the number of parallel lines is restricted the bandwidth is mostly increased by raising the clock frequency. But this technology also makes the processing elements faster. As a result the ratio between processing power and bus bandwidth is nearly constant. Next we present multistage switching networks which offer the possibility of simultaneously achieving higher bandwidth and moderate cost for switches.

Multistage switching networks. Multistage networks are used for the establishing parallel reconfigurable point-to-point connections. To reduce the hardware cost for the switches as much as possible 2x2 crossbars are very often used for the switches. Because in this case no broadcasting is realized the 2^2 necessary connections for this are reduced to 2! connections which can be achieved with a 1-bit address switching box. The base operation of such boxes is an exchange of the data or a bypass (Fig.14.8). A column of such boxes is called an exchange stage. Together with a static permutation stage, such a dynamic exchange stage is the basis of exchange networks. As we will see later the fixed permutation is of special importance for an optical realization of the node links.

It is necessary to use nearly $2(\log_2 N)$ stages with N/2 switches per stage which leads to a reduction of the cost for switches from N^2 in a crossbar to $N\log_2 N$. The price which has to be paid for that is an increase of the latency on the order of $\log_2 N$.

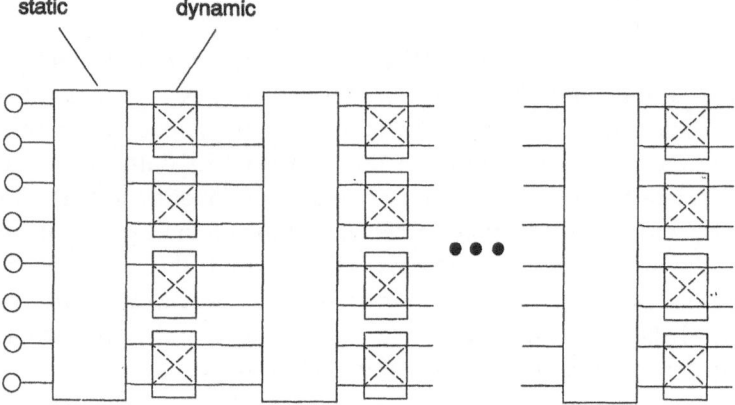

Fig. 14.8 Multi-stage networks

For two such networks with a minimum number of $2\log_2 N$-1 stages, efficient control algorithms are available [10]. But such networks need a central control unit which calculates the position of the switches previously for a current permutation. This calculation is very extensive for a large value of N and additional time is required for the positioning of the switches. If this has to be performed sequentially $N\log_2 N$ steps are necessary. In general the positioning of the switches has to be recalculated for every new permutation even if the changes are only small. In conclusion multistage networks with central control are only meaningful if the number of rearrangements is limited.

Therefore in systems with a large number of participants it is desirable to install multistage networks with distributed control. In such networks the switching boxes have internal logic to decide how the internal switch has to be positioned. For this decision binary address headers are attached in front of the data bit stream. From a control signal, a switching box is informed that the incoming data are address bits and that the internal switch position is to be calculated. Depending on these address bits the i.th switching stage is responsible for the i.th bit position[1]. Often the total data transfer is pipelined. This means that in one stage address bits are processed and at the same time in the previous stages data bits following the header bits are passed by. After the switches in the last stage are calculated it is guaranteed that the subsequent data stream reaches the destination. This distributed control problem is equivalent to a sorting problem in which an unordered sequence of numbers at the input has to be changed into a sequence of ordered numbers at the output. There are control algorithms for such self-routing networks with $(\log_2 N)^2$ stages [2],[16]. The price for the distributed control is again an increase in latency but this disadvantage can de diminished by pipelining. For example a switching stage can be reconfigured after a current data package has been passed by. These networks are especially applicable to telephone traffic.

Next we give some examples of frequently installed multistage networks in multiprocessor systems. All these networks are only slightly different because many of them are topologically equivalent. Their fundamental setup is a sequence of static and dynamic permutations which can differ in the function of the switching boxes and the exact connection topology. We start our presentation with the omega network which is a suitable representative to explain the characteristics of this class of switching network.

Omega networks. The omega network is an economic model because it has only one unique connection path between the dynamic permutations or the switching stages. Besides this advantage this feature has also further advantages and disadvantages as we will show.

Originally the omega network contained as switching boxes a 2x2 crossbar which had an upper and a lower broadcast (Fig.14.9) possibility additional to the exchange/bypass function [9]. The interconnection topologies between such switching stages are always perfect shuffles which are of special importance for optical solutions and will be discussed in the next chapter. The expression "perfect shuffle" refers to the shuffling process in card playing where the top half of a deck of cards is perfectly interleaved with the bottom half. The

---

1 Addresses have to be sent in reversed bit presentation (Most significant bit first)

corresponding address operation can be calculated by rotating the binary input address one bit position to the left.

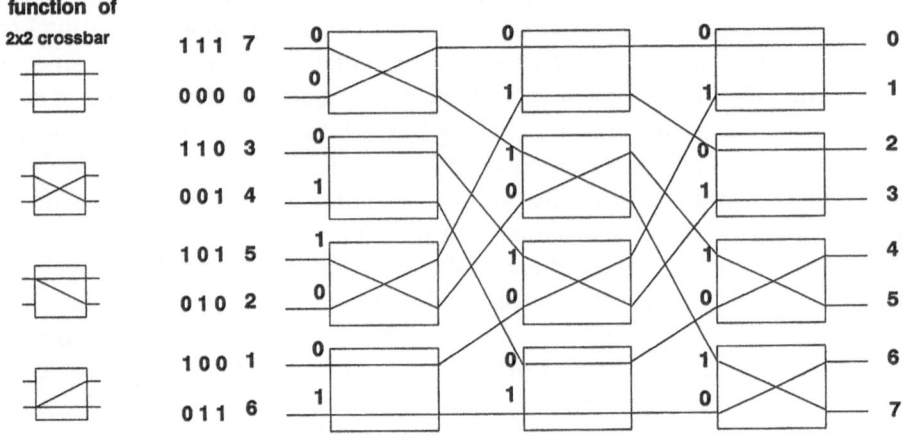

Fig. 14.9 Omega network

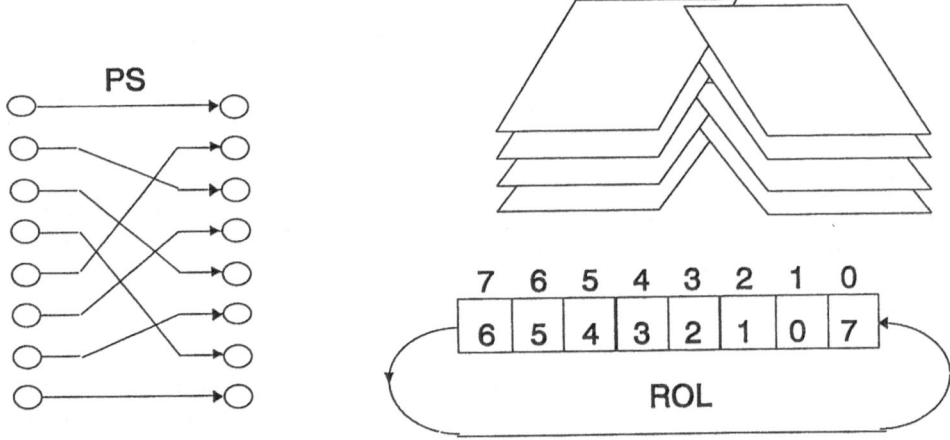

Fig. 14.10 A perfect shuffle

The switching method in an omega network is packet switching. The routing algorithm is performed by distributed control, as has already been described above. This allows a data packet, which has in front of it the reversed representation of its receiver address, to navigate itself to the destination (Fig.14.10). The simple configuration of the omega network is its advantage leading to low price, fast processing, and small size. For example the omega network could be realized as a single-stage because the functionality of all switching stages and the interconnection scheme between them are identical. The price for this further reduction in costs and size is the loss of pipelining. In general the disadvantage of the omega network is the high probability for blocking of the messages.

There are some solutions to this blocking problem. It has been proven that at least three passes through an omega network are required to realize any permutation. Whether two passes are sufficient is unproven as yet. Another way to make a network universal is to add further hardware to the switching boxes. The NYU (New York Ultracomputer) and the RP3 (Research Parallel processor prototype) both use omega networks in which the switching boxes can combine messages with the same address in queues. Another way to secure non-blocking data transfer is presented by the next network.

Benes networks. In contrast to an omega network a Benes network (Fig.14.11) is non-blocking exactly like the crossbar network. But the data transfer is not realized in one stage as in the crossbar, but in more stages. Therefore the Benes network could be defined as a member of an intermediate class between omega and crossbar networks which are called rearrangeable non-blocking networks. In a Benes network there are multiple ways to get from the source node to the destination node. It is necessary to know in advance the receiver addresses and then a central control unit can calculate the positioning of the switches in the exchange/bypass boxes. Therefore Benes networks are suitable for SIMD (Single Instruction Multiple Data) designs with circuit switching for interconnection.

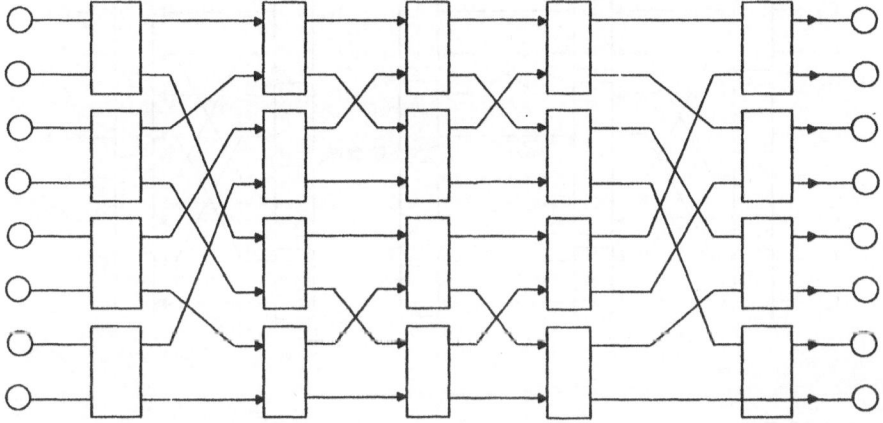

Fig. **14.11** Benes network

As can be seen in Fig.14.11 a Benes network looks similar to an omega network with additional switching stages in the middle. These further hardware elements are the price for the possibility of multiple paths between a sender/receiver pair and lead to an increase in costs, size and latency. The permutation process can be thought of as a sorting of odd and even addresses in the middle of the net. For universal permutations $2\log_2 N-1$ stages are necessary with $N/2$ switches in each stage. As a result we get $N\log_2 N - N/2$ binary switches if we use 2x2 crossbars. In the IBM GF11 a Benes network of 24x24 crossbars is used to interconnect 576 processing elements. The following networks are all topologically isomorphic to the Benes network. For specific permutations or also for optical implementations one of these networks has advantages over the others.

Banyan networks. Banyan networks (Fig.14.12) and the binary n-cube network used for realization of the hypercube interconnection topology have nearly the same static permutation pattern. The only difference is that an inverse perfect shuffle permutation terminates the binary n-Cube network which is not the case in the Banyan network. Both networks are not universal. Because of the irregular structure of such networks all stages have always to be built differently, in contrast to the previously described Omega network. In a hypercube every node has a number of direct connections that is equal to the dimension of the hypercube. For an interconnection to nodes which are not neighbours the latter have to work as intermediate stages. In transferring a message from a node in the (i-1)th stage to the ith, the ith bit is inverted. This inverting can be performed by the binary switching boxes in each stage. This process is repeated until the data has arrived at the destination address. When the address bits are all zeros then the message is at its destination node.

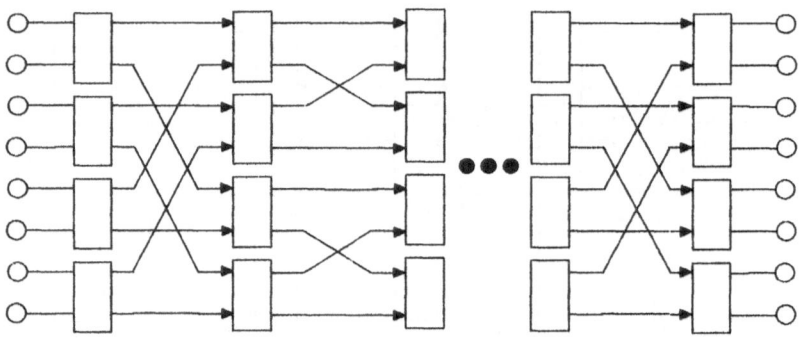

Fig. 14.12 Banyan network

Crossover networks. In the crossover network (Fig.14.13) the reflection is exploited as a static permutation which is easy to realize with optical components [8]. Every second address is reflected at the middle of the array. This is expressed by an conversion of the corresponding binary address. This address inverting is repeated successively on half of the array.

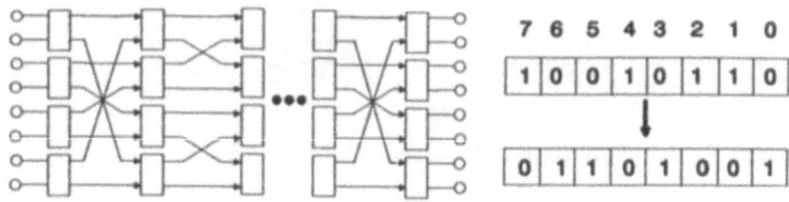

Fig. 14.13 Crossover network

Sorting networks. One subclass of sorting networks are shuffle-exchange networks, which employ a perfect shuffle as the static permutation (Fig.14.10) [18]. An important feature of the shuffle-exchange network for optics is that all permutation stages are identical. Recall that in the Benes network the static permutations in different stages are working on distinct large arrays.

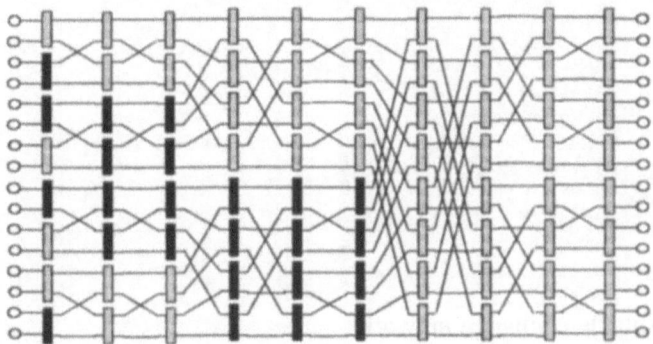

Fig. 14.14 Batchers parallel sorting network

In Fig.14.14 a sorting network is shown which is an implementation of Batcher's bitonic sorting algorithm [2]. This network is a non-blocking self-routing network. It needs $\log_2 N(\log_2 N+1)/2$ stages and $N/2$ switches in each stage. The function of the switching boxes is not the same in the whole array. There are two different boxes with different switching functions. The black boxes represent a switch in which the maximum of the two inputs is transferred to the upper and the minimum to the lower output. The gray box represents exactly the inverse functional behaviour (Fig.14.15). In order to realize the implementation with many identical components as possible Stone published a procedure [16] to substitute the different static permutations of Fig.14.14 with perfect shuffles (Fig.14.16). In this shuffle-exchange network $(\log_2 N-2)(\log_2 N-1)/2$ additional passive stages are needed leading to a total number of $\log_2 N(\log_2 N-1)+1$ stages. Because m perfect shuffle stages can be substituted by $\log_2 N$-m unshuffle stages this number can be diminished to $\log_2 N(3\log_2 N.2)/4$ which means a decrease by about 25%. In Fig.14.17 the first three shuffles of Fig.14.16 have been substituted by an unshuffle.

348

Fig. 14.15 Exchange/bypass and inverse exchange/bypass

These non-blocking shuffle-exchange networks are very interesting for optical implementation. Before we give reasons for that, we emphasize two further theoretical questions. How many switching elements are needed in a non-blocking sorting network and how can we solve conflicts generated by transfer requests of more than one sender to the same receiver?

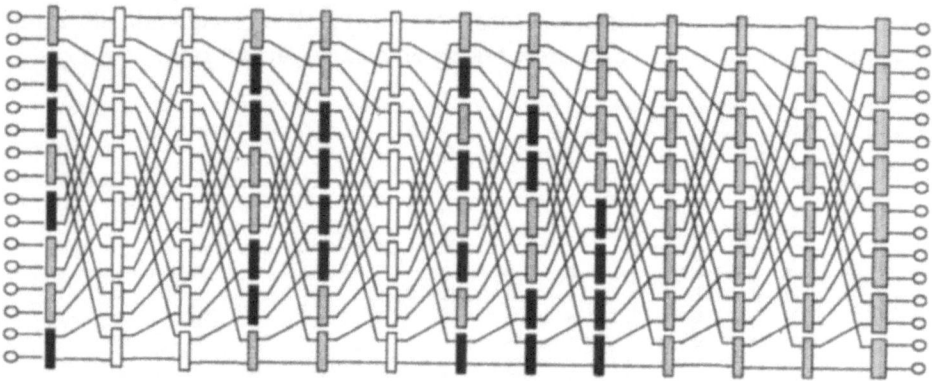

Fig. 14.16 Shuffle exchange network

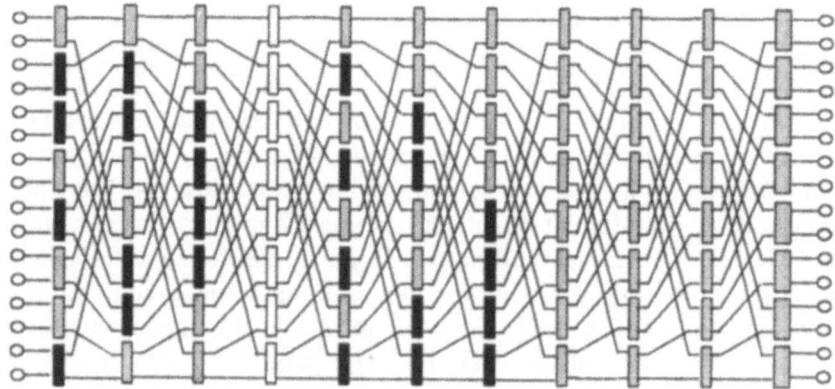

Fig. 14.17 Shuffle unshuffle network

14.1.4 Minimum number of switching elements in non-blocking networks

In order to realize any of NN permutations a multistage network with N inputs and N outputs with an (a)x(a) crossbar as switching element needs $\log_a N$ stages with N/a switching boxes in each stage. The total number of binary switches Z is then

$$Z = \log_a N \frac{N}{a} a^2 = aN \frac{\ln N}{\ln a} \qquad \left[\ln \equiv \log_e \right] \qquad (14.1)$$

A minimum value for Z is given for a=2.71... , the nearest integer number to e is 3. In a switching network with $\log_3 N$ stages the reduction of switches compared with the case of a=2 is only 5.3% . The reduction grows if networks with $(\log_a N)^2$ stages are used, because then 7 or 8 is the next nearest integer value to e2. An 8x8 switching primitive leads to 55% less switches. For optics a 4x4 switching matrix is of interest because this is the base element for 2-dimensional permutation networks which still brings a reduction of about 50% compared with 2x2 switching boxes. But as far as we know there is no control algorithm for bases greater than 2 for non-blocking networks with of the order of $\log_a N$ stages. Therefore it makes sense to substitute a 4x4 switching matrix by five 2x2 elements which still produces a reduction of 37.5% in the number of switching boxes. In addition to questions concerning the efficiency, like the minimum number of switches, questions concerning the effectiveness are also important. For example what should be done if address conflicts occur.

14.1.5 Arbitration in the case of address conflicts

In order to avoid conflicts in a sorting network it is necessary that the receiver addresses are disjunctive. This assumption is unrealistic both for a telephone network and for multiprocessor systems. In a crossbar network it is possible to solve this using a second network. Before the sender starts his transfer he can ask this second network if the receiver address is not occupied. If this is the case the sender polls as many times as necessary until the link is free.

In a parallel sorting network such a solution is not possible, all addresses are sorted in increasing order. This means that in the worst case only one sender is switched to the right destination address. An example of this is shown in Fig.14.18. The destination address is written on the left side of the input lines. Only address 0 has found its correct way, for the other addresses we can determine that they are all at a wrong position. In order to avoid such problems either a control unit or a more complicated algorithm has to be established. Such an algorithm can be found in [2]. The idea is to send back all sender requests which arrive at the wrong addresses. The effective algorithm requires that senders and receivers have to be at the same side of the network and each of them can transfer addresses into the network. For the receivers it is only allowed

350

to send their own address. In the following we denote the sender with S_i and the receiver with R_i. The addresses in the net are defined in such a way that the following condition is fulfilled.

$$S_i = <R_i < S_{i+1} \tag{14.2}$$

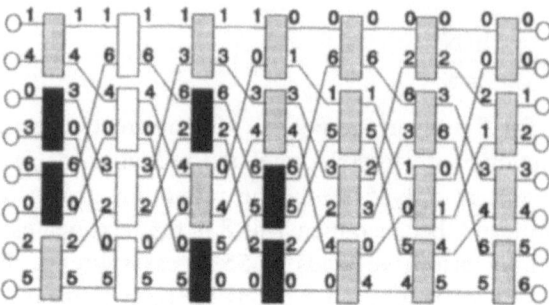

Fig. 14.18 Shuffle exchange network with address conflict

This means that, for example, all sender addresses have even numbers and all receivers have odd numbers. Additional senders always give the receiver address i+1 on the line if they wish to send a message to receiver i. In a data transfer without conflicts this leads to a situation in which every sender has a closed link to a place on the right side of the network which is exactly one position above its corresponding receiver (Fig.14.19). Thus neighbouring positions exchange their places. As result the address and the data of the sender are transferred to the correct receiver. This is shown for sender address 0 and receiver address 1 in the example of Fig.14.19 at the right side of the network at position 1 and 2. In the case of an address conflict two sender addresses are directly positioned above each other at the right end of the network architecture (for example in Fig.14.19 sender addresses 0 at the top of the right network side). All messages which did not find their correct address are simply reflected and will be transmitted back to the senders.

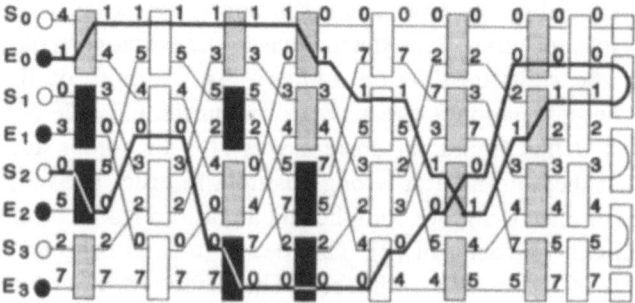

Fig. 14.19 Non-bloking network with arbitration

A condition for this described procedure is of course the existence of bidirectional paths. A pure straight forward solution would be desirable but the authors are not aware of an effective procedure. Finally we want to calculate the number of switches for Batcher's non-blocking self-routing network. In the previous section we pointed out that Batcher's network needs $\log_2 N(\log_2 N+1)/2$ switches. Here we have to substitute N by 2N because of the bidirectional paths. In addition two further stages are required at the end of the network. As a result we get

$$Z = N\frac{\text{ldN}(\text{ldN}+3)+7}{2}. \tag{14.3}$$

If a shuffle-unshuffle network is used the total number of switches is

$$Z = N(\text{ldN}+1)(3\text{ldN}-1). \tag{14.4}$$

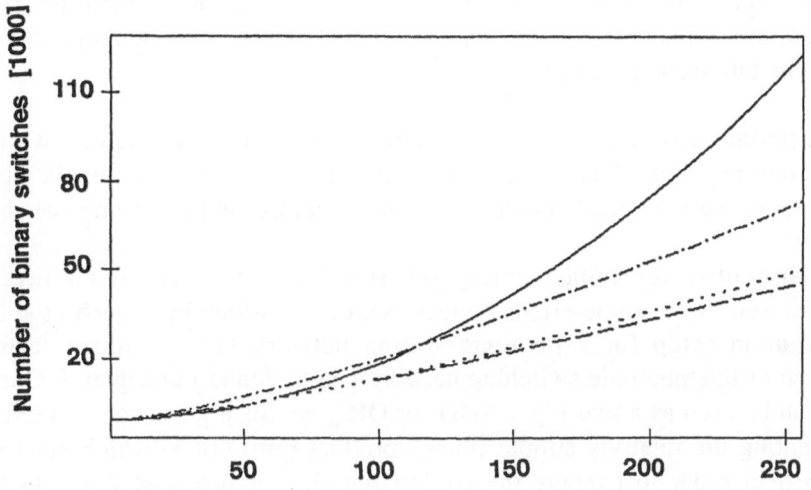

Fig. 14.20 Number of binary switches for Batcher (---), shuffle unshuffle (...), shuffle exchange (-·-·) and crossbar (——)

To conclude we present in Fig.14.20 a drawing which shows the number of switches for a crossbar, a shuffle-unshuffle and a shuffle-exchange network and Batcher's sorting network as functions of the number of communication participants. As a result we can see that for more than 100 participants the use of sorting networks reduces the number of binary switches.

14.2 Concepts and experiments for free-space optical interconnections

In this section we give examples for experimental setups for some of the schemes which have been presented in the previous section. The examples concern dynamic connection topologies like multistage networks and the single-stage crossbar networks.

14.2.1 Experiments for multistage networks

In section 2.2.3 it was shown that multistage networks consist of a dynamic exchange stage, which is an arrangement of 2x2 crossbars without broadcast function, and a static permutation stage, which can be implemented with passive optical components. In the following sections we treat the realization of an optoelectronic switching stage (3.1.1) and some different implementations of the static permutation stage (3.1.2).

Optoelectronic switches. In an optoelectronic switching network the switches for the routing of the data streams are realized as simple smart pixels which operate as simple 2x2-crossbar elements with optical I/O and electronic switching functions.

One possibility for implementing the switching stage are the arrays of symmetric self-electrooptic-effect devices (S-SEEDs) which have been used in a demonstration setup for a photonic sorting network [12]. A more detailed discussion of this photonic switching network can be found in chapter 8. One S-SEED can be used as a two-input AND- or OR-gate. Such gates can be used for implementing the relativly simple sum-of-product form (14.5) which has to be performed in order to execute the sorting function of one switching element (Fig.14.21). The variables s_0 and s_1 are two registers in which a detected switching state is stored.

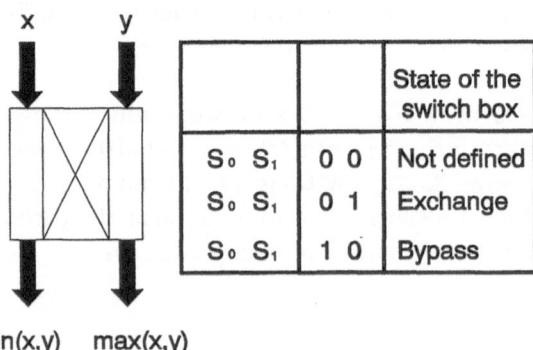

			State of the switch box
S_0 S_1	0	0	Not defined
S_0 S_1	0	1	Exchange
S_0 S_1	1	0	Bypass

Fig. 14.21 Scheme of a two-channel sorting node

In [13] a concept for a sorting network is presented which is based on S-SEED arrays with NOR-OR logic and regular optical crossover interconnects between the switching stages. This concept considers also the case of conflicts which occur if two senders intend to address the same receiver. Such a conflict is solved by the help of a waitingqueue. Messages which cannot be transmitted are fed back and have to reenter the whole network. A similar strategy was planned as a conflict solving aproach in the NYU Ultracomputer [5] by combining and buffering conflicting messages.

An alternative to SEEDs as smart pixel element is possible with photodetectors for optical inputs and light-emitting elements like a light-emitting diode (LED) or a laser diode for optical outputs. Between this optoelectronic I/O interfaces electronic logic circuits can be used to realize the exchange/bypass logic for the switching element and also a few bits of local memory to register a calculated switching state for example.

Another example of such a part of an optoelectronic switching stage is a so called OPTO-ASIC, which was published in [20]. This OPTO-ASIC is an optoelectronic integrated circuit which integrates the function of data receiving and the exchange/bypass logic on a silicon chip. The outputs from this device are electronic signals which have to be converted to optical signals externally. A more detailed description of the setup and experimental results of the OPTO-ASIC, including experimental results, are shown in chapter 15. With such an element it is in principle possible to implement the dynamic switching stages of self-routing Omega and sorting networks. Because the setting of the switching elements on one chip is controlled by an external electronic input signal this device can also be used as a switching stage in the centrally controlled, universal Benes network.

Static permutation stages. In the following section we give some examples for optical setups of a perfect shuffle connection, which is used in every permutation stage of a shuffle-exchange network, and a butterfly connection which is used as a permutation stage in Batcher's sorting network.

Space-invariant implementation of the perfect shuffle. Fig.14.22 shows a cross section of an optical setup using conventional optical elements [11]. By this setup eight parallel beams are mapped as a perfect shuffle permutation onto a switching stage, for example. At the input of the system the upper and the lower half of the beams are divided into two different groups by prisms. By a lens they are mapped onto two further prisms in the transform domain. Because the beams spread out in these prisms on different optical path lengths they obtain different phase shifts. This results in a physically seperation of the beams in space after they are mapped by a lens. With a second lens the two groups of beams are interleaved for the desired perfect shuffle interconnection.

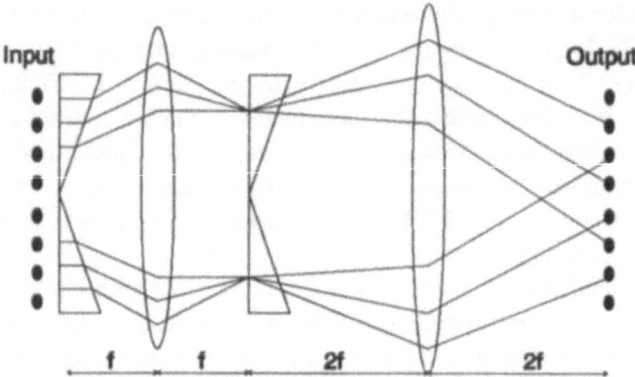

Fig. 14.22 Optical setup for a perfect shuffle permutation with separation in the input pupil, (left: input, right: output)

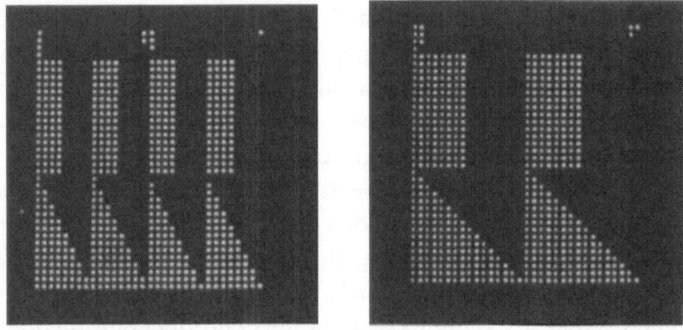

Fig. 14.23 Experimental results of an optical perfect shuffle with separation in the input pupil.

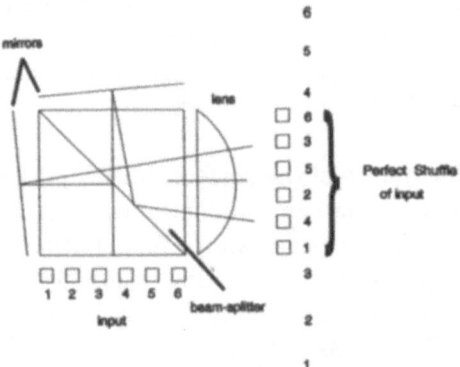

Fig. 14.24 Optical implementation, of a perfect shuffle with a beam splitter.

Fig. 14.23 shows experimental results for an array of 32x32 pixels. Each row of the input image, shown on the left side, is permutated as a one-dimensional perfect shuffle, the result is shown on the right side.

Fig.14.24 shows another concept for an optical perfect shuffle [6]. A perfecι shuffle is performed along the horizontal lines of a two-dimensional image. The input image is divided by a beam splitter into two identical copies. Each copy is mapped by a lens which performs a magnification by a factor of 2. The magnified copies are interlaced and the result in the middle of the output plane is the perfect shuffle of the input image.

A disadvantage of this perfect shuffle implementation is the loss of light caused by using only a part of the output plane. The loss of energy amounts to a factor of four per line. A factor of 2 is lost at the second pass through the beam splitter and a further factor of two arises by the separation of the exit pupil. For a two-dimensional input image the energy loss becomes a factor of 16. In contrast to the solution presented above, in this setup no astigmatism and distortion occurs because collimated beams are used which cross only plane surfaces. By the use of polarization beam splitters and half wave plates the energy loss can be reduced.

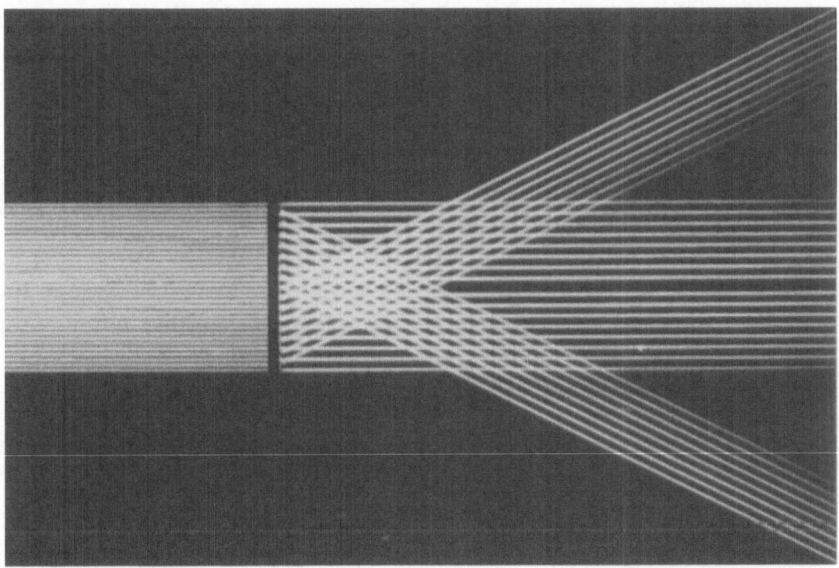

Fig. 14.25 Butterfly permutation realized with holographical optical elements.

Space-variant butterfly connection with holograms. Fig.14.25 shows a butterfly permutation for N=16 senders realized with volume holograms recorded in dichromated gelatin. The hologram consists of space-variant facetted holographic gratings. The light beams have been made visible by a flourescent water bath. The hologram deflects the incident beams into the desired direction. In experiments the amount of stray light was measured to be less than 1%. The

356

diffration efficiency was determined to be 80%, and with surface coating an improvement of over 90% is possible [17].

14.2.2 Optical implementations of a crossbar network

As already mentioned in section 14.2.1 a crossbar allows the simultaneous connection of N senders with N receivers. In the following section different concepts for an optical realization of crossbar networks are presented. The following concepts can be divided in two approaches. One approach intends to make use of the temporal bandwidth of the light. Wavelength multiplexing (3.2.1) is a technique which can be mentioned in this context in which coding of sender addresses uses different wavelengths. Another possibility to exploit the temporal bandwidth is coherence multiplexing. Here the time coherent function is applied as the information carrier. Address coding is performed by a linear combination of different spectral components. For a detailed discussion of this possibility we reference the work of Goedgebuer et al. [4]. In contrast to time division systems in space frequency systems the space or the propagation direction of the light beams is manipulated for the data transmission (3.2.2 and 3.2.3).

Optical crossbar as wavelength switch. Systems with wavelength multiplexing are an approach to exploiting the large temporal bandwidth of the light. The different participants of a communication system are coded by different spectral bands. Fig.14.26 shows the principle of such a wavelength switch system. Either the senders or the receivers can handle variable wavelengths. A lens system distributes the light from the sources relatively equally about the receivers. The receivers filter out the wavelength which is intended for them.

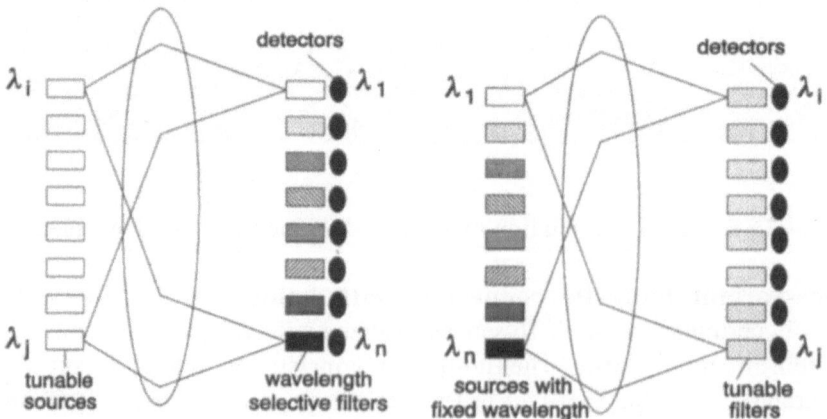

Fig. 14.26 Optical crossbar realized as wavelength switch.

Every transmitter source in the system of Fig.14.26 needs a fan-out of N. This leads to an energy loss of N because the light from one source has to be distributed equally among N receivers. The space-variant solution shown in Fig.14.27 avoids this disadvantage by the application of amici prisms. The deflection angle of such prisms depends on the wavelength. The price one has to pay for this is a loss of space bandwidth because space is needed for the prism arrays. In this setup it is necessary to use tunable wavelength at the sender side. From a system theoretical view point it is unimportant whether the wavelength coding takes place on the sender side or on the receiver side. But in practice it is difficult to produce light emitters with tunable wavelength. The wavelength of laser diodes could only be varied slightly in the range of a few nanometers [14],[15]. Therefore emitters with fixed wavelength should be preferred. Acoustooptic or electrooptic filters for example could be used as dynamically tunable wavelength filters for the receivers.

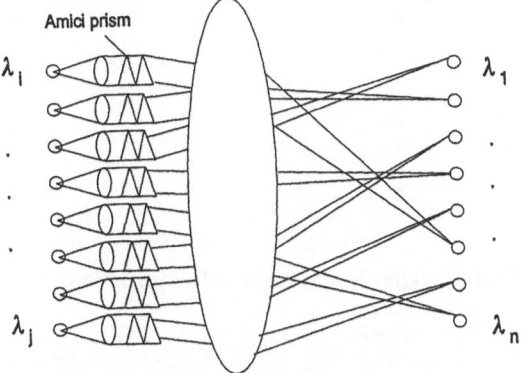

Fig. 14.27 Space-variant wavelength multiplexing.

Optical crossbar as space frequency systems. Another solution for implementing an optical crossbar is the use of switchable half wave plates and deflectors. Fig.14.28 shows the principle. Only a fan-out of 1 is necessary. But the deflectors need N different states. The deflectors need deflection angles of $\Delta\alpha = N \cdot B/F$, if B is the lateral extension of one source element, F is the focal length of the large lens and N is the number of senders and receivers respectively. Realistic values for B are 0.1mm (multi-mode fibre core, LED) and for F 1-5mm. This results in an angle range of $N°-2N°$.

As deflectors, galvanic mirrors, acoustooptic or electrooptic deflectors could be used in principle. With galvanic mirrors it would be possible to obtain an angle range of 10-40° but it takes a few milliseconds to switch between two different states [10]. In addition it is difficult to miniaturize them. Acoustooptic deflectors offer the possibility to switch 300-400 addressable points within microseconds but only small deflection angles are possible (2-3° at about 50MHz). A similar situation exists for electrooptical deflectors. Because of the small deflection angles only a small number of communication participants would be possible. Therefore a digital deflector [19] (Fig. 14.29) has to be used which consists of subsequent stages of deflectors. In front of each deflector stage an array of half wave plates is arranged. The linearly polarized light of the sources can be switched to the perpendicular polarization state. Depending on the polarization state light can be deflected by an angle $\pm\alpha$ through a wollaston prism. The deflection angles of the wollaston prism increase by a factor of 2 in every stage. With the polarization switches the sign of the deflection angle can be selected. With $\log_2 N$ stages N deflection angles and therefore N positions in the image plane are addressable.

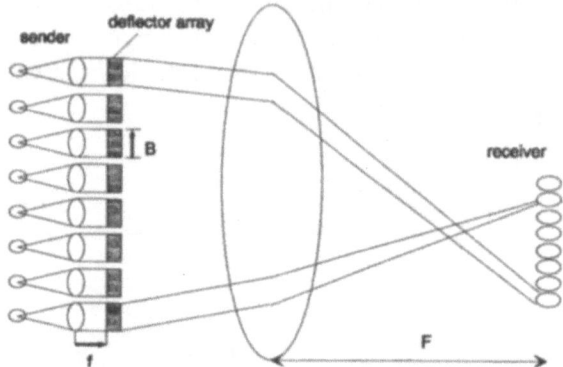

Fig. 14.28 Optical crossbar based on a deflector array.

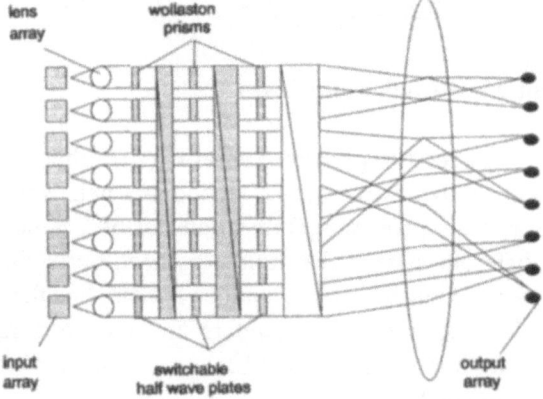

Fig. 14.29 Optical crossbar based on a digital deflector.

The advantageous features of this system are the minimum number $N.\log_2 N$ of binary switches and that only a fan-out of 1 is required. An additional advantage is the availability of cheap half wave plates realized in ferroelectric liquid crystals (FLC). These FLC are bistable elements with a switching time in the range of microseconds [7].

With Wollaston prisms divergence angles 2α up to $20°$ are possible. This implies a possible total deflection $\Delta\alpha$ about $30°$ in 4-5 stages. Projected in two dimensions this can be exploited to a maximum number of 256-1024 communication participants. Instead of comparative expensive wollaston prisms also cube beam splitters can be used which fulfill the same task. If grating beamsplitters with polraization-dependent properties are commercially available they can also be used instead of Wollaston prisms which would also reduce the cost.

Spatial light modulator crossbar switches. Goodman et al. presented a system which is appropriate for matrix-vector multiplication based on spatial light modulators (SLM). The principles of this system could also be used for the implementation of an optical crossbar. In such systems a spatial light modulator as input element consisting of N^2 elements operates as a transmission matrix. A row of separate controllable light sources is used which are arranged along the vertical direction so that every input row element is transmitted to a column of the transmission matrix. Depending on the desired connection one cell in the column of the transmission matrix is switched in the transmissive state. The light passing through this cell is mapped by a further lens pair to the desired receiver node. The receivers consist of a column of detectors which is arranged perpendicular to the light sources. A detailed explanation of such a crossbar based on an optical matrix-vector multiplier can be found in the contribution of A. Kirk and T. Hall in chapter 12 and A. Walker in chapter 8. There one can also find a description of a telecommunication crossbar switch for two-dimensional inputs and outputs which also uses a SLM.

The number of participants in this system is limited by the required fan-out. In addition to this the high absorption in the spatial light modulators has to be considered. Magneto-optical and PLZT modulators have a transmission coefficient in the range of 1-15%. Concerning this 20-30 channels at a data rate of 100MHz are regarded as realistic [3].

References

1. G.S. Almasi, A. Gottlieb, "Highly parallel computing", Benjamin/CummingsPublishing Company, Inc., 1989.
2. K.E. Batcher, "Sorting Networks and their Applications", in Proc. AFIPS Conf. 3071968. 3. A.R. Dias, R.F. Kalman, J.W. Goodman, A.A. Sawchuk, "Fiber-optic crossbar withbroadcast capability", Opt. Eng. **27** 955, 1988.
4. J.P. Goedgebuer, H.Porte, A.Hamel, "Electrooptic modulation of multilongitudinalmode laser diodes: demonstration at 850nm with

simultaneous data transmission by coherence multiplexing", IEEE Journ. of Quantum Electronics, **QE-23**, 1135, 1987. 5. A. Gottlieb, "An Overview of the NYU Ultracomputer Project", in J. Dongarra, ed.Experimental Parallel Computing Architectures, North-Holland, 1987.

6. K.H. Brenner, A.Huang, "Optical implementations of the perfect shuffleinterconnection", Applied Optics **27**, 135, 1988.

7. N.A. Clark, S.T. Lagerwall, "Submicrosecond bistable electro-optic switching inliquid crystals", Applied Physics Letters **36**, 899, 1980.

8. J.Jahns, M.J. Murdocca, "Crossover networks and their optical implementation",Appl. Opt. **27**, 3155, 1988.

9. D.H. Lawrie, "Access and Alignment of Data in an Array Processor", IEEETransactions on Computers **24**, no. 12, pp. 1145-1155, December 1975.

10. K.Y. Lee, "On the rearrangeability of 2Log2N-1 stage permutation networks",IEEE Trans. on Computers, **C-34**, 412 1985.

11. A. Lohmann, W.Stork, G.Stucke, "Optical perfect shuffle", Applied Optics **25**, 1530, 1986.

12. S.B. McCormick et. al., Optomecanics of a free-space switching fabric: Thesystem, SPIE Proceedings, Vol. 1553, 1991.

13. M. Murdocca, T. Cloonan, "Optical design of a digital switch", Applied Optics **28**, pp. 2505-2517, 1989.

14. S.Sampei, H.Tsuchida, M. Ohtsu, T. Tako, "Frequency stabilization of AlGaAs semiconductor lasers with external grating feedback," Jap. Journal of Applied Physics **22**, L258, 1983.

15. N.A. Olsson, W.T. Tsang, "An optical switching and routing system usingfrequency tunable cleaved-coupled-cavity semiconductor lasers," IEEE J. Quantum Electronics, **QE-20**, 332, 1984. 16. H.S. Stone, "Parallel Processing with the Perfect Shuffle", IEEE Transactions on Computers **C-20**, 153, 1971.

17. N. Streibl, "Diffractive optical elements for optical interconnects", Proc. SPIE,1574, International Colloquium on diffractive optical elements, 1991.

18. C. Wu, T. Feng, "The Universality of the Shuffle-Exchange Network", IEEE Transactions on Computers **C-30**, 324, 1981.

19. W. Stork, "Optical crossbar", Optik **76**, 173-157, 1987.

20. K. Zürl, N. Streibl, "Optoelectronic array interconnection", Optical and Quantum Electronics **24**, S405-S414, 1992.

Second section: Interconnection schemes and systems

Part 2.2 Limits

15 Limitations and scaling laws in parallel optoelectronic interconnections

K. Zürl
Physikalisches Institut der Universität Erlangen-Nürnberg, Angewandte Optik

For short range optoelectronic interconnects, both the concept of parallelism with moderate data rates on each channel and systems with a few high-data rate channels are possible. The transmission may be realized either with relatively high or low light energy per transmitted bit requiring simple or complicated (amplified) receivers respectively. In this chapter, limits and scaling laws are given for such systems, based on energy requirements of the complete system. Two examples are presented.

15.1 Introduction

Optical communication is in some cases superior to electronic communication even for short interconnection distances [1]. The data throughput of such shortrange optoelectronic data links is not limited by the optical setup (e.g. absorption, dispersion etc.) but by the electronic-optical and opto-electronical transducers.

Assuming a shortrange, parallel 2-dimensional data transmission, two parameters are applied to evaluate such systems: The data throughput per area which can be achieved and the required energy per bit.

In our approach, the entire data transmission system is considered, comprising the transmitters, optical channels and receivers. Limiting factors at the transmitter are the tolerable dissipated heat and the transmitter efficiency. The optical channel is characterized by the transmission and the crosstalk behavior and the receivers require a certain energy per bit for an operation with a tolerable bit error rate. Two receiver concepts are investigated: the simplest receiver without any electrical amplification and a highly amplified system operating at the noise limit.

15.2 Basic concepts

Fig.15.1 shows the basic model of the data interconnection system. This unidirectional point-to-point interconnection is suitable for the theoretical discussion; in section 15.8 realistic interconnection structures are presented. The data enters the system as electrical signals, and at the receiver they have to be reconverted into the same form as they entered (i.e. same number of channels and same data rate).

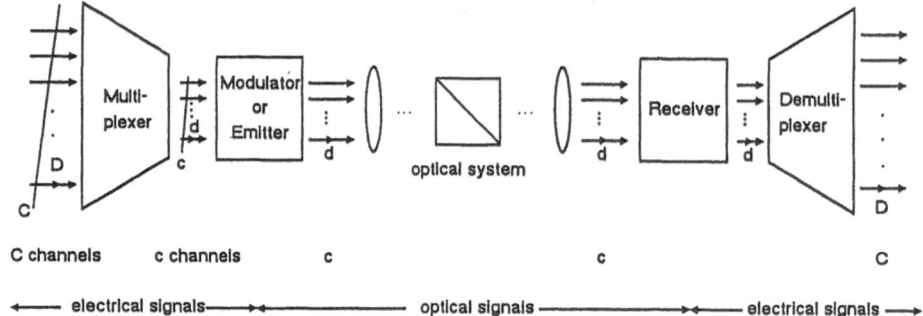

Fig. 15.1 Basic model of a data link.

Two classes of systems can be identified:
(a) Systems in which the data is present in one or a few channels C with high data rate D on each channel. Such systems are found for example in long distance communications and as shortrange optical interconnections within telephone switching networks. The data stream is highly multiplexed and we find data rates on the order of several Gbit/s. Optical links with a bandwidth of up to 12 GHz are currently commercially available [2].
(b) Systems with highly parallel data stream ($C>>1$) and a moderate data rate on each channel, for example in parts of a bus system in a computer with $C\approx100$ and $D=10...100$ Mbit/s. Since the number of pins of electronic chips is limited by current packaging technology to about 500, optical interconnections between chips or hybrids become increasingly interesting.

In both types of systems the overall data throughput $C\,D$ has the same order of magnitude. For the theoretical discussion it is sufficient to consider the one-channel-system (a) as a limiting case of the multi-channel-system (b) with $C=1$.

A basic limiting factor of shortrange optoelectronic interconnections is the dissipated heat which can be removed. Most applications for such interconnections are in compact systems like multiprocessors with already considerable heat problems, where the dissipated power of the interconnection system is a crucial factor. As all heat power has to be removed, independent of its originating location, a global view is applied, i.e. all particular heat power fractions are added up. This model is applicable for short range optoelectronic interconnections in power limited compact systems.

15.3 Data multiplexers/demultiplexers

Time-multiplexing is a common technique to optimize a system's performance, for example to save the cost of optical fibers and transducers. The overall data throughput is not changed by multiplexing ($c\,d = C\,D$), however the total number of optical channels c can be made smaller than the number of electrical channels C (Fig. 15.1).

Multiplexers and demultiplexers usually are cascaded systems. The energy required for each stage of the cascade depends on the technology which is applied, on the maximum data rate which is involved and on the number of input and output channels.

The energy per bit which is required for multiplexing was estimated for a 8:1 multiplexer stage with its adjacent counter (Fig. 15.2). Not included are the energies for the clock generation and for the load at the output O. Several commercially available multiplexers [3 - 9] had been investigated based on the data sheets (Fig. 15.3).

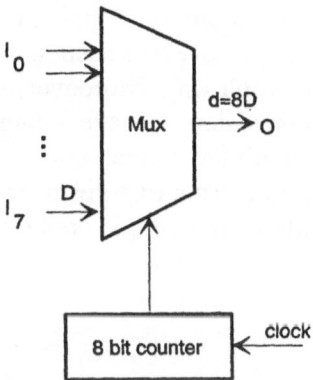

Fig. 15.2 A 8:1 multiplexer stage.

Bipolar ICs like TTL, FAST and the investigated GaAs multiplexer have a constant power consumption independent of the clock rate. Thus, the energy per bit decreases with increasing data rate[1]. In contrast, for CMOS devices, the

---

[1] If a highly capacitive load were involved, bipolar ICs at high data rates would also have a constant energy consumption per bit. In this section, the load is not considered and only the fundamental behavior is shown. Moreover, including this behavior would yield an even worse figure of energy consumption.

power consumption is proportional to the switching frequency, thus leading to horizontal bars in the diagram.

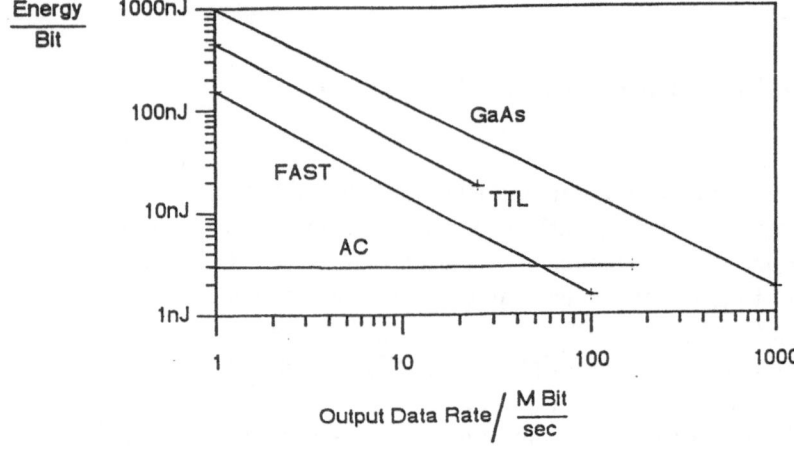

Fig. 15.3 Energy per bit required for multiplexing. AC = Advanced CMOS Logic, FAST = Fast Bipolar Logic, TTL = TTL Logic, GaAs = Gallium Arsenide Emitter Coupled Logic (ECL).

In the best case, the multiplexing energy which is required for *one* stage is around 1 nJ/bit, which is at least 3 orders of magnitude higher than the energy required to communicate a bit optically (see section 15.7). A similar amount of energy/bit is needed for demultiplexing. Moreover, most designers of digital circuits will find it most inconvenient to have a multiplexing/demultiplexing device on the board, as this circuit has a clock rate that is e.g. eight or 64 times higher than the system clock rate. This requires a lot of shielding, decoupling etc. to avoid interferences with the other logic on the board.

15.4 Data transmitters

In order to translate electronic signals into optical signals, either light modulator arrays or light emitter arrays (LEDs or laser diodes) are required. The data throughput per area of such devices is limited.

Limiting factors are

(a)   system parameters such as the quantum efficiency η_q and the radiation characteristic of a light source, or the transmission and reflection coefficients in the ON- and OFF-state of a modulator,

(b)   the required optical energy per bit which has to be emitted. It depends on the receiver sensitivity and the optical channel (including fan out, if necessary),

(c)   the heat dissipated per area P_{heat}/A that can be removed by a cooling system.

Both active emitter arrays and modulator arrays are treated in a common model (Fig. 15.4).

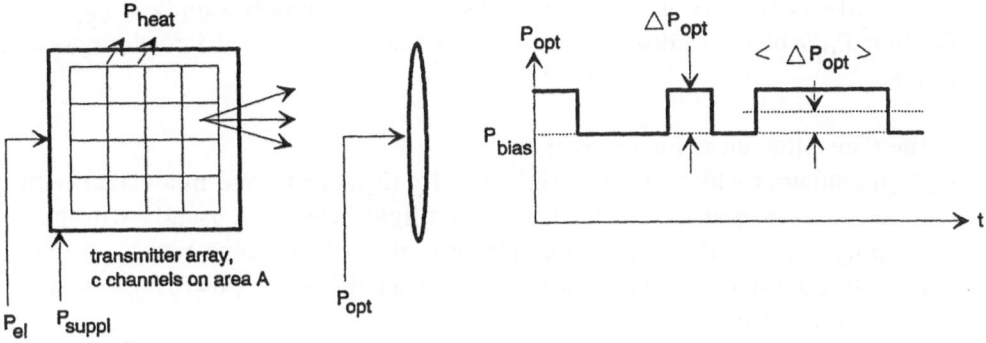

Fig. 15.4 Data transmitter.

An amplitude modulation of the light beams is assumed with a NRZ[2] code according to

$$P_{opt}=P_{bias}+S\,\Delta P_{opt}$$
with
S=1 for coding of a logical 'H'
S=0 for coding of a logical 'L'   (15.1)

Furthermore, we suppose that the probabilities p of logical 'H' and 'L' states are equal:

$$p(\,'H'\,)=p(\,'L'\,)=0.5 \qquad (15.2)$$

The light beams may be considered as a superposition of a bias fraction P_{bias} and a time variable fraction P_{opt} which carries information i.e. only this is the signal. With (15.2) follows that the time average[3] signal power $<P_{opt}>$ is $P_{op}/2$

---

[2]non-return-zero coding

[3]the signs <...> denote the time average of a variable

The modulation index m is defined as the ratio of the average signal amplitude to the time average of the light beam

$$m = \frac{\langle \Delta P_{opt} \rangle}{\langle P_{opt} \rangle} = \frac{\Delta P_{opt}/2}{P_{bias} + \Delta P_{opt}/2} \tag{15.3}$$

In an area A are c transmitter channels (Fig. 15.4). Power is fed to the array electrically (P_{el}) and, in some cases, by an optical power supply P_{suppl}. A fraction P_{opt} of the emitted light of each channel is coupled into the optical setup.

The transmitter array may consist of
(a) modulators which are powered optically ($P_{suppl} \neq 0$) and modulated with electrical signals ($P_{el} \neq 0$). Their output light beams are usually with bias ($P_{bias} \neq 0$) and the modulation index is in the order of some 10%,
(b) active emitters run in the small signal modulation mode (m=10%). Power is fed only electrically ($P_{suppl}=0$), or
(c) active emitters with a modulation index m=100% ($P_{suppl}=0$, $P_{bias}=0$).

The overall efficiency for a data transmitter array η_t is given by

$$\eta_t = \frac{\text{average of all emitted signal power fed into a setup}}{\text{average of all power fed into the device}}$$

$$= \frac{c\,\Delta P_{opt}/2}{\langle P_{el} + P_{suppl} \rangle} \tag{15.4}$$

and it consists of the product of the device efficiency η_d and the coupling efficiency η_c:

$$\eta_t = \eta_d\,\eta_c \tag{15.5}$$

η_d describes the efficiency of the transmitter array itself

$$\eta_d = \frac{\text{average of all emitted signal power}}{\text{average of all power fed into the device}} \tag{15.6}$$

and the coupling efficiency characterizes the coupling into the optical system:

$$\eta_c = \frac{\text{average of all emitted signal power fed into a setup}}{\text{average of all emitted signal power}} \qquad (15.7)$$

Both η_d and η_c will be evaluated separately for the different types of transmitters.

15.4.1 Active emitters

Active emitter arrays may have LEDs or laser diodes as emitters. In general, the radiation power of an emitter $P_e(A,\Omega)$ can be calculated by evaluating

$$P_e(A,\Omega) = \iint_{A,\Omega} P_{A,\Omega} dA\, d\Omega \qquad (15.8)$$

where A is the emitting area and Ω is the solid angle. $P_{A,\Omega}$ is the radiance which is the emitted light power per area and solid angle.

Laser diodes. For light sources with a cone shaped beam, eq. 15.8 can be evaluated using the approximation [10]

$$P_{A,\Omega} = P_{A,\Omega,0} \cos^k(\theta)\, \cos^l(\Phi) \qquad (15.9)$$

where θ and Φ are the radiation angles parallel and orthogonal to the laser mounting plane respectively (Fig. 15.5) and $P_{A,\Omega,0}$ is the radiance perpendicular to the surface.
The values of k and l are given by

$$k = \frac{\log 0.5}{\log(\cos\theta_0)} \quad \text{and} \quad l = \frac{\log 0.5}{\log(\cos\Phi_0)} \qquad (15.10)$$

The θ_0 and Φ_0 represent the width of the gaussian beam. Using a system of coordinates as shown in Fig. 15.5 and supposing an ideal alignment yields with eq. 15.9 a coupling efficiency of

$$\eta_c = \int_0^{\theta_0}\int_0^{\Phi_0} \cos^{k+1}(\theta)\cos^l(\Phi)d\theta\, d\Phi \Bigg/ \int_0^{\pi/2}\int_0^{\pi/2} \cos^{k+1}(\theta)\cos^l(\Phi)d\theta\, d\Phi \qquad (15.11)$$

Where θ_0 and Φ_0 are the arcsin of the N.A. (numerical aperture) of the incoupling setup. For a setup with radial symmetry, $\theta_0 = \Phi_0$. Fig 15.6a shows a plot of the coupling efficiency vs. the N.A. for this case and a conventional edge emitting laser diode.

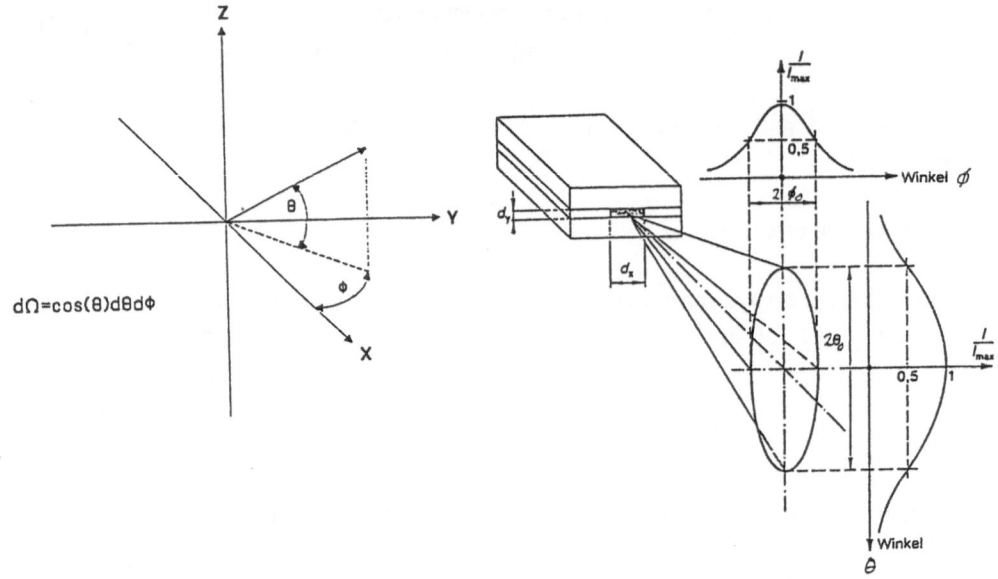

$$d\Omega = \cos(\theta)d\theta d\phi$$

Fig. 15.5 Radiation caracteristics of an edge emitting laser diode [11] and the coordinates used for the evaluation.

Frequently, multimode fibers are applied in the optical system to provide easy alignment (due to the large core diameter) and a mechanically flexible setup. In this case, coupling can be achieved either directly or by using a lens. Feeding the light directly into the fiber, the N.A. of the fiber (typically 0.2) limits the efficiency to around 60%. Using a lens as coupling element between laser diode and fiber, the N.A. of the lens is applied and this improves the efficiency to approx. 90% (for a typical lens with a N.A. of 0.5). This is possible due to the coherence of the laser diode light and thus the small emitting area with high intensity. The lens transforms the gaussian beam and benefits from the wider angle spectrum which is accepted by the fiber compared to the angle spectrum of the laser diode [12]).

The device efficiency η_d of laser diodes is of the order of 10%. So, an overall transmitter efficiency η_t of 5% to 10% is achievable.

Laser diodes have to be run above the laser threshold current which is presently around some milliamperes (thresholds lower than $100\mu A$ have been reported [13], [14]). In many cases, the laser diodes emit much too much light energy per bit (compared to the required energy/bit) if they are run during the full period duration of a bit. To save energy, they may be strobed. However, that requires faster laser driver electronics.

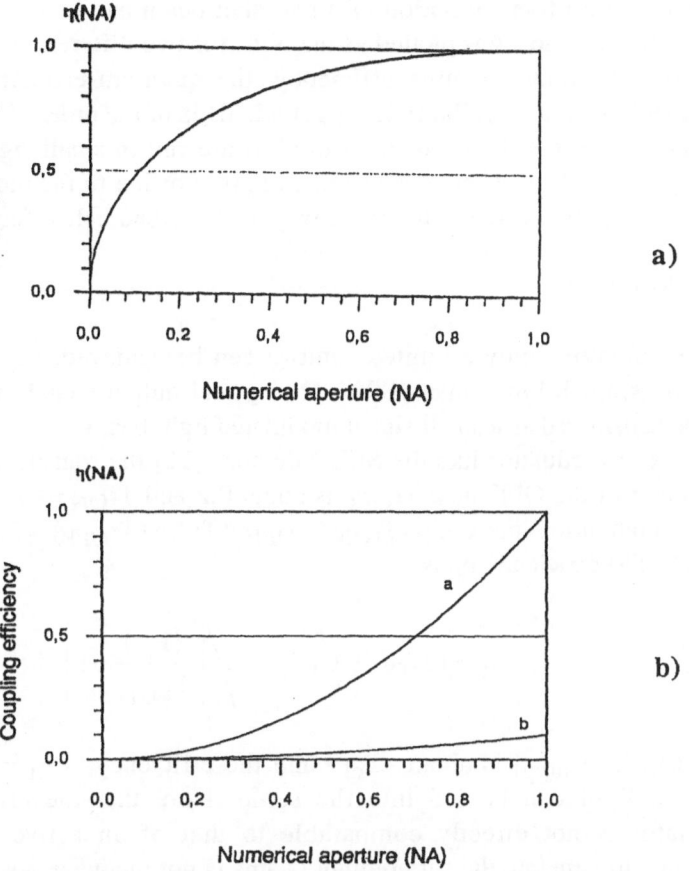

Fig. 15.6 Coupling efficiency into a numerical aperture N.A. (fibre or lens) a) of an edge emitting laser diode b) of a LED, curve a with a core area equal to the LED area and curve b with a core area equal to 10% of the LED area (reflexion losses are not included).

LEDs. LEDs are incoherent light sources with a lambertian emission characteristic. The coupling efficiency is given by [15]

$$\eta_c = \int_0^{r_e} r\, dr \int_0^{2\pi} d\Phi \int_0^{\theta_0} \sin\theta \cos\theta\, d\theta \Big/ \int_0^{r_{LED}} r\, dr \int_0^{2\pi} d\Phi \int_0^{\pi/2} \sin\theta \cos\theta\, d\theta$$

$$= \frac{A_{fiber}}{A_{LED}} NA^2$$

(15.12)

As LEDs have a relatively large emitting area A_{LED}, the ratio of it to the area of the fiber A_{fiber} is involved in eq.15.12. Applying a lens will not improve the efficiency as no transformation of a gaussian beam can be made. In Fig.15.6b the coupling efficiency is plotted vs the N.A. for two different area ratios.

Besides the poor coupling efficiency, the quantum efficiency is low, too (about 1%). So the overall efficiency for a LED is of the order of 10^{-3} to 10^{-4}.

If active emitters (LEDs or laser diodes) are run in small signal modulation mode (m < 1), their efficiency decreases in proportion to the modulation index. They then can be treated in the same way as described below for modulators.

15.4.2 Modulators

With modulators, only a limited contrast can be achieved, i.e. the modulation index m is much lower than 100%. The optical output signal of the modulator may be considered as a small signal modulated light beam.

For a fast modulator like the SEED devices [16] the transmission (reflexion) coefficient in the OFF-state Tr_{OFF} is about 0.2 and Tr_{ON}=0.4 in the ON-state. So, the modulation index $m = (Tr_{ON}-Tr_{OFF})/(Tr_{ON}+Tr_{OFF}) = 33\%$ according to eq.15.3 . The efficiency η_d is

$$\eta_d = (Tr_{ON} - Tr_{OFF})/2 \left/ \left[\frac{\langle P_{el} \rangle}{P_{sup\,pl}} + 1 \right] \right. \qquad (15.13)$$

For SEED's, P_{suppl} is about $\langle P_{el} \rangle$ and the efficiency is $\eta_t = 0.05$, if all the reflected light can be fed into the setup. Note that the efficiency of the modulator is not directly comparable to that of an active emitter as the efficiency to generate the supply light beams is not included. As the related heat is dissipated somewhere else it is not taken into account here. (The efficiency of the generation of the source light beams and its dissipated heat could be included applying the same calculation, but using only the electrical power fed to the system)

15.4.3 Conclusions for Transmitters

All dissipated heat, independent of the location of its origin, will be summarized as stated in section 15.2. Both, the fraction of the optical power which is *not* fed into the setup *and* the electrical losses[4], are ultimately dissipated as heat at the transmitter $P_{heat, transm}$:

$$P_{heat,transm} = \langle P_{el} + P_{suppl} \rangle - c \langle \Delta P_{opt} \rangle = (1-\eta_t)/\eta_t \; c \; \Delta P_{opt}/2 \qquad (15.14)$$

---

[4]The dissipated power in the driver electronics (for an active emitter in the same order of magnitude as the dissipated power of the emitter) may be included in by replacing the electrical power fed *to the emitters* by the electrical power fed *into the driver electronics*.

We assume that a certain quantity of energy per bit $<E_{em}>$ has to be sent into the setup for proper working of the receiver (for details see section 15.6). Data rate d, signal power and average signal energy per bit $<E_{em}>$ at the emitter are related by

$$\Delta P_{opt}/2 = <E_{em}> d \tag{15.15}$$

and with eq.15.2, the average energy per bit $<E_{em}> = E_{em}/2$. The heat $P_{heat,transm}$ which can be removed from a given area A is limited by the cooling capability provided for the system. According to eq.15.14, the heat power is related to the signal power. With eq. 15.15, this yields

$$\frac{cd}{A} = \frac{P_{heat,transm}}{A} \frac{\eta_t}{1-\eta_t} \frac{1}{<E_{em}>} \tag{15.16}$$

For transmitters, the data throughput per area $c\,d/A$ is limited by the removable heat power, the transmitter efficiency and the required energy per bit. A direct trade-off between the number of channels c per area and the data rate d of each channel is obtained.

15.5 The optical channel(s)

Any real optical system is affected by crosstalk and losses. Crosstalk describes the coupling of light power from one channel to another in a multi-channel system.

Losses can be characterized by a transmission Tr<1. (A fraction of this losses might appear as crosstalk on other channels.) If we have a system with fan-out F, e.g. a splitting of the transmitter power to F receivers, the average energy per bit at one receiver channel is

$$<E_{rec}> = (<E_{em}> Tr)/F + <E_{ct}> \tag{15.17}$$

The crosstalk energy per bit on a channel E_{ct} is a *deterministic* signal, it may not be treated as noise[5] but worst case considerations have to be applied. The crosstalk may be separated into two fractions:

---

[5]Due to the quantum character, any light causes photon shot noise at the receiver. However, for the light power involved in *technical* systems, the shot noise can be ignored compared to thermal noise of the receiver (see section 15.6).

(a) a background crosstalk energy which is homogeneous across the receiver field. A fraction of it might be external scattered light E_{ext} entering into the system, another, due to scattering effects etc., might be caused by the light which was coupled into the optical setup and will therefore be proportional to $(P_{bias} + P_{opt} /2)$ c Tr / (F d) (global crosstalk).
It is assumed that this background light is temporally and spatially constant.

(b) crosstalk of the neighboring channels due to scattered light, diffraction etc., which is proportional to the light energy impinging on each adjoining channel and is weighted by the distance inbetween (local crosstalk).

So, an expression for the crosstalk energy per bit is obtained (eq. 15.18):

$$E_{ct} = a_1 E_{ext} + a_2 \frac{\left(P_{bias} + \Delta P_{opt} /2\right) c\, Tr}{Fd} + \sum_i a_3 \frac{1}{x_i^2} \frac{Tr}{Fd}\left[P_{bias} + S_i(t)\Delta p_{opt}\right]$$

where a_1, a_2, a_3 are proportional factors, the x_i characterize the distance between channels and $S_i(t)$ are the momentary logic states of the channels according eq.15.1.
If the crosstalk energy E_{ct} would be temporally constant, it appeared simply as a constant bias light at the receiver which could be easily compensated by selecting an adequate threshold level. However, it is obvious that it varies with the bit clock rate depending on the $S_i(t)$ in the second term in the square brackets of eq.15.18. This fraction of crosstalk is generated by the signal light. For example, in one clock period all channels neighboring the observed channel are on (all $S_i(t)=1$, Fig.15.17a), in another clock period all are off (all $S_i(t)=0$, Fig.15.17b).

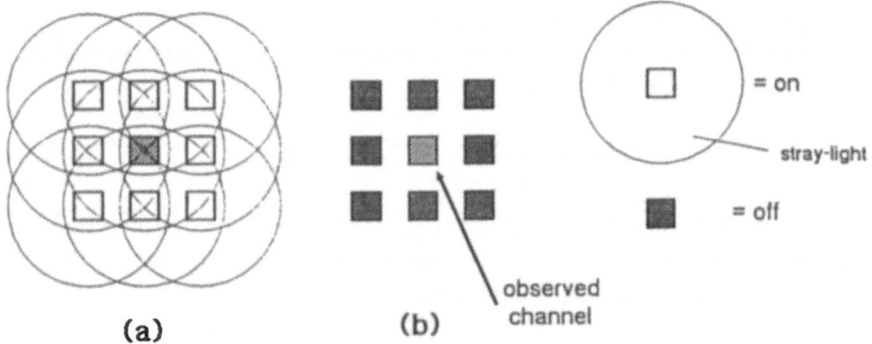

(a)   (b)

Fig. 15.7 The crosstalk of neighboring channels (no interference between the light sources) a) maximum crosstalk: all neighboring channels are 'on' and b) minimum crosstalk: all neighboring channels are 'off'.

The *temporal variation* of the crosstalk $\Delta E_{ct}(t)$ is the crucial parameter as it decreases the "quality" (bit error rate) of the received data:

$$\Delta E_{ct}(t) = a_3 \frac{\Delta P_{opt} Tr}{Fd} \sum_i \frac{1}{x_i^2} S_i(t) \qquad (15.19)$$

The factor a_3 characterizes the local crosstalk behavior of the optical system, due to diffraction, pitch of the channels etc. The minimum value for ΔE_{ct} is 0, the maximum $\Delta E_{ct,max}$ is obtained for all $S_i=1$. It has been assumed that a_3 is equal for all pixels of the array. (If the crosstalk were space variant, a_3 would have to be replaced by many proportional factors $a_{3,i}$.)

15.6 Data receivers

Data receivers consist of several building blocks (Fig.15.8): the photon sensing device (usually pin or avalanche photodiodes) with its biasing circuits, if necessary an amplifier and/or a filter (which may be included in the amplifier) and a decision unit (comparator) which reproduces a binary data stream from its analogue, noisy input signal.

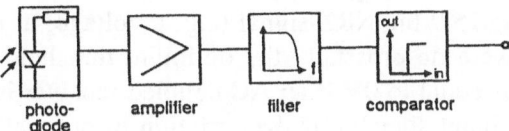

photo-diode   amplifier   filter   comparator

Fig.15.8 Typical stages of a data receiver.

The bit error rate (BER) is one of the critical parameters of every data link. It describes the ratio of faulty received bits to the number of all bits at the output of the receiver. To achieve acceptable values, the signal to noise ratio SNR at the input of the decision unit inside the receiver has to be high enough. The SNR is defined as rms (root mean square) ratio of signal to noise

$$SNR(\text{in dB}) = 20 \log_{10} \left(\frac{< i_{signal}(t)^2 >}{< i_{noise}(t)^2 >} \right)^{1/2} \qquad (15.20)$$

Note that for calculating the SNR from power ratios, $SNR = 10 \log(P_{sig}/P_{noise})$, as $P \propto i^2$. Supposed that the noise obeys gaussian statistics, SNR and BER are related for NRZ signals by the error function (erf) [17]:

$$BER = \frac{1}{2}(1 - erf(\sqrt{SNR/2})) = \frac{1}{2}erfc(\sqrt{SNR/2}) \qquad (15.21)$$

These equations are only valid for a transmission of a 'H' state, since the signal for the 'L' state is zero for the NRZ code. To calculate the BER the *SNR at the decision unit* has to be taken. Acceptable bit error rates for telecommunications are 10^{-9}. To achieve this, a SNR of 21.5dB [18] is required.

There are several noise sources in the system:
- noise of the amplifier (thermal and shot noise),
- noise generated in the photodiode and its biasing circuits (thermal and shot noise),
- photon noise (shot noise) due to the quantum character of the light (both signal light and crosstalk),
- perhaps noise due to fluctuations of the light power of the emitters.

These noise considerations have been examined in many publications (for example, [18-21]). Usually, in real systems, one of the above mentioned noise sources is dominant and can be treated as the only one. In most applications of *data receiving* systems at room temperature the thermal noise is the dominant value.

For multi-channel systems, the BER may increase not only with increasing noise, but also with decreasing of the *usable* signal power due to the variation of crosstalk. In Fig.15.9 the NRZ-signal (e.g. a voltage) at the decision unit is plotted. If there were no crosstalk, the optimum threshold level would be at $U_{th,0}$ (this may be equal to 0V if an AC coupled receiver is used, suppressing all constant (DC) input signals). If the variation of crosstalk is maximum, the threshold should be at $U_{th,max}$. Both the two extremes and discrete values inbetween occur with a certain probability which is determined by the geometry of the receiver array, a_3 and the S_i in eq.15.18. With eq.15.2, the values of ΔE_{ct} are equally distributed and thus the best value for U_{th} is $(U_{th,0} + U_{th,max})/2$.

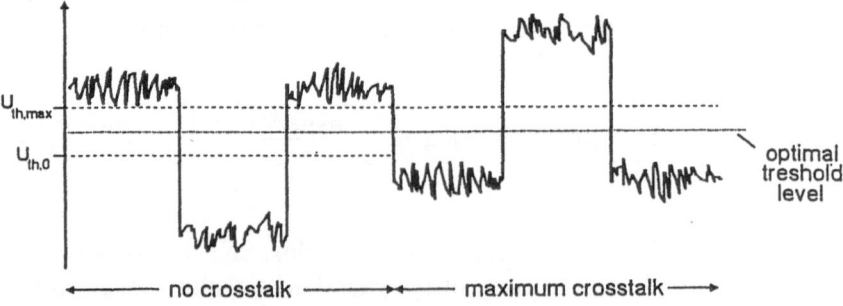

Fig.15.9 Optimal threshold level depending on the crosstalk.

Taking this optimum threshold level yields a decreasing signal-to-noise ratio for many cases, e.g. if the maximum or minimum crosstalk occurs, with respect to the crosstalk-free SNR. As the relation between SNR and BER is strongly nonlinear, worst case considerations have to be made. According to Fig.15.9, the amount of signal power above the threshold level is lowered by half the maximum variation of the crosstalk powers $\Delta E_{ct,max}$ d/2 and the usable signal power at the receiver $P_{sig,rec,ct}$ is

$$P_{sig,rec,ct} = \frac{\Delta P_{opt} \, Tr}{F} \left(1 - \frac{1}{2}a_3 \sum_i \frac{1}{x_i^2}\right) \tag{15.22}$$

The SNR with crosstalk SNR_{ct} is

$$SNR_{ct} = \frac{P_{sig,rec,ct}}{P_{noise}} = SNR_{no\,ct}\left(1 - \frac{a_3}{2} \sum_i \frac{1}{x_i^2}\right) \tag{15.23}$$

were $SNR_{no\,ct}$ is the SNR of a crosstalk free system. Obviously, a parallel data transmitting system can only operate if the crosstalk (represented by a_3) is small enough. With (15.23), the maximum tolerable crosstalk can be calculated for a given receiver SNR and the required BER of the system.

In Fig.15.10, the basic schematic of the front end of a photodiode receiver is shown. The resistor R is used to bias the photodiode PD which works in the photoconductive-mode [20]. C represents the junction capacity of the diode and all scattered capacities involved.

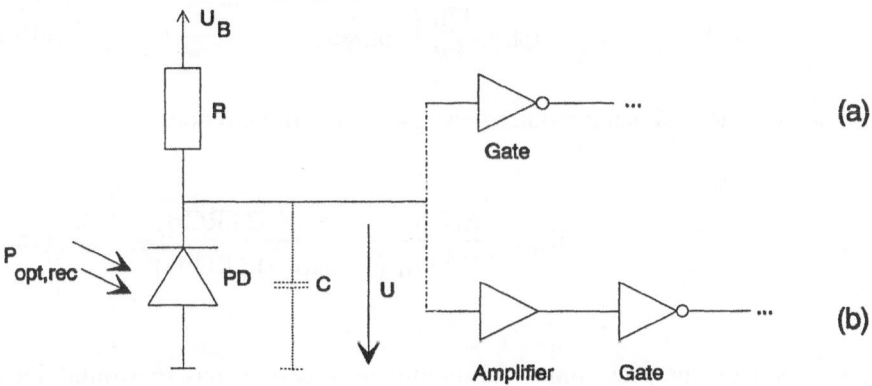

Fig. 15.10 Basic schematic of a simple receiver.

There are two ways to design optoelectronic data links:
(a) With high light levels, causing a high voltage swing U at the receiver diode, simple receivers are possible and no noise considerations for the receiver electronics are necessary. The voltage U is fed directly into the input of a gate.

(b) A low light concept with high sensitive, noise limited receivers. Here, low noise amplifiers are required before the signal may be fed into the input of a gate. However they will consume energy, too. This energy consumption is not included in the calculations here.

15.6.1 Receivers without amplification

For this type of receiver, the light power $P_{opt,rec}$ (Fig.15.10) has to be high enough to generate a desired voltage swing U at the output for a given data rate. Based on the equivalent circuit (Fig.15.11), this type of receiver can be evaluated.

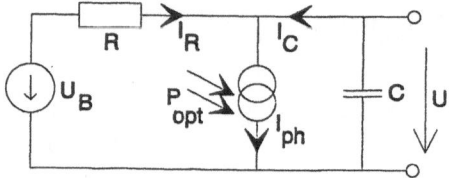

Fig. 15.11 Equivalent circuit for a non amplified receiver.

The output voltage U is

$$U(t) = (U_B - RI_{ph}) (1-\exp(-t/RC)) + U_0 \exp(-t/RC) \tag{15.24}$$

with the photocurrent I_{ph}

$$I_{ph} = \frac{e\eta_r}{h\nu} P_{opt,rec} \tag{15.25}$$

Solving eq.15.24 yields the required energy per bit at the receiver E_{rec}

$$E_{rec} = \frac{\Delta U}{Rd} \frac{h\nu}{e\eta_r} \frac{1-\exp(-2/RCd)}{(1-\exp(-1/RCd))^2} \tag{15.26}$$

ΔU is given by the electronics following the receiver and is around 1V for CMOS applications [22,23] in order to provide an acceptable BER. The optimum receiver design is for d = 1/ RC, providing the minimum required energy per bit $E_{ec,min}$:

$$E_{rec,min} = \Delta U\, C\, \frac{h\nu}{e\eta_r}\, 2.16 \tag{15.27}$$

15.6.2 Noise limited receivers

If used at ambient temperature, the sensitivity of the receivers is usually limited by thermal noise. The noise equivalent (optical) power for a photodiode P_n is [20]

$$P_n = \frac{2h\nu}{e\eta_r} \sqrt{\frac{kT\Delta f}{R}} \qquad (15.28)$$

with the conversion efficiency of the receiver $h\nu/e\eta_r$ ($h\nu$ is the photon energy, e the electron charge and η_r the receiver quantum efficiency) and where the thermal noise power $4kT\,\Delta f$ is produced by the resistor R.

The maximum channel data rate d is proportional to the channel bandwidth Δf. For a NRZ coded signal, the theoretical limit is

$$d = 2\,\Delta f \qquad (15.29)$$

If we assume that the thermal receiver noise is the only relevant noise source in the system, the required energy per bit E_{rec} at the receiver to receive a "H" information is

$$E_{rec} = \frac{SNR\,P_n}{d} = \frac{SNR\,2h\nu}{e\eta_r} \sqrt{\frac{kT}{2Rd}} \qquad (15.30)$$

The required energy per bit scales with the inverse square root of the channel data rate d.

If (like in the foregoing section) an optimum receiver design with $d=1/RC$ is considered, the E_{rec} may be written as

$$E_{rec} = \frac{SNR\,2h\nu}{e\eta_r} \sqrt{\frac{kT}{2}C} \qquad (15.31)$$

which means that the required energy/bit scales with the square root of the capacitance and is independent of the data rate. Note that the energy consumption of the required amplifier is not included in the calculations.

15.7 The optoelectronic data link

Assuming an optical path with no crosstalk, the energies per bit at the receiver and at the transmitter can be related according to eq.15.17

$$< E_{em} >= \frac{E_{rec}\, F}{2\, Tr} \tag{15.32}$$

15.7.1 Data link with a receiver without amplification

For the concept with high light power, we get, using eq.15.32, 15.26 and 15.16

$$\frac{c\,d}{A} = \frac{P_{heat,transm}}{A} \frac{2\,Tr}{F} \frac{\eta_t}{1-\eta_t} \frac{R\,d}{\Delta U} \frac{e\eta_r}{h\nu} \frac{(1-\exp(-1/RCd))^2}{1-\exp(-2/RCd)} \tag{15.33}$$

For discussion of eq.15.33, two regions can be identified: data rates small compared to the RC time constant ($1/d \gg RC$) and high data rates with $1/d \ll RC$.

For low data rates, the exponent expressions are negligible and we get

$$\frac{c}{A} = \frac{P_{heat,transm}}{A} \frac{2\,Tr}{F} \frac{\eta_t}{1-\eta_t} \frac{R}{\Delta U} \frac{e\eta_r}{h\nu} \tag{15.34}$$

which means that the number of channels per area is limited. Note, that the data rate on each channel is limited, too, by the RC time constant, which was the assumption for the derivation.

For high data rates, the series expansions of the exponent may be stopped after the linear term giving

$$\frac{c\,d}{A} = \frac{P_{heat,transm}}{A} \frac{Tr}{F} \frac{\eta_t}{1-\eta_t} \frac{1}{C\Delta U} \frac{e\eta_r}{h\nu} \tag{15.35}$$

The receiver behaves as a capacitive load upon the photodiode. For data rates above the RC time constant (supposed here), the energy per bit rises with the data rate, since charging and discharging of the capacitance can not be completed. There is a direct trade off between the number of channels and the data rate on each channel, e.g. the data throughput per area is limited by system parameters as expressed in eq.15.35.

For an optimum receiver ($d = 1/RC$), eq.15.27 may be applied instead of 15.26 and we get

$$\frac{cd}{A} = \frac{P_{heat,transm}}{A} \frac{2Tr}{F} \frac{\eta_t}{1-\eta_t} \frac{1}{\Delta U C 2.16} \frac{e\eta_r}{h\nu} \qquad (15.36)$$

which is nearly the same result as eq.15.35 (the difference is caused by the approximation in eq.15.35).

15.7.2 Data link with a noise limited receiver

Using the noise limited approach, substituting eq.15.32 and 15.30 in eq. 15.16 leads to

$$\frac{c\sqrt{d}}{A} = \frac{P_{heat,transm}}{A} \frac{Tr}{SNR} \frac{\eta_t}{F} \frac{e\eta_r}{1-\eta_t} \frac{e\eta_r}{h\nu} \sqrt{\frac{2R}{kT}} \qquad (15.37)$$

Supposing an optimum receiver with d=1/RC, eq.15.31 may be applied instead of eq.15.30 which yields to

$$\frac{cd}{A} = \frac{P_{heat,transm}}{A} \frac{Tr}{SNR} \frac{\eta_t}{F} \frac{e\eta_r}{1-\eta_t} \frac{e\eta_r}{h\nu} \sqrt{\frac{2}{kTC}} \qquad (15.38)$$

In this calculations, the heat generated *by the receiver amplifier* is not included. If we suppose that transmitters and receivers are located geometrically close to each other, in eq.15.38 the amplifier power has to be subtracted from the heat power which can be removed.

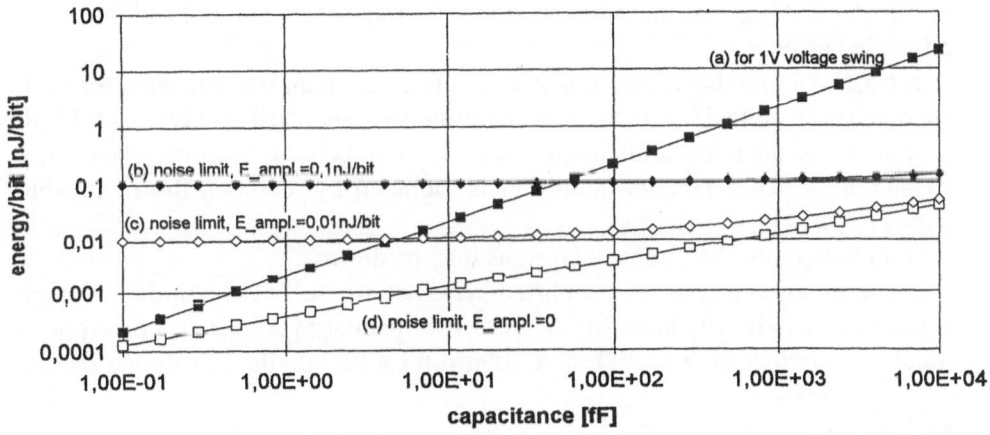

Fig. 15.12. Overall average energy/bit, including driver electronics, as a function of the receiver capacitance for a) non amplified receivers and b), c) and d) for noise limited receivers with different $E_{ampl.}$.

15.7.3 Comparison

For the further calculations, only optimum receivers (d = 1/RC) are considered. The overall required energy per bit E_{bit} is the sum of all power fed into the data link divided by the number of channels and bits/s (i.e. the data rate):

$$<E_{bit}> = \frac{\left\langle P_{el} + P_{sup\,pl}\right\rangle}{C\,d} \tag{15.39}$$

and using eq.15.4, 15.15 and 15.32, we get

$$<E_{bit}> = \frac{1}{2\eta_t}\frac{E_{rec}F}{Tr} \tag{15.40}$$

where E_{rec} is given by eq. 15.27 or 15.31. However, for the noise limited approach, the electrical energy consumption per bit of the receiver amplifier $E_{ampl.}$ has to be added:

$$<E_{bit}> = \frac{1}{\eta_t}\frac{F}{Tr}SNR\frac{h\nu}{e\eta_r}\sqrt{\frac{kTC}{2}} + E_{ampl.} \tag{15.41}$$

Depending on the receiver technology, the $E_{ampl.}$ is proportional to the inverse data rate (for bipolar amplifiers) or constant (for CMOS technology, and the power consumption is proportional to the data rate).

The overall energy per bit is plotted[6] as a function of the receiver capacitance in Fig.15.12. Note that the plots have logarithmic scales and only indicate the principle behavior.

In fig.15.13, the theoretical maximum data throughput per unit area for both approaches is plotted[7]. For the concept with non amplified receivers, eq.15.36 is applied, for the noise limited approach, eq.15.38 is used, however, the energy consumption of the receiver amplifier is included by reducing the removable power.

From both plots, the same conclusions may be drawn.

Low light energy systems (amplified systems) are the better solution, if large capacitances (>100fF) have to be used. A capacitance of 100fF corresponds roughly to a diode area of 200μm x 200μm for a PIN diode. A receiver energy

---

[6] The calculations for the plots were carried out using η_t (transmitter efficiency) = 1%, F (fan out) = 1, Tr (transmission) = 10%, SNR (signal to noise ratio) = 21.5dB, $h\nu/(e\eta_r)$ = 2V, T = 300K.

[7] For the plots, the same values stated in the previous footnote were used. Further, P_{heat}/A = 1W/cm2, U=1V.

consumption of 0,1nJ/bit (plot (b) in Fig.15.12 and 15.13) corresponds to 10mW for a channel data rate of 100 Mbit/s or to 100mW for 1Gbit/s, respectively, which are achievable values.

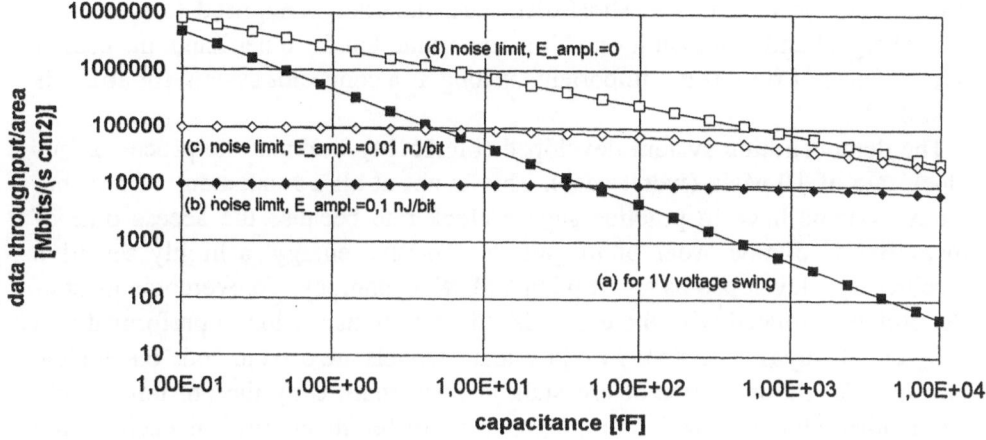

Fig. 15.13. Maximum data throughput per area for an optoelectronic system according to eq.15.38 (plot a), 15.33 (plot b) and 15.36 (plot c)

Besides its power consumption, the required areas for the amplifiers have also to be considered. As the capacitance of the receiver diode scales roughly with the area (for conventionally built diodes), the optical problems of focussing light onto small area diodes have to taken into account, too.

Comparing the two extreme concepts (at the noise limit, plot (d) and using non amplified receivers, plot (a)) it is important to note that, for the noise limited concept, the required energy/bit scales only with the square root of the capacitance, whereas a linear dependance exists for the non amplified receivers.

15.8 Examples

In this section, two example systems are presented: one, which provides many parallel channels with moderate channel data rate for use in a multiprocessor system, and another with a high channel data rate and a moderate number of channels, developed for telecommunication applications.

15.8.1 A parallel system at the noise limit

This system provides highly *parallel* optical channels with moderate channel data rates for parallel data transmission at the system clock rate. A typical application is the link between processors and shared memories within a multiprocessor system. For short distances, the energy required to create the optical signal and to transmit one bit is important. On the other hand, the cost of a few meters of fibre is not important, as long as a connector system for arrays is affordable.

The demonstration system developed (Fig.15.14) works at the processor bus clock rate of 10 MHz (worst case). This is not at all a fundamental limit, but many systems have in practice such a clock rate because the access time of memories is of this order of magnitude. To save energy, a highly sensitive receiver was developed as an ASIC in CMOS technology. To overcome most of the problems related with the crosstalk, the data transmission is performed in a differential way (Fig.15.15): Two physical channels are used to code one logical channel. So, the influence of crosstalk is very small. Only the portion of light from other channels which impinges with different energy on each of the receiver photodiode pair acts as the *variation of crosstalk* described in section 15.5 .

To avoid an external optical power source for the data conversion from electrically to optically coded data, active emitters (LEDs or laser diodes) are used. Since 2-D, parallel addressable monolithic laser diode-arrays are not yet commercially available (but under development, [13, 14]), a two-dimensional hybrid array of laser diodes was used for the first setup. Due to the high sensitivity of the receiver, the system concept also allows the use of 2-D LED-arrays.

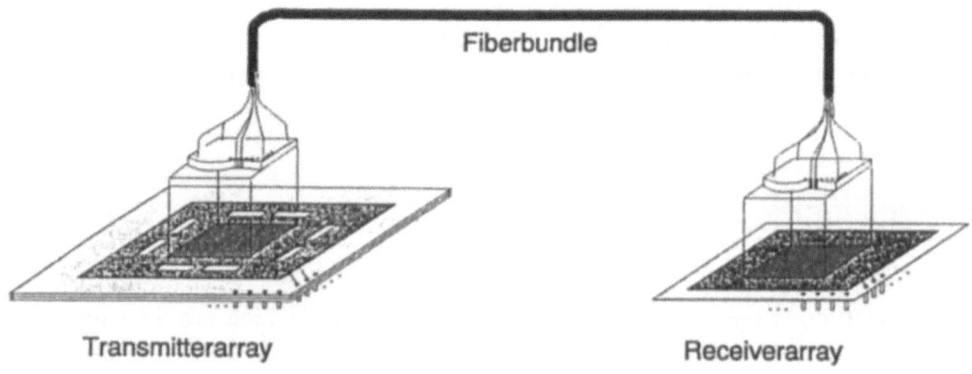

Fig. 15.14 A parallel optoelectronic interconnection system.

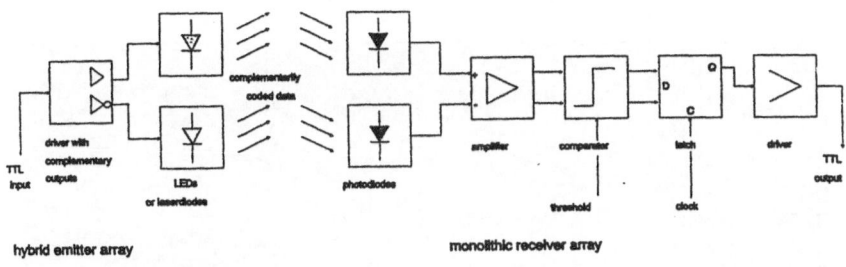

Fig. 15.15 Block diagram of one channel of the optoelectronic interconnection system.

For the optical system 2-D arrays of multimode fibers are used. The ends are held by a special mount. Such arrays can be produced by a drilling technique similar to the way SMA-fibre-connectors are produced or using a v-groove technology in silicon. To avoid problems with critical mechanical adjustments, multimode fibers with 100 μm diameter, a N.A. of 0.21 and a pitch of 420 μm are used.

Free space modules are used to provide more complex structures for example splitting onto several receivers (fan-out) or bidirectional communication through one fiber bundle.

The Optoelectronic Receiver IC (opto-ASIC). In Fig. 15.15, the basic functions of one logical channel of the receiver chip are depicted. On the final chip there will be an array of 8 x 16 logical channels each of which includes 2 photodiodes.

At a data rate of 10 MBit/s per channel (worst case), an overall data throughput of about 1 GBit/s is achieved. To provide high signal-to-noise margins, a differential concept was used both in the optical part and in the receiver. Scattered light impinging with the same intensity on both receiver diodes does in principle not affect the performance of the system. In addition to that, drift effects due to thermal shifts can largely be compensated in the receiver electronics. The receiver was designed to operate at the noise limit.

The complete receiver front end is fully custom designed. So it was possible to achieve a very high sensitivity. In worst-case simulations an optical energy of 10 fJ/bit was sufficient for a SNR of 23 dB yielding a bit error rate of better than 10^{-11} [18]. The received energy per bit corresponds to about 30.000 photons/bit.

The very high sensitivity was only made possible by the use of a completely symmetric, differential input stage and by providing an auto-zero adjustment at every clock cycle. Fig.15.16 shows the geometrical layout of two logical channels. The four large grey areas are the photodiodes (150 μm x 150 μm , pitch 420 μm, capacitance about 1pF). The receiver power consumption per channel is approx. 1mW, which yields to 0,1nJ/bit at 10 Mbit/s channel data rate.

Fig.15.16 Photograph of the layout of the prototype IC.

The data link, measurements. In the first experimental setup, commercially available laser diodes were used which are focused with an objective directly onto a prototype of the receiver ASIC. Fig.15.17 shows the setup for BER measurements. The data of the word generator which is fed to the laser diodes and the received data of the ASIC are compared and any errors are counted. As the data throughput is known, the BER can be calculated. With a light pulse duration of 50ns and a peak power of 0.21µW the timing parameters of the receiver were optimized and a BER better than 10^{-12} was achieved.

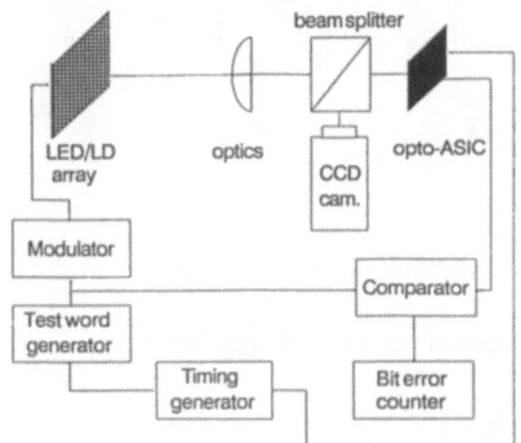

Fig. 15.17 Schematic of the experimental BER measurement setup.

Limits and Perspectives. For an optimum system with the opto-ASIC ($<E_{rec}>$ = 10fJ, $E_{ampl.}$= 0,1 nJ/bit, 10 Mbit/s per channel) using laser diodes as emitters (η_t=1%), a point to point interconnection (Tr/F=0.1) and a moderate heat power which can be removed (P_{heat}=1W/cm^2), the theoretical limit of the data throughput per area is, according to Fig.15.13, in the order of 10^{10} bit/(s cm^2). With a channel data rate of 10^7 bit/s, 10^3 channels should be implemented on one cm^2, which corresponds to a channel area of approx. 300µm x 300µm. A somewhat lower channel density has been used to simplify the alignment of the multimode fibers.

15.8.2 A high-speed system with high light levels

The second example is a prototype of an optoelectronic interconnection system which is suitable as a backplane for switching systems in telecommunications. As the data in such systems are already highly multiplexed, a concept with high channel data rate and a moderate number of channels is used. Holographic coupling elements mounted on a light-guiding plate provide highly efficient data transmission allowing different configurations such as point-to-point and broadcast links. Data transmission was demonstrated with several parallel channels at a data rate of more than 500 Mbit/s per channel [24].

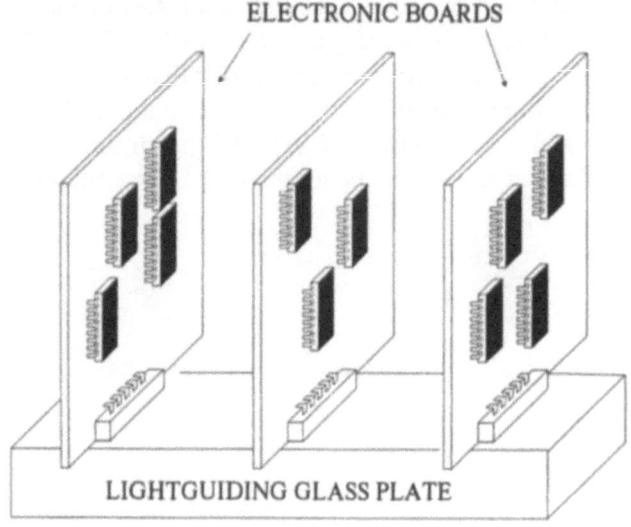

Fig. 15.18 Optical backplane with light guiding plate.

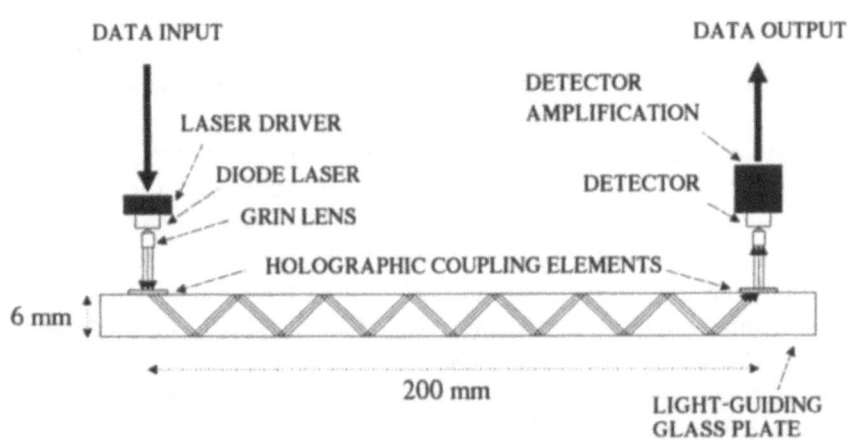

Fig. 15.19 Illustration of the principle of the light transmission inside the glass plate.

System concept. The concept for a backplane using optoelectronic interconnections is shown in Fig.15.18. Every electronic board is equipped with optoelectronic converters (laser diodes and photodiodes) to send and detect light pulses. The collimated light is coupled into a plane-parallel glass plate using holographic coupling elements. The deflection angle is 45°, the light is guided inside the glass plate by total internal reflection (Fig.15.19).

Corresponding holograms perform the coupling of the light from the glass plate to the detector.

Since the glass plate is unstructured, this concept can be regarded as a free-space interconnection, also allowing fan-in, fan-out or data permutation. In this case, the 'free space' is filled with glass which provides mechanical stability as well as protection against disturbing influences e.g. air turbulence.

Experimental setup. To demonstrate an optical backplane a point-to-point link was realized [24]. The light emitted by a semiconductor laser was collimated by a graded-index (GRIN) lens and was coupled into a glass plate with a thickness of 6 mm.

The coupling elements were realized as volume holograms in dichromated gelatin (DCG). The diffraction efficiency of a coupling element at the desired 1st order is higher than 90%, but it is also important to achieve a low amount of energy in the undesired -1st order (in that case well below 1%), since this light is also guided inside the glass-plate and may cause crosstalk.

As detector a silicon PIN-diode was used. The distance between the emitter and the detector was 11 cm. The channels were implemented with a pitch of 1 cm, due to the size of the discrete lasers and detector elements.

The overall optical losses caused by multiple reflection and absorption inside the glass are around 45% (Tr=0.55). The emitted light power is 3mW (cw). As in this setup no lens was used for focussing on the detector, only 220 µW of light power enters the detector. For calculating the performance of the system, the coupling efficiency at the receiver can be included into the coupling efficiency η_c, and with a typical laserdiode efficiency η_d =10%, an η_t of about 1% is obtained. Since bipolar laser drivers and receiver amplifiers are used for such high-speed systems, the power consumption of these devices is high and independent of the data rate. In the setup, the laser driver consumes approx. 0.6W, and the receiver amplifier 170mW. These values are not included in the efficiency η_t. The theoretical bandwidth of the system is 800MHz, and thus the maximum data rate is 1.6 GBit/s.

The noise equivalent power (eq.15.28) is 1,3 µW[8] yielding to a SNR of approx. 22,3 dB. As this value provides for a BER of approx. 10^{-10} the system operates at the noise limit.

The theoretical maximum data throughput per area is limited by the power dissipated of the laser driver and receiver amplifier circuits. With a cooling system which can remove 1W/cm^2 of heat power, the maximum channel density is 1.3cm^{-2}, which corresponds to a theoretical maximum data throughput of 1.6 GBit/s*1.3 = 2 Gbit/(s cm^2) which is not a bad number for a technical realized system (compare Fig.15.13).

---

[8]where λ =785nm, η_r=0.8, Δf=800MHz (the bandwidth of the receiver front end due to the RC cutoff time constant) R=100Ω and ambient temperature were applied.

15.9 Conclusions

Interconnection systems may be classified in two groups:
(a) For long distances, the price of the transportation media and the repeaters, amplifiers etc. is important.

(b) For short distances, e.g. board to board interconnections in the range of 10cm to 10m, it is more important to establish communications with low energy, to avoid too much heat dissipation in the system and to allow high packing densities.

For this short distance point-to-point interconnection, (1) the maximum overall data throughput per unit area c d / A at (2) a minimum energy per bit usually is the design goal. With a limited cooling power, both features are linked.

Two concepts may be applied to reach that goal: Interconnection systems with receivers which operate at the noise limit or systems with high light power at the receiver in order to avoid amplification.

For both concepts, the receiver capacitance is a key value, however, for the noise limited approach, the energy consumption of the receiver amplifier dominates the performance.

For both concepts, a direct trade off between the number of parallel channels and the channel data rate is obtained, if optimum designed receivers are applied. However, taking into account that the energy required for multiplexing is three orders of magnitude higher than the communication energy, data should be transmitted over short distances in the format it is in: same data rate, same parallelism.

Moreover, if we want to perform data *processing* optically, the implementation of certain operations on parallel data planes may prove advantageous, for example data shifting, data combination ("wired OR-function"), or data permutations. In these cases the design goal is not only high data throughput, but taking other advantage of the inherent parallelism of optics.

References

1. D.A.B. Miller, "Optics for low-energy communication inside digital processors: quantum detectors, sources, and modulators as efficient impedance converters", Opt. Letters, vol.14, p.146 (1989).

391

2. Ortel corporation, "12 GHz Laser Module", 12 GHz Photodiode Module", data sheets, Alhambra, California, 1989

3. Texas Instruments,"Pocketguide Band 1: Integrierte Schaltungen", BRD, 1983

4. National Semiconductor,"FAST advanced Schottky TTL logic databook", USA, 1988

5. National Semiconductor,"F100K ECL logic databook and design guide, USA, 1989

6. Giga Bit Logic, "GaAs IC Data Book & Designer's Guide", Newbury Park, CA, 1989

7. Philips,"High-speed CMOS 74HC/HCT/HCU Logic family", Niederlande, 1991

8. National Semiconductor,"FACT advanced CMOS logic databook", USA, 1989

9. Philips,"ECL 10K and 100K logic families-Book IC08", Niederlande, 1986

10. E. Grimm, W. Nowak, "Lichtwellenleitertechnik", Berlin, 1988

11. H. Bauer, "Lasertechnik", Würzburg, 1991

12. E. Hecht: "Optics", Sydney, 1987

13. T. Sakaguchi, F. Koyama, K. Iga, "Vertical cavity surface-emitting laser with AlGaAs/AlAs Bragg reflector", El. Letters, 24, pp. 928-929

14. J. Jewell, A. Scherer, S.L. McCall, Y.H. Lee, S. Walker, J.P. Harbison, L.T. Florez, "Low threshold electrically pumped vertical-cavity surface-emitting microlasers", El. Letters, 25, No.17, p.1123

15. W. Bludau, H. M. Gündner, M. Kaiser, "Systemgrundlagen und Meßtechnik in der optischen Übertragungstechnik", Stuttgart, 1985

16. 4000-Series SEED Optical Arrays, Preliminary Data Sheet, AT&T Bell Laboratories

17. K. Zürl, "Optoelektronische Feldverbinder", Thesis, Univ. of Erlangen, 1992

18. A. Yariv:"Introduction to Optical Electronics", New York, 1976

19. H. Kressel (editor):"Semiconductor devices for optical communication", Berlin, 1987

20. T. E. Jenkins: "Optical sensing techniques and signal processing", London, 1987

21. R. Miller, "Rauschen", Berlin, 1979

22. M. R. Feldmann, S. C. Esener, C. C. Guest, S. H. Lee, "Critical issues in free space intrachip optical interconnect technology", SPIE Vol. 836 Optoelectronic Materials, Devices, Packaging, and Interconnects (1987)

23. L. A. Glasser, D. W. Dobberpuhl, "The design and analysis of VLSI circuits", 1985

24. H.-J. Haumann, H. Kobolla, F. Sauer, J. Schmidt, J. Schwider, W. Stork, N. Streibl, R. Völkel: "Optoelectronic interconnection based on a light-guiding plate with holographic coupling elements", Opt. Eng. 1991, p.1620

16 Comparison between electrical and optical interconnects

Th. Maurin
Institut d'Electronique Fondamentale, CNRS, Orsay

Ph. Lalanne and P. Chavel
Institut d'Optique, CNRS, Orsay

16.1 Introduction

One attractive feature of optics is its ability to communicate large amounts of information in parallel. Another, already commonly used, advantage of optical interconnects is to provide high data flow rate in communication. The latter superiority has already allowed optical fibers to enter the two first stages of the usual five stage communication hierarchy in computers:
  - module to module, i.e. between processors located some distance, typically at least meters and up to many kilometers between each other;
  - backplane to backplane, i.e. between cabinets forming together a single machine;
  - board to board, i.e. between boards inside a given cabinet, each board being conventionally accessed through a backplane connector;
  - chip to chip, i.e. inside a board;
  - gate to gate inside a chip, which is the last stage of the classification .
The three last stages are schematically depicted in Fig. 16.1.

The main purpose of this chapter is to explore how far down this hierarchy optics will penetrate rather than to study the possible means by which optics might contribute to the solution for the different levels of communication. The interest of this classification, as was pointed out by Goodman et al [14], is to indicate an almost continuous evolution between the upper stage requiring a relatively small number of long and high bandwidth channels, and the gates level requiring a rather large number of short and non-time multiplexed channels. In fact, the two extreme cases of the hierarchy are completely different. For long distances, little parallelism is possible: because of the cost of fibers (or electrical wire), a single link containing one or a small number of fibers is preferable and because of the modal dispersion, single mode fibers are often the only suitable. Moreover, optical device technology is bound to use the telecommunication wavelengths 1.3 and 1.55 μm to minimise attenuations and material dispersions. On the contrary, for intra-computers distances, both multimode fibers and multichannel systems are of interest and no argument allows to choose a given wavelength at the present time. This significant difference between the two extreme levels of the hierarchy implies that the optical intra-computer communication is a new problem and that the knowledge learned from long distance communications is not always valid for this case.

Best optical switching energies of present optoelectronic logic gates are on the order of femtojoules [1] per micrometer squared and are now comparable with CMOS technologies switching energies. This is not surprising as the two techniques rely on the same basic concept: loading some capacity to induce an electron migration. However, while smooth scaling down of microelectronics is expected down to feature size of 0.1 μm, the integration density of optoelectronic chips is limited by diffraction, by alignment constraints or by sensitivity to laser power fluctuations (the symmetric SEED's [1] consist of two quantum well devices operating in a differential way). So, while pure electronics integration leads to millions of logic elements operating at hundreds of MHz, optoelectronic chips cannot presently expect reaching such massive integration. Moreover, optical connections suffer from the penalty in the power budget due to electron to photon and photon to electron conversions in spite of the many improvements achieved during the last decade. This difficulty in implementing good optical generators or modulators is obviously a drawback for a deep penetration of optics in the hierarchy if we consider that an entire connection includes end components and a pure communication link and that, as we go down the hierarchy, the share of the end components in the total power (or bandwidth or even design) budget considerably increases and even becomes predominant. For instance, at the last stage of the classification, nearest gate to gate interconnects can be considered as simple electrical nodes (no capacitance or resistive effect) for bandwidth, speed or energy consumption considerations.

As we go up the hierarchy, limitations of electrical interconnects induce a quick decrease of the interconnection density or connectivity. Reaching 10,000 cm/cm^2 for 1μm CMOS chips, the interconnectivity falls down to 800 for MultiChipModules and depending on the technology [5,28] drops to 50 or 90 (per level) for Printed Circuit Boards.

Although this whole chapter is more based on fundamental arguments rather than on a technical description of the means by which optical links may be handled for intracomputer communication, the following section describes different practical systems under progress in laboratories. After a brief overview of emitters and detectors, section 2 will introduce the more commonly selected approaches for implementing the connections at the different levels of the hierarchy. Section 3 gives an overview of interconnect limitations in VLSI systems. The last section makes a comparison between optical and electrical interconnects based on speed and power considerations at the different levels of the hierarchy. It takes advantage of the almost continuous evolution of the power budget, the available bandwidth and the interconnectivity between the upper and the lower stage of the hierarchy to infer a threshold above which optics has little interest. Surely, this comparison relies on the present technology. Many communications indicate that non-linear optics is an extremely active area of research and that new and improved concepts for logic elements arise at a quick pace; emitters such as laser diodes have received much attention and improvements during the past decade. Also and above all, technical and physical limits of such parameters as modulation frequency are still out of sight[32]. Consequently, the validity of our concluding remarks may have a limited lifetime.

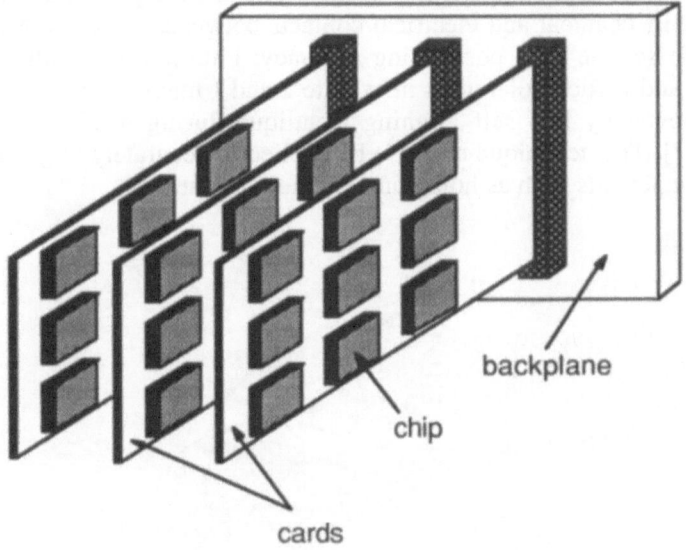

Fig. 16.1 Illustration of chips, board and backplane organisation for digital computers.

16.2 Optical intracomputer communication systems

16.2.1 Sources and detectors

The source can be a quite broad band light emitting diode or a directly modulated semiconductor laser or an external powerful cw laser coupled to several local modulators. As was pointed out in the previous chapter, this last possibility reduces the local power consumption since no threshold currents have to be considered. On the counterpart, as it does not provide any amplification, the cascadability is not ensured. Moreover, the architectures based on local modulators are often less manageable than those using semiconductor lasers. Modulators or laser diodes are generally integrated on GaAs substrates for wavelengths about 0.8 micrometer. So, they are not directly compatible with classical silicon computers. Consequently, in most cases, the first stage of an optoelectronic communication linking two points A and B begins with an electrical wire from A to an intermediate GaAs chip located some distance near A. This electrical wire can be a classical pad to pad printed link or, as was first proposed by IBM, a solder bump. Figure 16-2 shows such a technique that offer a natural way for integrating different substrates such as gallium arsenide optoelectronics chips and conventional silicon ones. Figure 16-3 shows the positioning technique. After a preliminary placement, a reflow operation, during which the solder bump is melted, is applied. The melting procedure provide at the same time mechanical and electrical contacts between the two substrates and is responsible for the high positioning accuracy: Lateral (i.e. parallel to plane of substrate) and vertical tolerances inferior to 2 and 1 micrometer respectively have been achieved by this self aligning technique during the Esprit II OLIVES Project [27]. This technique can also be applied to accurately align emitters with optical components such as holograms or microlens arrays.

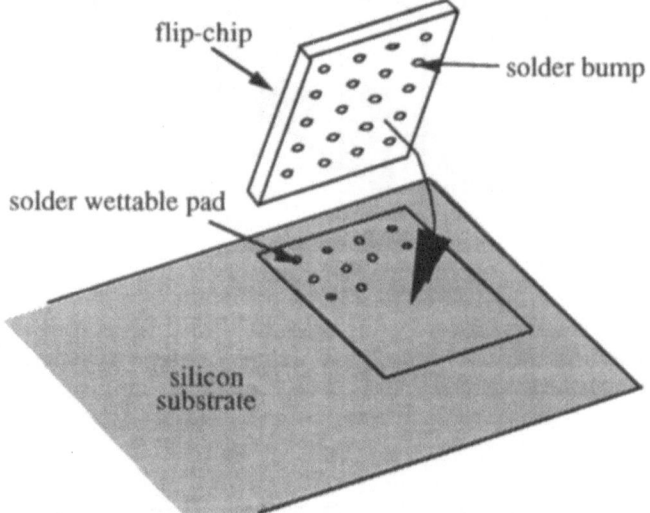

Fig. 16.2 Flip chip bonding in flip chip bonding.

As was pointed out in chapter 2, laser diodes or modulators can be also grown directly on silicon substrate. Although major problems such as interface stresses owing to a lattice mismatch and variations in thermal conductibility occur and that this approach has already a long history of failure, recent improvements offering an increasing operating life have been demonstrated [7] and nothing presently allows to disregard this powerful step toward "optical computing". While classical laser diodes are more tractable when used with waveguides circuits, vertical cavity structure arrays offer massive parallel free space interconnects.

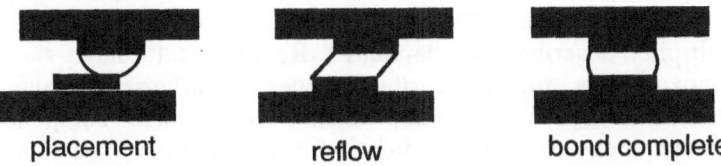

placement   reflow   bond complete

Fig. 16.3 Self-alignment technique.

At the end of the optical interconnect, the receiver is generally a photodiode which can be integrated on a classical Si substrate. The photodiode can be connected to an amplifier and a comparator to recover the logic level in order to deliver a global standardised voltage level for all links. To reduce the circuitry, not only efficient electron to photon and photon to electron converters are needed but also small losses in the optical interconnection are required to ensure that most of the emitted photons are received by the photodiode. This partly justifies the efforts made in the design and fabrications of the high efficiency diffraction holograms presented in the previous chapter.

16.2.2 Optical channels

A lot of very different optical channels can be implemented for communications in digital systems. According to volume constraint specificity at each level of the hierarchy, waveguides technology is generally preferred at the interchip level, planar confined free space interconnect for the backplane interconnects while the compatibility of optical fiber technology is tested at the upper level of the hierarchy. Little literature is available for implementations at the intrachip level; Because it covers all the interconnect levels of the hierarchy and because it is quite similar with the other projects under progress, the ESPRIT II OLIVES programme [27], that involved many industries and universities of the Economic European Community, will mainly be used in the following, to illustrate the different approaches chosen for intracomputer interconnects.

Waveguide circuits. Optical fibers are well suited for high data motor ways linking cabinets [11]. The technology developed in that system allows low bit error rates (typically 10^{-15}) and reconfigurability for long distance communication of several kilometres without the use of repeaters. The interconnection technique enhances flexibility in the placement of the processors while increasing the reliability of the overall system.

At the backplane level, works are presently under progress for implementing a multifiber bus [6,27]. In the ESPRIT II OLIVES Project [27], several nodes consisting of 8 lasers and receivers are connected by ribbon fibers through an array of passive star couplers. With a data rate of 1 Gbit per second and per channel and a fanout equal to 32, the system implemented has a global data rate of 32 Gbit per second.

A combination of multimode optical fibers and free space interconnects is involved in the MILORD project [14] which consists in an optical crossbar network providing the required interconnections for a parallel processor consisting of 128 transputers. The optical crossbar is similar to the optical vector matrix multiplier described in chapter 12. Recent results have shown that a 128x128 network operating at 150 Mbits/s with a reconfiguration delay of 10 ms (ferroelectric liquid crystal) can be implemented with available components.

Optical waveguides are good candidates for interchip interconnects. They are compatible with planar cards constraints, offer an increased bandwidth compared to their electrical counterpart. Because of the very small crosstalk existing between two adjacent lines, the interconnectivity is highly increased and can reach 500 cm/cm^2 (per level). Different technical approaches are under tests: Honeywell [4] and ATT Bell [18] consider polymer waveguides for chip to chip interconnects while the Si-substrate of wafers is structured by wet etching in the OLIVES project [20].

Attenuations of straight lines vary between 1 and 3 dB/cm according to the technology. Curvature losses are very important and increase proportionally to the inverse of the curvature radius. On the contrary, attenuations in a 90° intersection is about 3 dB with a negligible crosstalk [4] (about 40 dB). The main constraint comes from the precise alignment at the laser to waveguide coupling location. The lateral alignment is quite tolerant whereas transversal constraints can be submicronic in some cases [24]. At the last stage of the interconnect, the waveguide to photodiode coupling is less critical. The photodiode surface can be extended to relax alignment constraints. Figure 16.4 sketches one possible concept based on this technology and developed in the OLIVES project [20] for connections between chips. Let us consider a classical card connecting several chips. To implement new interconnects through optics, the project associates an overlay card. This overlay silicon card is structured with waveguides and contains laser diode and receiver arrays directly coupled by the waveguides. When mounted on the classical multichip card, wire bonds ensure the local electrical connections between each mother chip and its corresponding laser diode array on one side, and between each daughterchip and its corresponding photodiode arrays on the other side. This scheme implemented with 8 waveguide channels of 125 micrometers pitch has shown an improved interconnection density with higher bit rates than classical electronics.

Another approach mixing optical and electrical channels has been proposed by Mac Donald et al. [23] to fulfil the density and propagation delay needs for submicron silicon devices operating in the range of 100 MHz and for digital GaAs operation in the GHz range. This concept includes two separate submodules allowing electrical and optical interconnects in the same package. Figure 16.5 illustrates the cross section of this structure. An advanced MultiChip Module technology, with electrical wiring, is used for short distance interconnections between neighbouring integrated circuits (IC). Electrical solder bumps and vias are suitable for frequency up to 40 GHz. A photonic module including passive waveguides and active optical devices (modulators, detectors and transmitters) is used as an add-on module electrically connected to the MultiChipModule by means of solder bumps. This submodule allows optical interconnects between different parts of the package. The photonic module has also optical I/Os used for communicating with other modules. These optical interconnects between photonic modules are done via optical fibers. For this concept, the challenge will still be to align the optical components to the waveguides.

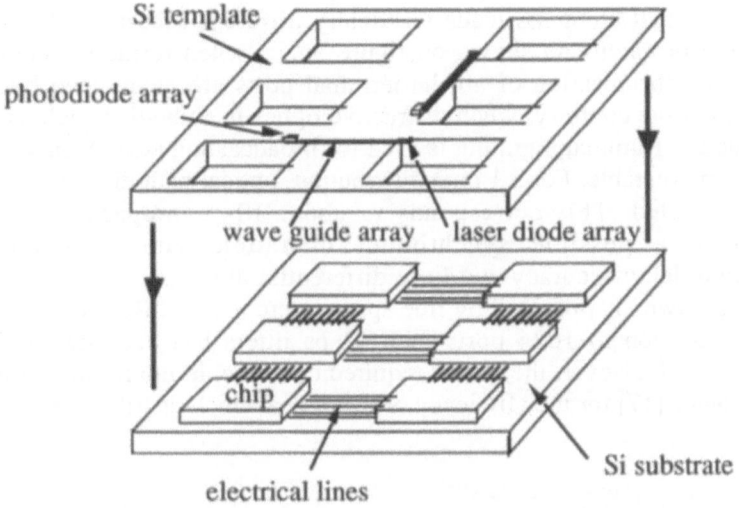

Fig. 16.4 Intra-card optical interconnections in the OLIVES II project.

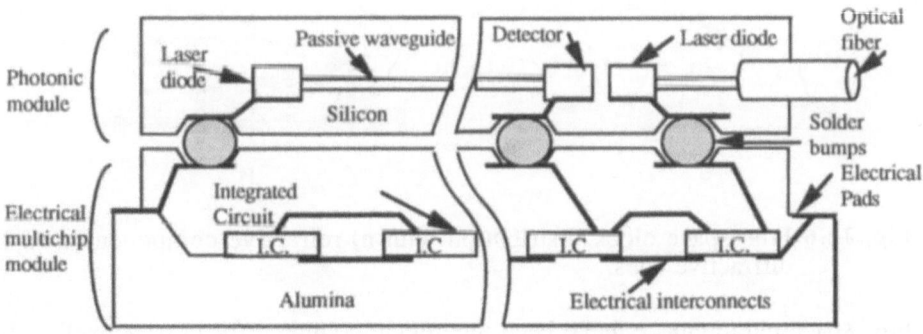

Fig. 16.5 Cross section of the Electro-Optic MultiChipModule concept based on High Density Interconnect. After reference 23.

Free space interconnects. Free space interconnects offer two additional flexibilities: freedom from planar or quasi-planar constraints and 2D parallelism. Historically, they have been first introduced by J.W. Goodman et al. [15] for optical clock distributions. Fibers or integrated optical waveguides or both can be used such as in the OLIVES project where an intercard clock distribution is implemented by adjusting delays in fibers and by using a Htree waveguide distribution at the board level. Array illuminators, realised by means of holographic optical elements or lenslet array, also provide a powerful way to distribute an optical clock over an arrays of photodetectors embedded in the integrated circuit [8] (see chapter 12). Let us note that the use of diffractive or refractive optical elements leads to slightly different accuracies for the clock distribution problem. As shown on figure 16.6a, when refractive elements are used, the synchronisation of all the terminal ports are guaranteed by Fermat's principle. On the contrary, when diffractive optics is embodied such as in figure 16.6b where a Dammann grating is used for broadcasting (see chapter 12), clock skews are predictable. For a 1 cm^2 distribution, optical path differences of 3 mm can be expected. This corresponds to about 10 ps propagation delays. In addition to a possible synchronous operation with picosecond-range propagation delay accuracy and little differential attenuation, fault-tolerance to local breakdowns is provided by free space interconnects. Recently, a 300 MHz clock distribution to 1024 ports with 12 ps jitters has been demonstrated at Bellcore [17]. However, the power required could be, in the future, an important limiting factor [17] for the efficiency of the optical clock distribution.

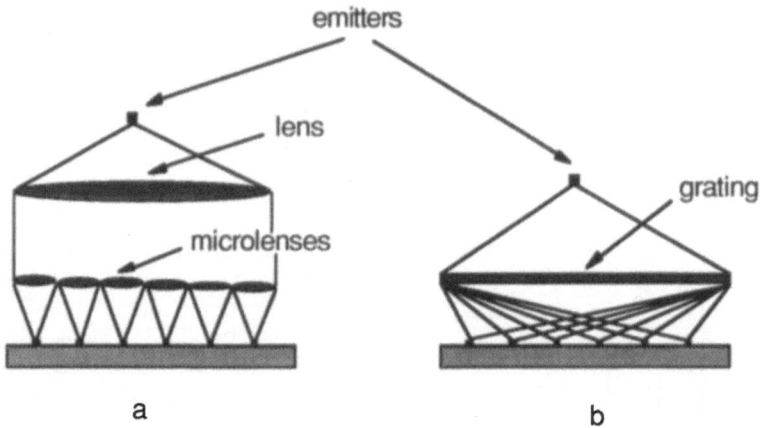

Fig. 16.6 Free space clock distribution with **a)** refractive components and **b)** diffractive ones.

Free space interconnects have been theoretically studied for inter as well as for intrachip communications with respect to their interconnect density capabilities. At the interchip level which requires space-variant interconnections, computer generated holograms have received much attention [3,12,16] because they are very flexible during their fabrication (see chapter 5 and 12). As shown on figure.

16.7a, the simplest scheme consists in associating one hologram per emitter. This hologram deviates the beam and, acting as a lens, focuses it onto the receiver. However, if we take care of crosstalk by aperture diffraction and of the divergence angle of light sources, this approach results in a weak interconnect density. According to reference 16, only 20 arbitrary interconnects can be achieved per cm^2. Compared to the purely electrical interconnectivity at the intracard level, this number is in the same range. It results that this approach presents little interest.

Various other possibilities have been studied in literature. Among them, the scheme sketched in figure 16.7b seems quite promising. Compared to the previous single hologram scheme, architectures employing a double pass hologram and a mirror increase the connectivity and allow larger interconnect distances[3,12]. As was shown in chapter 2, up to 500 arbitrary space variant interconnection can be implemented per cm^2. However, this approach still complicates the alignment procedure since one more component has to be aligned. The following section will discuss a recent technique simplifying the alignment.

Because they can be fabricated in an efficient highly reliable manner using procedures compatible with existing integrated circuit fabrication process, computer generated holograms have received much attention [39]. Various possible interconnection schemes have been studied in order to take into account the fault-tolerance to misalignment, crosstalk by aperture diffraction, the numerical aperture of lenses and divergence angle of light sources [3].

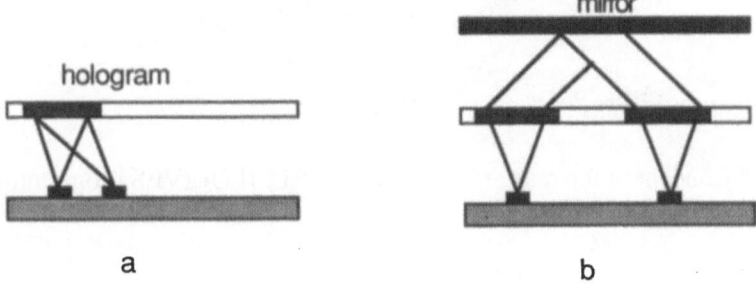

Fig. 16.7 a) Single and b) double pass hologram interconnection.

Planar confined free space interconnects. A compact copy [29] of the double-pass technique has been demonstrated by using the total internal reflection in the glass plate of holographic gratings fabricated in dichromated gelatine. In this approach, the glass face without gelatine is used as the mirror of the previous double-pass structure and the tilted propagation operates inside the glass plate. The result is that free space is confined in the thickness of the glass plate which structures the wave on only one of its faces. More recently, the combination of the latter free space planar approach and of electron beam written computer generated hologram techniques has been proposed for integrating many different optical components onto one single substrate with flexibility (see chapter 2). By providing stability with accurate alignment, this kind of approach is more likely to reach racks tolerances than the pure free space one. A Htree distribution using this concept is actually under study at Bell Laboratories. They succeeded [36] in the implementation of beamsplitter gratings with 98% efficiency and beam deflector gratings with 90% efficiency. It could result in an overall efficiency of 75% for 1x4 distribution system.

Besides, this concept has been studied in the OLIVES programme [27] at the backplane interconnects level. Figure 16.8 sketches the so called mastercard demonstrator. The mastercard is fabricated from a quartz glass mask plate and is etched with holograms on one side. As shown on figure 16.9, a collimated light from a laser diode located on a mother board is first incoupled, then driven by several internal reflections and finally outcoupled toward different photodiodes located on daughterboards and mounted with an additional microlens fixed to the standard electronic package. A clock distribution with a fanout of 4 has already been implemented [27] with a clock skew of 100 ps and a BER of 10^{-11} (see also reference 30 for more details).

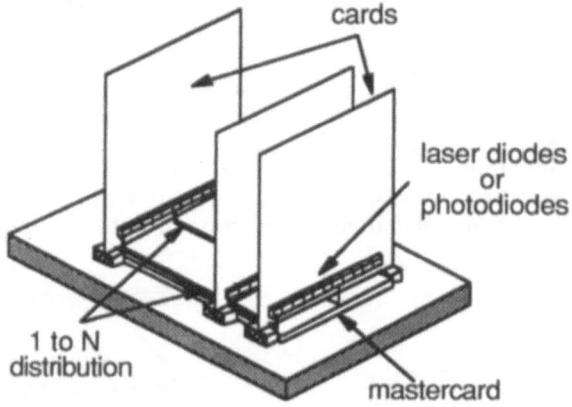

Fig. 16.8 Concept of the mastercard in the ESPRIT II OLIVES programme.

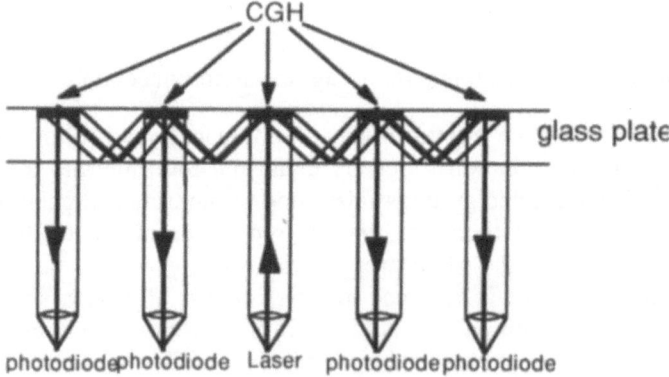

Fig. 16.9 Mastercard distribution: cross section.

This concept has also been applied to provide full interconnectivity as a backplane between 9 boards (81 links). A prototype of such a computer is shown on figure 16.10. The pitch between each hologram is 6.35 mm and the data rate implemented is 1.4 Gbits per second within the 1 dm^3 computer. The BER was measured to less than 10^{-10}.

Fig. 16.10 Prototype of the 1 dm^3 computer. (Photograph provided by the courtesy of B. Houssay, Thomson CSF/RCM)

Stacked planar optics. Vertical cavity laser diodes arrays combined with microlens arrays can be directly used for intercard communications. Figure 16.11 shows the concept. A classical mother chip is connected with classical electrical wires to a laser diode array surmounted by a technologically compatible lenslet array. On the adjacent card, the same symmetrical lenslet array conjugates the laser diodes with a set of photodiodes directly integrated onto the daughter chip. In this particular case, very high connectivity can be achieved. This concept has been validated by Tsang [34] who has demonstrated a 19% differential current efficiency in a direct source to detector optical link operating at one-gigabit per second. This approach is also under study with LED arrays in the OLIVES project [27]. The practical limitation comes from the weak tolerance in the misalignment between cards.

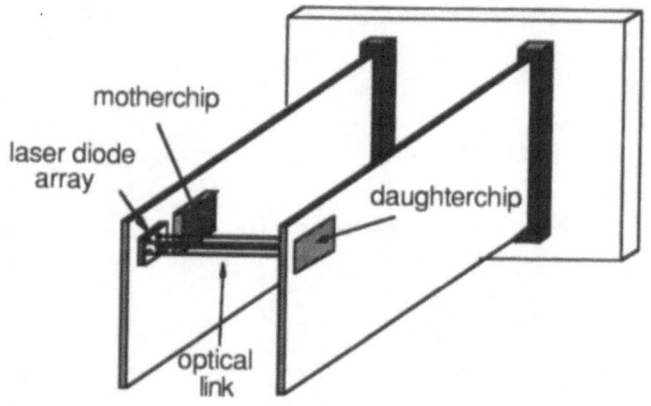

Fig. 16.11 Adjacent card optical connector.

16.3 Analysis of interconnect limitations in VLSI systems

16.3.1 Introduction

The present trends and advances in the domain of microelectronics can be divided into two fields : technology and architecture. On the one hand, silicon devices continue to shrink and chip complexity, size, IO count, operating speed to increase. On the other hand, parallel architectures try to overcome the intrinsic limitations of conventional sequential architectures. These technological and architectural trends continuously increase the communication needs inside and between chips, thus, the interconnection crisis becomes a crucial problem for on board and intrachip interconnections.

16.3.2 Technological constraints in VLSI

The main features of recent advance in integrated circuits technology can be characterised by the decreasing feature size and the increasing chip size. On the one hand, submicronic processes are presently in use and, on the other hand, processes are able to reach, with a good yield, a chip area larger than 1 cm^2. Obviously, these features are leading to more and more powerful integrated circuits. Thus, nowadays, more than 1 million transistors can be integrated on a chip, the CMOS inverter delay time can be less than 0.5 ns, and new 32 bits VLSI processors are not far from 100.10^6 clock cycles per second.

If $1/\alpha$ is the scaling factor ($\alpha > 1$) of the MOS transistor feature size induced by a CMOS process size shrink, (for instance, due to this process improvement, the width of a transistor is reduced from W to W/α) and if the very simple first order "constant field" model is used, the gate delay time and gate area have respectively a scaling factor of $1/\alpha$ and $1/\alpha^2$ [37]. If moreover the chip size expansion factor is β then the factor of merit (Number-of-gate)/(gate-delay) grows as $\beta^2\alpha^3$.

But some drawbacks also arise from these improvements, drawbacks which can be seen as major limitation factors.

Speed. One of the major issues is the propagation delay time. This overall delay, through gate and line, is roughly the sum of the gate delay and the line propagation delay. The former scales with $1/\alpha$, but the latter depends on the distributed resistance and capacitance of the line and is approximately proportional to the RC product of the wire. The first order model shows that the RC time constant must stay unchanged by scaling. Then, the propagation delay remains unchanged while the gate delay decreases. So, due to the decreasing feature size, the main limiting factor in overall delay becomes the RC line response time. Moreover, a more accurate model shows that the RC product does not remain constant but increases as the feature size decreases. This is due to fringing capacitance, which also produces a crosstalk effect, and even more importantly to the average line length that does not scale with $1/\alpha$ but, on the contrary, follows an (area)$^{1/2}$ dependence, that is to say a β scaling [31].

This issue is more noticeable with some signals which are distributed all over the chip. The length of the clock signals, for instance, does not scale down at all and the clock line response time RC is affected by a factor $\beta^2\alpha^2$. The chip crossing delay is affected by the same factor. This problem is fundamental for clock signal distribution and synchronisation over the die. To fully take advantage of reducing the gate delay, higher operating frequency have to be used. This fact imposes to use more and more clocking schemes based on edge triggered latches. The use of this type of latches demands higher precision in clock generation. In a downscaled CMOS technology a typical gate delay is less than 1 ns, then clock skew must be less than 1 ns which is not an easy design objective to achieve.

I/O number and complexity. Another important difficulty, magnified by the technological advances is the increasing I/O number on a chip. So, input and output data are becoming a major problem. Rent's rule estimates the I/O number on a chip [24]. This number is given by : $N = k . N_g^p$, where N_g is the number of gates in the chip, and p is Rent's exponent which is tied up with the complexity level of the circuit and falls within the range of 0.4 to 0.6. [13]. Obviously the critical value of p is 0.5 because if it is higher than 0.5, the number of I/O grows with a factor $(\beta\alpha)^{2p}$ which is larger than $\beta\alpha$, the scaling factor of the chip perimeter. So, if peripheral pads (I.C. connections) are used, the chip perimeter may become insufficient.

We can note that the greater the complexity, the greater the value of p. Moreover, the increasing number of components integrated on one chip may also lead to an important communication crisis within the die. Thus, if the maximum length of interconnections follow a β factor, if the number of connections grows like the I/O count (that is to say : follow Rent's rule) then the scaling factor of the connection density within the die can be estimated at $(\beta\alpha)^{2p-1}$. So, data bus connections between different functional blocks may become wider and wider and take more and more area as the complexity of the chip increases. This problem is becoming crucial because of the crossing limitation due to the limited number of connection levels within the die (usually 2 or 3 metal levels and 1 polysilicon level), even if this number tends to increase.

Power. For chips, boards and modules, power dissipation constraints include all parameters involved in the conduction and convection phenomena used to remove heat energy from the transistors to the air or cooling fluid. For instance maximum junction temperature, ambient temperature, thermal resistance of the different components, cooling fluid parameters, heat sinks geometrical parameters are important. Limiting power dissipation will limit performance, so power management becomes a key to performance in large systems design [25].

According to the first order model, the switching capacitance C scales with $1/\alpha$ and the maximum switching frequency f varying like 1/(gate delay) scales with α. To maintain the electric field constant, we need a supply voltage V_{dd} scaling factor of $1/\alpha$. So, the power dissipated by switching capacitance or dynamic power (CV_{dd}^2f) decreases with $1/\alpha^2$. Thus, the dynamic power density is technology independent. In fact, without supply voltage scaling, this parameter increases by α^3. The maximum junction temperature imposes a maximum power density dissipation due to the thermal resistance of the package and then limits the maximum clock frequency or the percentage of gates allowed to switch within the same clock cycle. For instance a 1 W, 1 cm^2 2μm CMOS with 400,000 transistors where only 10% of the gates are active on the same cycle, is limited at 25 MHz [5]. The constraint on the number of simultaneously switched line drivers cannot be overcome even with optical interconnections.

Other related problems. The Fan-out remains an important problem as well. Gates or I/O pads able to drive a large number of gates require powerful transistors with large dimensions. This has the effect of increasing the total capacitance and then the power required. For instance, standard cell and gate arrays rely on gates which can drive large loads in order to reduce the sensitivity to interconnect capacitance which is unknown and depends on customer specification. Thus, the transistor are much larger than necessary and the performance are roughly 5 times less than a full custom design.

Within the chip, reducing the line metallization width brings new problems because these lines must carry a higher current density which follow a α factor. Thus, electron migration appears and becomes a major factor to consider. New metallization schemes will be required to accommodate the higher current density and new materials have to be used (Molybdenum....)

The crosstalk is strongly influenced by the geometrical parameters of the interconnection lines and by the rise/fall time of the input signal as well. For instance, the peak coupling noise on an adjacent inactive line increases when the rise time of the input signal decreases. Thus, increasing the switching frequency and reducing the clock skew implies reducing the rise time of the clock signal and therefore increasing the crosstalk which could become an important limiting factor. The crosstalk can be reduced by using a shield realised by interleaving ground and signal metallization, but this solution increases the layout complexity area and price [2,21,38].

16.3.3 Architectural constraints

On the other hand, advances in architecture have been stimulated by the intrinsic limitations of conventional architectures. For instance, the well known Von Neumann memory-CPU bottleneck is still presently an important factor in limiting performances of standard sequential processors. Thus, new concepts in architecture have been developed to overcome these limitations and to increase performances of general and dedicated applications.

Conventional architecture. Without fundamental change in architecture, some new structures have improved results. For instance, the Reduced Instruction Set Computer (RISC) architecture can be seen as a means to reduce complexity in order to enhance overall capability. This kind of architecture reducing the set of computer instructions allows a shorter cycle time and a simpler instruction decoder which lead to a more efficient computer. The silicon area allocated to the sequencing unit is drastically reduced so that area used as register and memory can be increased significantly. This point is important for power consideration as well. In memory only a few transistor are involved in an access, so the percentage of transistor switching in the same clock cycle is small. A RISC architecture using a more significant part of its area for memory has a percentage of transistor switching in the same clock reduced in comparison with the conventional architecture. As seen previously this is an important advantage.

We can also note the present trend towards 64 bits address and data buses. Thus, each chip with separate instruction and data buses (Harvard architecture) will require more than 256 I/O. RISC is nowadays a very useful concept used in many computers from the SUN Sparc to the IBM R6000 [9]. But this new concept always includes the intrinsic limitation of a sequential treatment.

Parallel architecture. Presently, researchers explore new concepts oriented towards parallel architectures. Parallelism, with coarse or fine grain, is the new way to respond to the performances challenge. For instance, systolic arrays, cellular architectures, automata arrays, neural networks are presently studied in many electronic laboratories all over the word. All these new architectures can be sorted into different classes like: SIMD or MIMD, synchronous or asynchronous and so on... Whatever it is, the main advantages of these new structures are tied up with high level of parallelism. However, from the communication point of view these new parallel architectures are not always the best solutions in terms of integration and design because they need more and more interconnections. For instance, the Connection Machine includes 50000 chips and new interconnection technologies have already been realised and tested [22].

We can also note that, in parallel architecture, the very frequent use of repetitive structures is a very useful feature especially for CAD testability and reorganisation.

But some problems must be solved to fully take advantage of high parallelism architectures. Among these problems, the communication problem: input connections, output connections, signal distributions and synchronisation are crucially amplified by the new parallel architectures arising.

New interconnect technologies for systems. Wafer Scale Integration (WSI) is a new technological and architectural approach to overcome the I/O terminals growth issue and to integrate devices at the wafer level in order to enhance hugely useful silicon area. In a WSI environment, clock synchronisation over all the chips on the substrate is very important, so, this objective is a new challenge for circuit designers [26]. Also required by this technique is the possibility of reorganisation of the connections between devices to bypass faulty ones.

The last decade, with the introduction of SMT (surface mount technology) and high I/O integrated circuits devices has seen the proliferation of packaging formats and technologies or "Packaging Big bang". All the new technologies aim to increase density and performances of the modules. Among them, multichip module (MCM) using a high connectivity substrate (silicon-on-silicon) provides an interesting solution as a next generation interconnection and packaging technology beyond SMT for high performance and high reliability systems using VLSI devices. The incorporation of optoelectronic devices within this technology is conceivable for offmodule and perhaps onmodule interconnection as seen previously in figure 5 [5,26,28].

To summarise, we can say that present technological problems can be seen as the search for an optimum in a sea of constraints. In particular, the point has now been reached where silicon designs can no longer be considered in isolation from interconnections and also from packaging problems [28].

16.4 Optical and electrical interconnects

As seen previously, advances in VLSI technologies are often limited by the power dissipation, the connectivity density and by the propagation delay time through gates and interconnections. For these three main points, optical interconnection brings valuable solutions.

16.4.1 Connectivity density considerations

The few basic architectures described in section 2 offer at least the same connectivity densities as electronics ones. For intracard interconnects using waveguides, the connectivity density is in the range of that of electronics; optics allows smaller crosstalk while electronic interconnects can be easily implemented on different layers. Considering now planar confined free space techniques, the density is largely increased. As was mentioned in chapter 3.1, about 500 arbitrary interconnects per cm^2 can be fairly expected (see also chapter 11). This number reaches that of electrical multichip modules in case of quasi parallel point to point wires. It surpasses the present density realised for intracard communications.

However, this number must greatly be moderated. The main practical problem with free space interconnects comes from alignment tolerances. For instance, for the mastercard configuration shown on figure 16.9, collimating lenses have been integrated at the board level with end components. This approach has shown improved alignment tolerances compared to other schemes implementing both collimator and deflector gratings at the glass slab level. 100 instead of 10 µm alignment tolerance have so been obtained.

In the direct board to board interconnection of figure 16.11, the pitch of the laser diode array is rather determined by alignment constraints including crosstalk than by heat dissipation limits.

Classical electronics packaging techniques also suffer from (in some sense) misalignment problems with regards to connectivity density. The pitches between the pins at the chip and board levels is determined by plug-in facility and testability considerations and not by fundamentals limitations. So, an accurate discussion of optics possibilities to overcome pin connector bottleneck limitations of electronics would necessary take into account packaging. Optical packaging solutions that present a clear supremacy over existing electronic techniques have still to be developed.

16.4.2 Bandwidth consideration

In a given technology, the length-bandwidth product is the main limiting factor of electrical interconnects. From this point of view, we can, according to Tsang [33], define regions of interest for electrical and optical interconnections based on both frequency and distance. For instance, transmission lines like stripline technologies are limited in the higher frequency domain by parasitic effects such as metal skin-effect. For the distances encountered in intracomputer interconnects, for modulations in the GHz range, optics does not suffer from length-bandwidth product limitations. Even with holograms or multimode guided propagation, optical interconnects present very little dispersion and no phenomenon such as skin effect or crosstalk have to be taken into account. For instance, even a multimode silicon fiber interconnecting two points whose distance is 1m (1 m is a quite long distance for an intracomputer communication) can propagate signal modulations up to about 10 GHz. Identically, a 1 cm diameter holographic lens working with a numerical aperture of 0.3 introduces a maximal propagation delay of 10 ps allowing modulation speeds larger than 10 GHz.

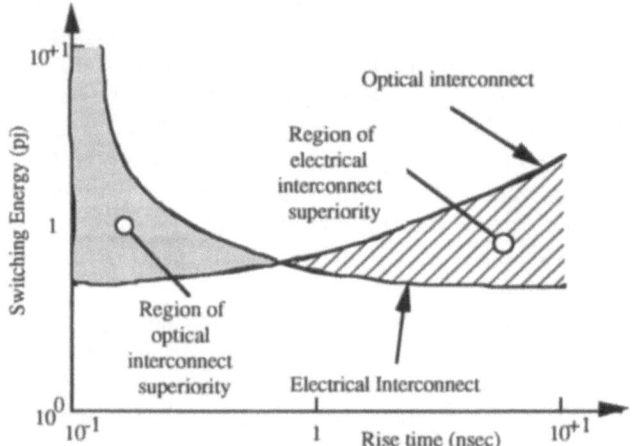

Fig 16.12 Regions of interest for electrical and optical interconnections. After reference 32.

16.4.3 Power consideration

Electrical lines techniques are well known and studied, it is possible to make a detailed energy budget for all the interconnects of the hierarchy. For optical techniques, since the technology is not mature, the energy balance is more difficult to establish accurately and depends on approaches investigated. A precise comparison can be found in the paper of Feldman et al. [11] and in the previous chapter. Figure 16.13 shows, for instance, the respective superiority domains of electrical and optical interconnects for a 1 mm intrachip interconnect with a fan-out of one and based on 3 μm CMOS process. The comparison is made between an 1 mm aluminium line and a 10 μm square detector driven by a 1 mW laser diode through a 9% efficiency optical line. From the energy point of view, optical interconnect is more efficient for signal rise time less than 0.5 ns. So, in this case, optical interconnects are well suited for frequency larger than 1 GHz.

Due to the high capacitance of an I/O pad (about 0.8 pF for a 4.10^4 mm^2) and due to the unscaling voltage supply at this level (V_{dd} =5 V), the switching energy involved in a connection through a pad reaches about 10 pJ. This value is more than one order of magnitude larger than the intrachip one. So, there is a large energy penalty in getting out of the chip. For an optical connection, a 1 mA threshold laser diode dissipates at least 1 mW per link. This power corresponds to electrical interchip communications in the range of 100 MHz which is the order of magnitude of present maximum chip frequencies. Thus, this rough comparison sets the present position of the optical techniques in the hierarchy from the power point of view : optical and electrical interconnect for interchip to intermodule connection seems, to date, to be equivalent. As seen previously, for intrachip connections electrical interconnects seems to be winner for frequency up to 1 GHz.

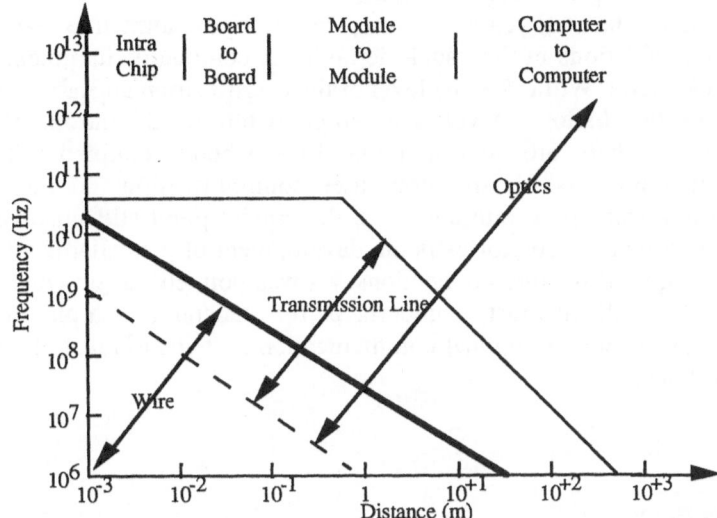

Fig 16.13 Switching energy versus rise time for 1 mm aluminium line and for a 9% efficient optical interconnect with a 1 mW laser diode. After reference 10.

However, the power budget cannot be obtained so straightfordwardly: it depends on the optical architecture. As was previously pointed out in the section about density considerations, alignment precision that must be established and maintained to assure that the light spots are striking the appropriate photodiodes influences the power budget. For instance, defocusing techniques can be applied to alleviate misalignment requirements with, on the counterpart a decreasing efficiency. So again, an accurate comparison is not still possible. Let us just say that, in the two systems shown on figures 16.8 and 16.11, energy requirements for optics and electronics are in the same range.

In conclusion, for an equivalent power budget, optics is expected to improve an important figure of merit of conventional computers: the bit rate per area.

16.4.4 Discussion and conclusion

The evolution of optical interconnections will be shaped by future advances in many component technologies. These technologies include monolithic optoelectronic integrated circuits, hybrid integration, laser arrays, surface-emitting laser, heteroepitaxy of GaAs on Si [33], micro-optics, fiber coupling techniques, holograms, laser amplifiers, modulators, spatial light modulators. A fundamental trump for free-space and guided optical interconnections is the availability of high density, low power budget in some cases and large length-bandwidth product through photons.

After more than ten years of efforts for specifying attractive features of optics for communications in the usual hierarchy of computers, its potentiality is now well understood. While the last level of the classification appears still out of sight for optics, the shallower levels seem to be mature for the introduction of optical techniques. There are no doubt that fiber ribbons coupled with multifiber connectors appears for intracomputer communications in the near future. Concerning free space interconnects, the crucial point still unsolved but under progress in many laboratories is the development of a standardisation of optical devices required for interconnections. As was pointed out by Huang at a recent meeting [19], efforts must be undertaken for realising now a platform for all the components needed in optical communication systems to reach the versatility of electrical lines.

References

1. Lentine, A.L., Chirovsky, L.M.F., D'Asaro, L.A., Tu, C.W. and Miller, D.A.B.: Energy scaling and subnanosecond switching of Symmetric Self-Electrooptic Effect Devices, vol.1, 1989, pp 129-131.
2. Abuelma'atti M. T.: The waveform degradation in VLSI interconnections. IEEE Journal of solid-state circuits, vol. 25, 1990, pp 1014-1016.
3. Brenner, K.H., Sauer, F.: Diffractive-reflective optical interconnects. Appl. Opt., vol 27, 1988, pp 4251-54.
4. Bristow, J.P.G., Sullivan, C.T., Guha, A., Ehramjian, J., Husain, A.: Polymer Waveguide-Based Optical Backplane for Fine-Grained Computing, SPIE vol.1178, Opt. Interconn. in the Computer Environment, 1989, pp103-114.
5. Burr, J.B., Burnham, J.R., Petreson, A.M.: System-wide energy optimization in MCM environment. Proc. Multichip Module Workshop, Santa Cruz, California, March 1991, pp 66-83.
6. Chiarulli, D.M., Levitan, S.P., Melhem, R.G.: Demonstration of an All Optical Addressing Circuit. Proc. Optical Computing, 1991, pp 235-238.

7. Choi, A.K., Turner, G.W., Windhorn, T.H., Tsaur, B.Y.: Monolithic integration of GaAs/AlGaAs double heterostructure LEDs and Si MOSFETs. IEE Electron Dev. Lett. EDL-7, 1986, pp 500-502.
8. Clymer, B.D., Goodman, J.W.: Timing uncertainty for receivers in optical clock distribution for VLSI. Optical engineering vol. 27, 1988, pp 944-954.
9. Cocke, J., Markstein V.: The evolution of RISC technology at IBM. IBM Journal of research and development, vol 34, 1990, pp 4 -11.
10. DeCusatis, C., Huffman A., DeMario, G., Stigliani D.: Fiber optic interconnects for the IBM System/390. Optics and Photon. news Vol.2, 1991, p 33.
11. Feldman, M., Esener, S., Guest, C., Lee,S.: Comparison between optical and electrical interconnects based on power and speed considerations. Appl. Optics, vol. 27,1988, pp 1742-1751.
12. Feldman, M., Guest, C.: Interconnect density capabilities of CGH for optical interconnections of VLSI circuits. Appl. Opt., vol 28, 1989, pp 3134-37.
13. Ferry, D.K.: Interconnections lengths and VLSI. IEEE Cir. Dev. Mag., juil. 1985, pp 39-42.
14. Fracès, M., Bouzinac, J.P., Bodin,E., Comte, D., Siron, P.: The optical crossbar network MILORD machine: 1st developments and results. SPIE, vol. 128, Optical Interconnections and Networks, 1990, pp 66-78.
15. Goodman, J.W., Leonberger, F., Kung, S.Y., Athale, R.: Optical Interconnections for VLSI systems. Proc. IEEE , vol 72, 1984, pp 850-866.
16. Goodman, J.W.: ICO 15, SPIE Vol. 1319, oral communication, Garmisch-Partenkirchen, August 5-10, 1990.
17. Hartman, D.H., et al.: Optical clock distribution using a mode-locked laser diode system. submitted to conference on Opt. Fiber Comm. (1991).
18. Hornak,L.A., Tewksbury, S.J., Weidman, T.W., Kwock, E.W.: Wafer-level Optical Interconnection Network Layout. SPIE, vol.1281, Optical Interconnections and Network, 1990, pp 16-22.
19. Huang, A.: Towards a digital Optics platform. SPIE Vol. 1319, Optics in Complex Systems, 1990, pp 156-160.
20. Karstensen, H., Schneider, H.W., Straudt, A., Zarschizky, H., Gerndt, C., Klement, E., Tischer H.: Optical multichannel parallel chip to chip data distribution. SPIE, vol. 1281, Optical Interconnections and Network, 1990, pp 23-32.
21. Kaupp, H.R.: Waveform degradation in VLSI interconnections. IEEE Journal of solid-state circuits, Vol.24, 1989, pp1150-1153.
22. Lane, T.A., Quam, J.A., Kahle, B.O., Parish E.C.: Gigabit optical interconnects for the connection machine. SPIE, Vol. 1178, Optical interconnects in computer environment, 1989, pp 24-35.
23. MacDonald, J.F., Vlannes, N.P., Lu, T., Wnek,G.E., Boden, E.P., Ghezzo, M., Steward, K.R., Yakymyshyn, C.: Electro-optic multichip modules with non-linear organic Waveguides. Proc. Multichip Module Workshop, Santa Cruz, California, March 1991, pp 93-100.
24. Mentzer, M.A.: Principles of Optical Circuit Engineering. Marcel Dekker, INC/Newyork Basel, 1990.

25. Moresco, L.L.: Electronic system packaging : the search for manufacturing the optimum in a sea of constraints. IEEE Transactions on components, hybrid, and manufacturing technology, Vol. 13, 1990, pp 494-508.
26. Neugebauer, C.A., Carlson, R.O.: Comparison of wafer scale integration with VLSI packaging approaches. IEEE Transactions on components, hybrid, and manufacturing technology, Vol. CHMT-10, 1987, pp 184-189.
27. Parker, J.W.: Optical interconnection for advanced processor systems: a review of the ESPRIT II OLIVES program. Journ. of Lightwave Techn., Vol.9, 1991 pp 1764-73.
    Final report on ESPRIT II project 2289 OLIVES, mai.1992.
28. Pedder, D.J.: Interconnection and packaging of solid-state circuits. IEEE Journal of solid-state circuits, Vol.24, 1989, pp 698-703.
29. Sauer, F.: Fabrication of diffractive-reflective optical interconnects for infrared operation based on total internal reflection. Appl. Opt., vol 28, 1989, pp 386-388.
30. Sebillotte, C.: Holographic optical backplane for board interconnections", SPIE vol.1389 on Advances in Interconnection and Packaging, 1990, pp 600-611.
31. Shozi, M.: CMOS Digital Circuit Technology. Prentice Hall, 1988.
32. Smith, P.W.: On the physical limits of Digital Optics Switching and Logic Elements. The Bell System Technical Journal, vol. 61, 1982, pp 1975-1993.
33. Tsang, D.Z.: Optical interconnections in digital systems status and prospects. Optics and Photonics news, October 1990, pp 23-29.
34. Tsang, D.Z.: One-gigabit per second free space optical interconnection. Appl. Opt., vol 29, 1990, pp 2034-37.
35. Vergnolle, C., Houssay, B.: Interconnection requirements in avionic systems. SPIE vol.1389 on Advances in Interconnection and Packaging, 1990, pp 648-658.
36. Walker, S.J., Jahns, J., Ailawadi, N.K., Mansfield W.M., Mulgrew, P.P., Pastalon, J.Z., Roberts, C.W., Ternant, D.M.: Efficient beam splitters and beam deflectors for integrated planar micro-optics. Diffractive Optics: design, fabrications and Applications (OSA, Washington, D.C., 1992) vol.9, pp 82-84.
37. Weste, N., Eshraghian, K.: Principles of CMOS VLSI Design. Addison and Wesley, 1985.
38. You, H., Soma, M.: Crosstalk analysis of interconnection lines and packages in high-speed integrated circuits. IEEE Transactions on circuits and systems, Vol. 37, 1990, pp 1019-1026.
39. See chapter 5 and also the technical digest on Diffractive Optics: design, fabrications and Applications (OSA, Washington, D.C., 1992) vol.9.

Index

Contributors

JARI TURUNEN, ANDREW C. WALKER   Heriot-Watt University, Department of Physics, Riccarton, Edinburgh EH14 4AS, United Kingdom.

RENE DÄNDLIKER, HANS PETER HERZIG Institute of Microtechnology, University of Neuchâtel, Bréguet 2, CH-2000 Neuchâtel, Switzerland.

MIKE THOMAS GALE, HANS WILLY LEHMANN, RUDOLF MORF,   Paul Scherrer Institut, Badenerstr. 569, CH-8048, Zürich, Switzerland.

DIETMAR FEY, NORBERT STREIBL, WILHELM STORK, KONRAD ZÜRL Physikalisches Institut der Universität Erlangen-Nürnberg, Staudstr. 7, D-91058 Erlangen, Federal Republic of Germany.

THIERRY MAURIN   Institut d'Electronique Fondamentale, Université Paris-Sud, Bât. 220, CNRS, F-91 405 Orsay Cedex, France.

TREVOR HALL, ANDREW KIRK   King's College London, University of London, Department of Physics, The Strand, London WC2R 2LS, United Kingdom.

SUZANNE BONNEFONT, FRANÇOISE LOZES-DUPUY, HENRI MARTINOT L.A.A.S., CNRS, 7 Av. du Colonel Roche, F-31 077 Toulouse Cedex, France.

DANIEL BIZE   C.E.R.N., Département d'Optique, 2 Avenue E. Belin, BP 4025, F-31 055 Toulouse, France.

PIERRE CHAVEL, PHILIPPE LALANNE, GILLES PAULIAT, GERALD ROOSEN   Institut d'Optique Théorique et Appliquée, CNRS, BP 147, F-91 403 Orsay Cedex, France.

ESPRIT Basic Research Series

J. W. Lloyd (Ed.): **Computational Logic.** Symposium Proceedings, Brussels, November 1990. XI, 211 pages. 1990

E. Klein, F. Veltman (Eds.): **Natural Language and Speech.** Symposium Proceedings, Brussels, November 1991. VIII, 192 pages. 1991

G. Gambosi, M. Scholl, H.-W. Six (Eds.): **Geographic Database Management System.** Workshop Proceedings, Capri, May 1991. VIII, 320 pages. 1992

R. Kassing (Ed.): **Scanning Microscopy.** Symposium Proceedings, Wetzlar, October 1990. X, 207 pages. 1992

G. A. Orban, H.-H. Nagel (Eds.): **Artificial and Biological Vision Systems.** XII, 389 pages. 1992

S. D. Smith, R. F. Neale (Eds.): **Optical Information Technology.** State-of-the-Art Report. XIV, 369 pages. 1993

Ph. Lalanne, P. Chavel (Eds.): **Perspectives for Parallel Optical Interconnects.** XIII, 417 pages. 1993